THIRD EDITION

Interpersonal Process in Psychotherapy

A Relational Approach

THIRD EDITION

Interpersonal Process in Psychotherapy

A Relational Approach

Edward Teyber

California State University, San Bernardino

Brooks/Cole Publishing Company

I(T)P® An International Thomson Publishing Company

Pacific Grove • Albany • Belmont • Bonn • Boston • Cincinnati • Detroit • Johannesburg • London
Madrid • Melbourne • Mexico City • New York • Paris • Singapore • Tokyo • Toronto • Washington

 A CLAIREMONT BOOK

Sponsoring Editor: *Eileen Murphy*
Marketing Team: *Jean Vevers Thompson and Margaret Parks*
Editorial Assistant: *Lisa Blanton*
Production Editor: *Laurel Jackson*
Manuscript Editor: *Bernard Gilbert*
Interior Design: *Lisa Thompson*

Interior Illustration: *Lisa Torri*
Cover Design: *Vernon T. Boes*
Cover Photo: *FPG International*
Indexers: *Diane Anthony and Dayle Hill*
Typesetting: *Joan Mueller Cochrane*
Cover Printing: *Phoenix Color Corporation*
Printing and Binding: *Quebecor Printing Fairfield*

For more information, contact:

BROOKS/COLE PUBLISHING COMPANY
511 Forest Lodge Road
Pacific Grove, CA 93950
USA

International Thomson Publishing Europe
Berkshire House 168-173
High Holborn
London WC1V 7AA
England

Thomas Nelson Australia
102 Dodds Street
South Melbourne, 3205
Victoria, Australia

Nelson Canada
1120 Birchmount Road
Scarborough, Ontario
Canada M1K 5G4

International Thomson Editores
Seneca 53
Col. Polanco
México D. F. México
C.P. 11560

International Thomson Publishing GmbH
Königswinterer Strasse 418
53227 Bonn
Germany

International Thomson Publishing Asia
221 Henderson Road
#05-10 Henderson Building
Singapore 0315

International Thomson Publishing Japan
Hirakawacho Kyowa Building, 3F
2-2-1 Hirakawacho
Chiyoda-ku, Tokyo 102
Japan

Printed in the United States of America

10 9 8 7 6 5

Beginning February 22, 1999, you can request permission to use material from this text through the following phone and fax numbers: Phone: 1-800-730-2214; Fax: 1-800-730-2215.

Library of Congress Cataloging-in-Publication Data

Teyber, Edward.
    Interpersonal process in psychotherapy : a relational approach /
    Edward Teyber.—3rd ed.
        p.   cm.
    Includes bibliographical references and index.
    ISBN 0-534-34020-2 (alk. paper)
    1. Psychotherapist and patient. 2. Psychotherapy. I. Title.
RC480.8.T38   1997
616.89'14023—dc20   96-41346
                                                        CIP

*Dedicated to those
who are struggling
to change*

ABOUT THE AUTHOR

Edward Teyber is a professor of psychology and Director of the Psychology Clinic at California State University, San Bernardino. Dr. Teyber received his Ph.D. in clinical psychology from Michigan State University. He is also the author of the popular-press book *Helping Children Cope with Divorce* and co-editor with Dr. Faith McClure of the clinical training text *Child and Adolescent Therapy: A Multicultural-Relational Approach.* He has published research articles on the effects of marital and family relations on child adjustment and has contributed articles on parenting and postdivorce family relations to popular-press newspapers and magazines. Dr. Teyber is also interested in supervision and clinical training, and he maintains a part-time private practice.

CONTENTS

CHAPTER THREE
Honoring the Client's Resistance   62

CHAPTER FOUR
An Internal Focus for Change   92

CHAPTER FIVE
Responding to Conflicted Emotions   123

CHAPTER EIGHT
Current Interpersonal Factors   226

PART FOUR
Resolution and Change   259

CHAPTER NINE
An Interpersonal Solution   261

CHAPTER TEN
Working Through and Termination   296

PREFACE

This compelling, multiculturally sensitive textbook elucidates each of the major issues that arise in treatment and shows how theory leads to practice. It examines actual clinical processes with clarity and immediacy and captures beginning therapists' questions and concerns. The book presents a practical relational framework for conceptualizing cases and developing treatment plans. This framework fosters therapists' flexibility by allowing them to draw from a variety of theoretical perspectives to maximize treatment effectiveness. Programmatic intervention guidelines, provided throughout each chapter, are illustrated with numerous therapist-client dialogues, clinical vignettes, and detailed case studies. Clinically authentic and personally engaging, this text will help students understand the therapeutic process and how change occurs. It is ideal both for graduate student therapists in practicum courses and for students in upper-division undergraduate courses that provide a rigorous introduction to counseling and therapy.

For all theoretical approaches, no matter whether treatment is short-term or long-term, the therapist-client relationship is the foundation of therapeutic change. Few texts provide specific guidelines for understanding and intervening in the process that takes place in the therapeutic relationship; this omission leaves beginning therapists uncertain as to how to proceed with their clients. Furthermore, in preparation for actual clinical work, students take courses on counseling theories and interpersonal skills development. Although useful, this preparation is too general to give students who are seeing clients the in-depth understanding needed for direct clinical practice. Fully cognizant of their limited experience and knowledge, these clinical trainees are often painfully aware that they do not really know what to do or how to help their clients. Many feel inadequate and worry about making mistakes and doing something wrong that would hurt their clients. As a result, beginning therapists are eager for practical instruction that will provide specific guidelines to help them understand and respond to their clients' problems. The purpose of this text is to meet this need by replacing the nebulous ideas that students often have about treatment with a well-developed framework for understanding how change occurs and what role the therapist-client relationship plays in that process.

This text helps beginning therapists to conceptualize what is happening in their real-life relationships with clients and to use this understanding to guide their treatment plans and intervention strategies. In addition, it teaches beginning therapists how to use themselves and their relationships with their clients to help the clients change.

I have developed several topics more fully in this edition. In every chapter, new case studies foster multicultural awareness by showing how the cultural element enhances the interpersonal process approach. Extensive links to other treatment modalities—short-term therapy, couples therapy, group therapy, and family therapy—also appear throughout the book. The interpersonal process approach provides an organizing framework that integrates cognitive-behavioral, family systems, and interpersonal-dynamic theories, and this relational meta-perspective is more clearly highlighted than before. This edition also includes standardized formats for writing structured case conceptualizations, treatment plans, and process notes within the interpersonal process framework. Finally, discussion of the role of attachment and attachment-related affects—especially shame—has been expanded so that students can better understand clients' presenting symptoms and problems.

Also new to the third edition, an instructor's manual accompanies the text. The manual contains an instructor's lecture outline, in-class student exercises, essay questions, and multiple-choice questions for every chapter.

Acknowledgments

Many colleagues, friends, and students have given much to me while I was preparing the third edition. First, I would like to thank the following reviewers of this edition for their valuable comments and suggestions: Myrna Friedlander, State University of New York at Albany; Keren M. Humphrey, Western Illinois University; Dana Schneider, Sonoma State University; and David Van Nuys, Sonoma State University.

From start to finish, it has been a pleasure to work with Eileen Murphy and the Brooks/Cole staff. My colleague Dr. Faith McClure has imbued the text with her unique multicultural sensitivity. Her compassion and emotional presence with clients are a continuing inspiration to me. My friends Drs. Robert Hilton, Sam Plyler, and Keith Valone have also helped with this revision in many ways. Several former graduate students at CSUSB—now my colleagues—have met with me weekly for the past two years and expressed eloquently the joys and anxieties of becoming a therapist. I was touched by them, have often laughed with them, and have learned much from them. In addition, I appreciate the significant contributions made by Diane Anthony, Dayle Hill, George Renshaw, and Nancy Wolfe. This book is graced with the very dear memory of our friend Joy Kirka.

Edward Teyber

PART ONE

An Interpersonal Process Approach

CHAPTER ONE
Introduction and Overview

CHAPTER ONE

Introduction
and Overview

Paul, a first-year practicum student, was about to see his first client. He had long and eagerly awaited this event. Like many of his classmates, Paul had decided to become a therapist while working on his undergraduate degree. Psychotherapy had always been intrinsically interesting to him. For Paul, being a therapist meant far more than just a "good job"; it was the fulfillment of a dream. How exciting, he thought, to make a living by talking to people about the most important concerns in their lives.

At this moment, though, Paul felt the real test was at hand: His first client would be arriving in a few minutes. Worries raced through his mind: What will we do for fifty minutes? How should I start? What if she doesn't show up? What if she doesn't come back next time? Paul also wondered if his client—a 45-year-old African American woman—would have difficulty relating to him, a white man in his early 20s. Needless to say, Paul was intent upon finding a way to help this client with her problems. He had learned something about therapy in his undergraduate psychology classes, and a good deal more from his volunteer experience with callers on the local crisis hotline. But even with these experiences and a supervisor to guide him, Paul was painfully aware of his novice status and of the fact that he didn't know very much about actually doing therapy.

Although some of Paul's classmates thought he was a little too idealistic, they did share his excitement about becoming a therapist. Many of them were older than Paul and far more experienced in life. Some had raised children; others had already had careers as teachers, nurses, and businesspeople. These new therapists were often departing from life roles in which they had felt competent and confident. For these more seasoned classmates as well, a counseling career held new hopes, but also evoked anxiety about their ability to meet new career demands. With each successive step—completing undergraduate requirements, being accepted into a graduate program, and now starting their first practicum—they were approaching their goals. Like Paul, however, they knew that realizing their dreams of a new and rewarding career would also depend on their ability to help others. And with the arrival of their first clients, that ability was about to be tested.

The Need for a Conceptual Framework

Usually, beginning therapists' anxieties about their ability to help are not neurotic insecurity or obsessive worrying. It is realistic to be concerned about one's performance in a new, complex, and ambiguous arena. Paul and his classmates, however, bring many strengths to their new profession. Most students who are selected for clinical training already possess sensitivity, intelligence, and a genuine concern for others. Such personal assets, as well as their own life experiences, will prove enormously helpful to their future clients.

Personal experience, common sense, and intuition are useful. However, in order for a therapist to be effective with a wide range of people and problems, these valuable human qualities must be wed to a conceptual framework. Therapists must understand very specifically what they are trying to do in therapy—where they are going and why—in order to be consistently helpful to clients. Without a conceptual framework as a guide, decisions about intervention strategies and case management are too arbitrary to be trustworthy, and the therapist's confidence in the therapeutic enterprise lessens as a result.

Many therapists become exceedingly anxious in their early counseling work because they lack a coherent and practical conceptual framework. Uncertainty over how to conceptualize client problems and ambiguous guidelines for how to proceed in the therapy session heighten their insecurity and preoccupation with their counseling performance. As a result, their effectiveness with clients is impaired.

Finally, the lack of a conceptual framework hinders novice therapists as they try to find their own professional identities. If therapists are to feel secure in their abilities, they will need a theoretical framework that is applicable across clients yet congruent with their own values and life experiences. The purpose of this book is to provide one such conceptual framework for understanding the therapist-client relationship and the developmental course of therapy. Students will be able to refine this framework to fit their own personalities and therapeutic styles as they develop professionally. The principles suggested here may be applied by therapists who have different personality styles, have grown up in a range of different cultural contexts, and work within varying theoretical approaches. With support from supervisors and instructors, beginning therapists will be able to personalize this model, use it flexibly, and make it their own over time.

The Interpersonal Process Approach
Theoretical and Historical Context

The purpose of this applied text is to help practicum students understand their clients' problems, and use this conceptualization to formulate treatment plans and intervention strategies that guide what they say and do with clients. The therapeutic model presented here will be referred to as the *interpersonal process approach*. This is not a new theory of psychotherapy but a synthesis of ideas from three rich clinical traditions: the collection of interpersonal psychodynamic theorists known variously as Sullivanians, neo-Freudians, culturalists, or inter-personalists; the relational branch of object relations theory (as opposed to the traditional psychoanalytic school of drive theory) and the attachment literature that has developed from it; and family systems theory. Although these approaches have a different focus—on interpersonal, intrapsychic, and intrafamilial domains, respectively—they share a *relational* emphasis and are fundamentally similar. This text highlights the relational dimension in these approaches and links these concepts to the practice of individual psychotherapy. These theories are briefly introduced here, so as to provide a theoretical context for the clinical approach outlined in the remainder of the book. There is a wealth of practical information in these three clinical theories, but far more complexity than beginning therapists can assimilate in a short time. However, the clinical guidelines in Chapters 2–10 are derived from these sources, and readers will want to explore these conceptual domains further as they continue their training.

Interpersonal theory   The interpersonal dimension in psychotherapy was origi-nally highlighted by Harry Stack Sullivan and is articulated and studied today in varying forms by Hans Strupp, Michael Kahn, Peter Sifneos, William Kiesler, Gerald Klerman, Myrna Wasserman, Lorna Benjamin, Judith Jordan, Irene Stiver, Stephen Mitchell, and many others. Sullivan (1968) first brought the relational focus to psychotherapy in the 1940s; he remains an enormously influential but insufficiently recognized figure. A maverick, Sullivan radically broke with Freud's biologically based libido theory and was one of the first major theorists to argue that the basic premises of Freud's drive theory were inaccurate with regard to human motivation, the nature of experience, and clinical tech-nique. Rather than Freud's focus on fantasy and intrapsychic processes, Sullivan emphasized interpersonal relations and the child's actual experience with parents. He argued that every major tenet of Freud's theory could be understood better through interpersonal and social processes. Sullivan developed an elaborate developmental theory of psychopathology that emphasized what people do to avoid or manage anxiety in interpersonal relations; he viewed anxiety as the central motivating force in human behavior. He described a painful and pervasive core anxiety rooted in dreaded expectations of derogation and rejection by parents and others, and later by oneself. Following from this, Sullivan conceptu-alized personality as the collection of interpersonal strategies that the individual employs to avoid or minimize anxiety, ward off disapproval, and maintain self-esteem.

According to Sullivan, the child develops this personality, or *self-system*, through repetitive interactions with his or her parents. More specifically, the child organizes a self around certain basic relational configurations that characterize important patterns of parent-child interaction. These repetitive interactions with parents, which may be painful or anxiety-arousing, are structured into the child's personality through a collection of complementary self-other relational patterns. For example, children may develop internal images of themselves as helpless or insignificant and expectations of parents and others as demanding or critical. Alternatively, children may evolve images of themselves as special and expectations of others as admiring.

Having learned these self-other relational patterns, people systematically behave in ways that avoid or minimize the experience of anxiety. For example, suppose that particular aspects of the child, such as sadness or crying, consistently result in parental rejection or ridicule. The child learns that these more vulnerable modalities represent a "bad self," and these anxiety-arousing aspects of the self are split off or disowned. The child also develops interpersonal coping styles (for example, a macho manner) that preclude being subjected to anxiety-arousing rejection again. These coping styles, or *security operations*, are interpersonal defenses that originally were necessary to protect the self in the early parent-child relationships. Unfortunately, these interpersonal defenses are overgeneralized to other relationships in adulthood and become "characterological" as the child, now grown to an adult, anticipates that new experiences with others will repeat the relational patterns of the past.

Sullivan developed his rich, if sometimes elusive, interpersonal theory through his clinical work with schizophrenic patients. He was a gifted therapist and had great respect for the personal dignity of very disturbed patients. His sensitivity to patients, compassion, and genius for understanding the intricate workings of anxiety inspired a new direction in American psychiatry. Sullivan worked with, and influenced, Erich Fromm (1982), Karen Horney (1970), Frieda Fromm-Reichmann (1960), and other seminal thinkers in the interpersonal field. A single integrated theory was never developed, and therapists working within this approach were never identified by a particular group name. Nevertheless, these theorists shared basic concepts and values and made enormously rich contributions to the theory and practice of counseling and psychotherapy.

Sullivan also influenced many other interpersonal dynamic theorists such as Erik Erikson (1968) and Hans Strupp (Strupp & Binder, 1984); client-centered theorists such as Rogers (1951) and Carkhuff (1969); and self theorists such as Kohut (1971). He even prefigured family systems theory in his study of close relational systems. In fact, two of the founding fathers in the family therapy field, Don Jackson and Murray Bowen, trained under Sullivan. In this textbook, we will also be examining a host of contemporary short-term and time-limited dynamic approaches to treatment—developed by Peter Sifneos, Lester Luborsky, Henry Davanloo, and others—that are based on this interpersonal orientation. Many of the basic concepts in the interpersonal process approach can be traced back in some form to Sullivan, Horney, Fromm-Reichmann, and the other interpersonal theorists.

In the chapters that follow, we will return repeatedly to two basic concepts from these interpersonal theorists. First, we will be exploring the broad *interpersonal coping strategies* or personality styles that clients employ with therapists and others to avoid anxiety and maintain interpersonal safety and self-esteem at the expense of change. Second, in order to conceptualize client dynamics and better understand the interaction or *process* between the therapist and client, later chapters will also examine the *maladaptive relational patterns* or *core relational conflicts* that clients repetitively reenact in close relationships with therapists and others.

Object relations theory   In the present context, *objects* are people or, more precisely, internal representations of them, and object relations theory is about interpersonal relationships—especially those between parents and young children. In particular, object relations theorists try to account for the meaning of attachment and the emotional need for relationships with others in the development of a self. Sullivan's theory centers on relationships, but the question that drove his work was very pragmatic: What do people do to avoid anxiety in interpersonal relations? In contrast, the question for object relations theorists is more abstract and inferential. Object relation theorists are interested in understanding how formative interactions between parents and children become *internalized* by the child and serve as cognitive schemas, relational templates, or internal working models that guide and shape how children establish subsequent relationships with others. These cognitive models provide basic expectations or road maps for what will transpire in relationships. Although these early templates for relationships will become more complex and change over time, object relation theorists believe that these formative parent-child interaction schemas provide the basic structure for developing a sense of self, organizing the interpersonal world, and shaping subsequent relational patterns.

Object relations therapists use these schemas or templates to conceptualize clients' problems and better understand the problematic relational patterns that clients are enacting with others and in the therapeutic relationship. For example, in order to understand these templates or faulty relational expectations, object relations therapists will ask themselves questions such as these: (a) What does the client want from others? (The client might want to be cared for or taken seriously, for example.) (b) What does the client expect from others? (The client might expect to be diminished or exploited, for example.) (c) What is the client's experience of self in relationship to others? (It might be burdensome or disconnected, for example.) (d) What are the emotional reactions that keep recurring? (The recurring reactions might be shame or affection, for example.) As a result of these core beliefs or relational templates, what are the client's recurrent interpersonal strategies for coping with their relational problems? (Possible strategies include complying with others, withdrawing from others, and trying to control others.) Finally, what do these interpersonal styles tend to elicit from the therapist and others?

Object relations theorists also believe that the primary motive of human development is to establish and maintain emotional ties to parental caregivers (object relatedness). The greatest conflicts in life are threats to, or disruptions of, these basic ties (separation anxieties and abandonment terrors). In this regard, anxiety is a *signal* that emotional ties are being threatened. If parents are emotionally available and reliably responsive to a child's attachment needs, the child is secure in his or her ties. Through the course of development, such children can gradually *internalize* their parents' availability and come to hold the same constant loving affect toward themselves that their parents originally demonstrated toward them. In other words, cognitive development increasingly allows young children to internally hold a mental representation of their parent when they are away; if securely attached, they are increasingly able to experience or sustain the same reliably loving affect toward themselves that their parents originally provided for them. As these children progress toward maturity, they become increasingly able to comfort and soothe themselves, function for increasingly longer periods without emotional refueling, and effectively elicit support when necessary. *Object constancy* develops as their ability to comfort and nurture themselves becomes the source of their own self-esteem and secure identity as capable, love-worthy persons. Further, they possess the internal working models or relational templates necessary to establish parallel new relationships with others that possess this same affirming affective valence.

But what if the child cannot establish and maintain emotional ties to the primary caregivers? In some families, parents are preoccupied, depressed, rejecting, or even abusive. Children of such parents are biologically organized to have their attachment longings fulfilled, but their emotional security is compromised by inconsistent or unresponsive parents. These children are trapped in an unsolvable dilemma: They cannot succeed in evoking a reliable response from their parents, nor can they escape or forsake their need for attachment. In a word, they feel *anguish*. What do they do? Object relations theory tries to account for the different solutions and adaptations the child must make to this poignant and all-too-human dilemma.

Object relations theorists propose complex (and competing) developmental theories to account for the intrapsychic and interpersonal mechanisms that children employ to protect against the painful separation anxieties aroused when they cannot maintain secure emotional ties. These theorists propose that all children mature through stages of development that progressively allow for a more integrated, stable view of self and others. However, if development is problematic, children remain "stuck" or else regress under stress to less integrated stages. In particular, when parents are excessively inconsistent or unresponsive, the child must resort to *splitting defenses* to maintain ties to an idealized loving parent; that is, the child separates and internalizes the "bad" (threatening or rejecting) aspects of the parent from the "good" (loving or responsive) aspects of the parent. Thus, in object relations theory, in contrast to psychoanalytic theories, it is not unacceptable sexual and aggressive impulses that are repressed, but what has been painful or threatening in children's relationships with their caretakers. In healthier parent-child interactions that provide secure attachments, however, the young child can tolerate increasing levels of frustration and disappointment as he or she matures. Gradually, the child will be able to increasingly integrate his or her ambivalent feelings toward the sometimes frustrating and sometimes responsive parent into a stable and affirming relational template of self and other, which becomes the basis of self-esteem and identity.

Under greater attachment threats, however, the child cannot do this without resorting to splitting defenses. These splitting defenses preserve the necessary image of an idealized, "all good," responsive parent with whom the child is internally connected. But the price is high: Reality is distorted; the self is fragmented; and the child becomes the one who is "bad." The frustrating parent is no longer "bad," which allows the child to view the external world as safe. The price, however, is inner conflict: The child believes that, if only he or she were different, parental love would be forthcoming. This situation is illustrated most poignantly by the bruised and battered child who nevertheless continues to idealize and defend the abusive parent, by maintaining that he or she is bad and deserves the punishment. The child's belief that he or she is bad maintains the illusion that the world is orderly and just. This self-negating distortion enables the child to cope—to feel that he or she has some control over events and is not shamefully helpless, ineffectual, or vulnerable.

Why is this complex and abstract theory relevant to interpersonally oriented therapists? Realizing that the child must find some way to maintain ties with a good, loving parent helps therapists to understand why clients persist in maladaptive behavior and self-defeating relationships. When clients relinquish symptoms, succeed, or become healthier, they often feel anxious, guilty, or depressed. Loyalty and allegiance to symptoms—and the bad objects they represent—are not maladaptive to insecurely attached children, because such loyalty preserves the ties to the "good" or loving parent; that is, clients maintain their childhood fear that, if they disengage from these internal objects (or the symptoms that represent them), they will find themselves again helpless, alone, and frightened. The interpersonal process approach is designed to resolve these dysfunctional cognitive schemas and painful interpersonal conflicts by expanding the client's relational templates and expectations of what can occur in close personal relationships. We will see later how therapists can help to do this by providing clients with the real-life experience that some relationships can be different, at least some of the time.

Certain ideas in this book are drawn from the *relational* branch of object relations, especially the writings of R. D. Fairbairn, W. D. Winnicott, H. Guntrip, and J. Bowlby, as opposed to the traditional psychoanalytic "drive" branch of object relations represented by S. Freud, M. Klein, M. Mahler, and O. Kernberg (Greenberg & Mitchell, 1983). These relational theorists have also influenced contemporary clinicians, such as J. Masterson's (1976) important contributions on treating borderline personality disorders and H. Kohut's (1971, 1977) work on self psychology and narcissistic conditions. Unfortunately, it is exasperating to many beginning therapists to find that much of object relations theory is presented in obtuse, inaccessible jargon. Paradoxically, object relations theorists address the most sensitive and personal dimensions of human experience in impersonal, distancing terms. However, we will be able to use certain aspects of this theory to understand clients better, relate more empathically to them, and guide our therapeutic interventions. In particular, we will be returning to the concepts of *relational templates* or *internal working models* throughout the text, in order to understand and intervene with the maladaptive relational patterns that clients continue to reenact. We will also be exploring *object ties* further, in order to better understand why clients may resist change—that is, to see how anxiety and guilt are evoked for some clients as their attachment ties to internalized parental figures are threatened by change.

Family systems theory   Not only must children adapt to their attachment figures, they must also adapt to their familial and cultural systems. Beginning therapists can glean a wealth of information about their adult clients from family systems theory, which provides an understanding of the familial rules and roles that shaped the clients' development. We become who we are in our families of origin, and family process researchers have learned much about multigenerational family relations.

In the late 1950s, Gregory Bateson, Virginia Satir, Jay Haley, and others began studying communication patterns in families (Goldenberg & Goldenberg, 1996). They found that intrafamilial communication followed definitive but often unspoken rules that determined who spoke to whom, about what, and when. Adult clients often continue faulty communication patterns learned in their families of origin (for example, addressing conflicts through a third party rather than speaking directly to the person involved; allowing others to speak for them and say what they are thinking or feeling; never making "I" statements or saying directly what they want).

Early family researchers also learned that children are often scripted into familial roles—such as the responsible or good child; the problem or bad child; and the hero, rescuer, or invisible child. Following object relations theory, the most common role is the good child/bad child role split that occurs in so many dysfunctional families. Most adult clients are still performing these childhood roles, which distort and deny many important aspects of who they really are. In the process of adapting in their families, these children placed their families' demands and expectations on themselves. Many clients who have played the good-child role in their families are still perfectionistically demanding of themselves, guilty about doing anything for themselves, worried about the needs of others, afraid of their anger, and confused by their sad and empty feelings. Whereas the bad child usually acts out personal conflicts externally (for example, through substance abuse or promiscuity), the good child often seeks treatment and enters therapy, presenting with symptoms of anxiety or depression.

Families also have unspoken rules about how, in late adolescence, children can leave home. For example, the oldest daughter may not be allowed to grow up and leave home successfully on her own. Often, she will seek emancipation through pregnancy, only to find herself even more dependent on her parents than before and forced to live at home again. A son may only be able to leave home through rejecting confrontations with parents, and then may live in another part of the country and have little or no continuing contact with the parents. Family rules, faulty communication patterns, and roles all serve to maintain family myths that, in turn, function to avoid anxiety-arousing conflicts. For example, these are some common family myths: Dad doesn't have a drinking problem; nobody is sad in our family; Mom and Dad never fight and are very happily married. In sum, family myths and all of these other family characteristics are rule-bound *homeostatic mechanisms* that govern family relations and establish repetitive, predictable patterns of family interaction. They are homeostatic in the sense that they tend to maintain the stability of the family system.

Salvadore Minuchin (1974) and other family researchers have also explored the alliances, coalitions, and subgroups that make up the structure of family relations. In some families, for example, the maternal grandmother, mother, and eldest daughter are allied together, and the father is the outsider. This structural road map for reading family relations becomes even more illuminating when these family dynamics are examined in a three-generational perspective (Bowen, 1966; Boszormenyi-Nagy & Spark, 1973). It is fascinating—and sometimes disturbing—to see how these same family rules, roles, myths, and structural relationships can be reenacted across three or four generations in a highly patterned, rule-governed system. Adding to the complexity of this tapestry, many of these family rules, roles, and communication patterns operate within a broader cultural context. (For example, the oldest son in a traditional Chinese family or the oldest daughter in a traditional Latino family may need to assume a responsible role.) It is useful to establish the extent to which the patterns are culturally sanctioned.

The interpersonal process approach draws on these basic family systems concepts, as well as something else the family therapy movement has contributed— the effects of parental relations and familial experience on clients' problems.

Therapists need to help clients make realistic assessments of the strengths and problems that actually existed in their families of origin. For many clients and therapists, however, it is culturally taboo to speak critically of parents; Alice Miller (1984) observes how this even breaks the Fourth Commandment: Honor thy parents. Herein, though, is one of the great strengths of family systems theory. The family systems therapist is trying to *understand* hurtful familial interactions, not to *place blame* or scapegoat any family member. The family therapist is concerned about the well-being of *every* family member, parents as well as children. In parallel, *as clients realize that their individual therapist wants to understand—rather than to blame, reject, or demonize—the internalized parent, clients are able to explore highly threatening material and make significant gains in therapy.* When therapists' own splitting defenses lead them to blame and reject parental figures (by saying, for example, "I can't believe how mean your mother was to you!") rather than to understand and empathize with their client's experience (by saying, for example, "I'm sorry you were hurt so much when she did that"), however, the client ultimately will not be able to progress as far in treatment, out of loyalty to the internalized parent.

According to object relations theory, there are some good things in almost every family, even in very troubled or abusive families. There are also limitations and conflicts in the healthiest families. When clients continue to idealize parents and deny real problems that existed, they are complying with binding family rules and protecting their parents at their own expense. However, clients cannot simply reject hurtful parents and emotionally cut them off, or they will tend to reenact these same problematic relations with others. This means that clients must find a way to keep parents (or healthier aspects of them) alive inside themselves as partial identifications. How can clients accomplish this?

In order to get better, clients must be able to come to terms with both the good news and the bad news in their families. As already noted, however, clients cannot achieve this integration and resolution if the therapist also employs his or her own splitting defenses. In other words, the therapist does not want to identify with and idealize the wounded child that exists within some clients and to reject the hurtful parent. Neither does the therapist want to support clients' continuing denial of familial problems and idealization of parents. Instead, the therapist's appropriate role is to try to understand what actually occurred in clients' development and to help them come to terms with the good and the bad in their experience. The therapist also helps the client distinguish between past experience and current reality.

As we see, family systems concepts inform and guide the interpersonal process approach. In particular, we will be returning to two family systems concepts throughout this text. First, in concert with the other two theoretical approaches, family systems theory helps us clarify the problematic interaction patterns that clients are enacting with the therapist and others. Second, it helps both therapists and clients to understand the developmental background of many current relational problems. With this awareness, clients do not need to continue reenacting or to angrily break off important but problematic familial relationships.

Basic Premises

What are people's problems about? Where do they come from and how can they be resolved? Every beginning therapist grapples with such questions and will continue to do so throughout his or her life. One of the enduring satisfactions of a counseling career is in continuing to develop one's own ideas about the nature of people and their problems.

This section introduces some ideas and assumptions that are central to the interpersonal process approach:

1. Problems are interpersonal in nature.
2. Familial experience is the most important source of learning about ourselves and others.
3. The therapist-client relationship can be used to resolve problems.

In other words, this overview examines how problems are expressed, how they have developed, and how they can be resolved.

The nature of problems: A separateness-relatedness dialectic   Most client conflicts arise in close relationships, and clients often enter therapy when they are beginning or ending a primary relationship. One useful way to conceptualize many of these interpersonal problems is along a continuum of separateness and relatedness. This will now be examined from a family systems perspective.

Most family systems theorists agree that the basic dimension of family life is a developmental continuum of separateness-relatedness. Although there will be great variation in the specific ways these basic family functions are carried out in different cultural contexts, all families must nurture young children and provide them with a secure sense of acceptance and belonging (relatedness). As children grow older, the developmental task of the family shifts from nurturing children to training them for independence and preparing them for a successful emancipation (separateness) that encompasses continuing ties and familial responsibilities. In other words, separateness does not mean disconnectedness; it simply means that the child has internalized and integrated various aspects of the parents, can function more independently, and is centered in his or her authentic self. Thus, in order to function well, the family must have the flexibility and interpersonal range to meet offspring's changing needs for both attachment and closeness in early childhood and for greater autonomy in later years. As we will see, healthy families achieve the complex, interdependent balance between the poles of separateness and relatedness and encompass aspects of *both* dimensions at the same time.

If there are problems in forming an attachment bond in infancy or in meeting the strong dependency needs of young children, family members will be disconnected from one another. Children will develop neither a shared sense of family identity nor loyalty to each other. Most importantly, children will not be motivated to comply with parents' socialization demands unless an affectional bond exists. Young children fail to relinquish their own wishes and adopt more mature and responsible behavior if they are not concerned about maintaining parental approval. As a result, children from such disengaged families do not internalize social rules and self-controls adequately. They do not feel responsibility toward others or the culture at large and are likely to develop acting-out symptoms and to externalize blame. These offspring are more likely to come in contact with police and judges than with mental health professionals.

Problems also arise at the other end of the continuum. When there is too much closeness and family members are enmeshed, parents cannot tolerate the individuation or emancipation of their offspring. An example of this can be seen when enmeshed families stifle individuation by making anger unacceptable and by avoiding conflict rather than addressing and resolving it. Individuation is also blurred by ineffective communication patterns, as when family members express what another is thinking or speak to each other through a third person rather than directly.

Enmeshed families also inhibit emancipation by discouraging developmentally appropriate steps toward independence and autonomy. For example, these families do not give offspring in late adolescence permission to leave home psychologically and succeed on their own. Such offspring often feel a binding guilt as they try to become independent adults and make commitments to love relationships and career choices (Teyber, 1983).

In contrast to offspring in disengaged and enmeshed families, offspring in families that function well develop a balance between these two extremes and integrate the relatedness and separateness dimensions. These offspring develop a sense of self that allows them to feel *emotionally connected to others while still remaining separate from them or individuated* (Jordan, Kaplan, Miller, Stiver, & Surrey, 1991; Stierlin, 1972; Mahler, Pine, & Bergman, 1975; Boszormenyi-Nagy & Spark, 1973).

The same dialectic of separateness and relatedness continues into adulthood. Here again, the challenge of successful relationships is to balance intimacy and autonomy. On the one hand, we need to have a sense of belonging and being cared about. It is important that others acknowledge our experience and respond to us in caring and confirming ways. At the same time, however, we must also be separate and independent. All of us have the right to our own feelings and perceptions, even if they differ from those of significant others. We need to be able to express our own preferences and beliefs and to pursue our own interests and goals. Healthy individuals can accommodate the polarities of separateness and relatedness; that is, they can set limits, have personal boundaries, and pursue their own interests, while still being committed to others in affirming and caring ways.

The problems that most clients present reflect an inability to achieve an integrated sense of self as both an independent and competent person and an emotionally available, responsive, and committed person. Thus, several unifying themes may be identified in clients' varied problems. Client conflicts will often embody separateness issues of autonomy, initiative, and power, on the one hand, or relatedness issues of intimacy, trust, and commitment, on the other.

The source of problems: Affective learning in the family of origin   Social learning in the family of origin is the most important source of long-standing personality strengths and conflicts. Family homeostasis requires that many patterns of inter-action and communication become repetitive and rule-bound. These transactional patterns are strongly maintained in the family of origin and can be highly resistant to change in adulthood.

There are two reasons why familial experience has such a powerful, long-term impact on the individual. The first is the sheer repetition of family transactional patterns. The same types of affect-laden interchanges are reenacted thousands of times in daily family life. Suppose, for example, that a parent has difficulty responding positively to a child's successes. When the young child enthusiasti-cally seeks the parent's approval for an accomplishment, the parent might ignore the child and change the topic, compare the child unfavorably with a sibling's greater accomplishment, turn away and look vaguely sad or hurt, or take the success away from the child by making it his or her own.

The same type of parental response usually occurs when the child shows the parent a favorite drawing, makes a new friend, wins a race at school, or earns a star from his or her teacher. This transactional pattern becomes a powerful source of learning when it continues over a period of years and even decades. As an adult, the child is likely to feel conflicted about completing his or her educational degree or about taking pleasure in a promotion earned at work. Thus, the most impor-tant problems in people's lives are the generic conflicts that develop from these characterological or habitual response patterns. Contrary to popular belief and portrayals in Hollywood's movies, long-standing personality problems are shaped more by repetitive family transactional patterns (*strain* trauma) than by isolated traumatic events or time-limited stressors (*shock* trauma) (Wenar, 1994).

Second, the impact of these repetitive transactions is magnified because of the intensity of the affect involved. Parents are the pillars of the child's universe, and children depend on them with a life-and-death intensity. These repetitive transactional patterns have been reenacted in highly charged affective relationships with the most important people in one's life. The child in the preceding example may well feel a desperate need somehow to win the parent's approval, yet simultaneously feel increasing anxiety about trying to succeed or about approaching the rejecting parent. Thus, our sense of self in relation to others is learned in the family of origin and, in many ways, will carry over to adulthood.

In summary, family interaction patterns may be hurtful and frustrating for the individual, or they may be validating and encouraging. For better or worse, these repetitive patterns of family interaction, roles, and relationships are internalized and become the foundation of our sense of self and the social world. Of course, other factors are influential as well. Familial relationships are molded within a cultural context that can affirm or repudiate parental behavior, and familial experience will not be central to many situational problems that clients present. Familial experience does, however, provide our first and most long-lasting model for what goes on in close relationships. It will figure significantly in the individual's choice of marital partner and career, in how adult offspring will, in turn, parent their own children, and in many of the other enduring problems and satisfactions of adult life.

Although the influence of the family of origin is profound, ineffective and painful transactional patterns can be relearned. Change is indeed possible, and occurs in part through a relational process of affective relearning, which we consider next.

Resolving problems: It is the relationship that heals   The relationship between the therapist and the client is the foundation of the therapeutic enterprise. The nature of this relationship is the therapist's *most important* means of effecting client change; it determines the success or failure of therapy. In order to utilize the therapeutic relationship systematically as a vehicle for change, however, therapists must understand very specifically the meaning of their dynamic interactions with their clients. Thus, the interpersonal process approach focuses on understanding what goes on between the therapist and client in terms of their interaction or *process*.

The therapist-client relationship is complex and multifaceted; different levels of communication occur simultaneously. For example, there is a subtle but important distinction between the overtly *spoken content* of what is discussed and the *process* dimension of how the therapist and client interact. In order to work with the process dimension, the therapist must make a perceptual shift away from the overt content of what is discussed and begin to track the relational process of how two people are interacting as well. This means that, at times, the therapist must be able to step beyond the usual social norms and take the risk of talking to the client about their current interaction. This will be challenging for many beginning therapists, because it conflicts with their familial and cultural prohibitions, and it may seem impolite or intrusive to speak forthrightly about "you and me." In this text, however, we will see how powerful it can be when the therapist talks directly with the client about what is going on between them. Further, we will also see how the therapist can use this metacommunication in a sensitive and respectful way that does not make the client feel uncomfortable. Consider the following example.

During the first session, the client tells the therapist that he resents his wife because she is always telling him what to do. He explains that he has always had trouble making decisions on his own and, as a result, his wife has often simply made decisions for him. Even though he felt that she was just trying to help him with his indecisiveness, he resented her "bossiness" and "know-it-all attitude." After describing the presenting problem in this way, the client asked what the therapist thought he should do.

Let us look at the process dimension of their interaction. Suppose the therapist complied with the client's request and said, "I think that, the next time your wife tells you what to do, you should. . . ." If therapy continues in this didactic vein, the therapist and client will begin to reenact in their relationship the same conflict that led the client to seek treatment; the therapist will be telling the client what to do, just as his wife has done. The client has certainly invited this response, and will probably welcome the therapist's suggestions at first. In the long run, however, he will probably come to resent the therapist's directives just as much as he resents his wife's, and will ultimately find the therapist's suggestions to be of as little help as hers.

Alternatively, the therapist might respond to the other side of the client's conflict and say, "I don't think I would really be helping you if I just told you what to do. I believe that clients need to find their own solutions to their problems." Frustrated, the client responds, "But I told you I don't know what to do! It's hard for me to make decisions. Aren't you the expert who is supposed to know what to do about these things?"

This response throws the client back on the other side of his conflict and leaves him stuck in his own inability to make decisions. If this mode of interaction continues and comes to characterize their relationship, their process will recapitulate the other side of the client's problem; his inability to initiate, make his own decisions, and be responsible for his own actions will immobilize him in therapy, just as it has in other areas of his life.

What we see here is that clients do not just talk with therapists about their problems in an abstract manner; rather, they actually recreate and act out in their relationship with the therapist the same conflicts that have led them to seek treatment. *This recapitulation of the client's problem is a regular and predictable phenomenon that will occur in most therapeutic relationships.* We will return to this scenario later in the chapter and explore better therapeutic responses.

To resolve problems, clients must experience in their real-life relationship with the therapist a new and more satisfying solution to their problems. When clients terminate prematurely or therapy reaches an impasse, the therapist and client are usually reenacting in their relationship subtle aspects of the same conflict that the client has been struggling with in other relationships, although neither of them may be aware of this reenactment. For example, clients may feel that they are being controlled by the therapist and have to do everything the therapist's way, or that they have to take care of the therapist and meet his or her needs, or that they must please the therapist and win his or her approval, or that they have to compete with the therapist—just as they are doing with others in their lives. When this occurs, the therapeutic process has metaphorically repeated the conflicted interaction that clients have not been able to resolve in other relationships and that they have often experienced in earlier formative relationships in the family of origin.

As clients begin to play out with the therapist the same relational patterns or interpersonal problems that originally brought them to therapy, the therapist must be able to respond in a new and more effective way that enables clients to resolve the conflict within their relationship. This concept may seem easy to understand but is very difficult to put into practice in affect-laden relationships with clients. Accordingly, this basic theme will be repeated in different ways and illustrated by varying case examples throughout the text.

In successful therapy, clients experience a new kind of relationship. They participate in a relationship in which their old conflicts or relational patterns are intensely aroused, but the therapist does not respond in the same problematic way that others have in the past. Following their relational templates (their ingrained relational responses and expectations), clients routinely try to elicit or reenact the same interpersonal conflicts with the therapist that they have been experiencing with others—often along the process dimension or way in which they interact. However, rather than responding in the problematic way that clients expect, the therapist is able to provide a different and more satisfying response to the old relational scenario. When the therapist and client are able to address and resolve these expectable conflicts or reenactments that occur in most therapeutic relationships, change has begun to occur.

In this process, clients begin to distinguish between past and present reality and learn that they do not have to respond in their old ways or always receive the familiar responses. Thus, clients begin to develop a greater repertoire of behaviors and expectations and become more flexible in their relationship with the therapist. As we will see, it is often relatively easy to help the client *generalize* this experience of change with the therapist and adopt new and more adaptive responses with others outside the therapy setting. This *corrective emotional experience* is the fundamental premise of interpersonal process psychotherapy and the basic mechanism of therapeutic progress and change. Let's examine more closely what occurs in these pivotal interactions.

Clients believe actions, not words. Accurate interpretations, relevant educational inputs, and effective cognitive restructuring will be useful in almost every therapy, but they are not the primary seat of action in the interpersonal process approach. This is a performance-based or experiential learning model: Clients change when they live through emotionally painful and ingrained relational scenarios with the therapist, and the therapeutic relationship gives rise to outcomes different from those expected, anticipated, or feared (Strupp, 1980). When the client reenacts important aspects of his or her conflict with the therapist, and the therapist's response does not fit the old relational templates, schemas, or expectations, the client has the *experience* that relationships can be another way. It is powerful and enlivening to find that, at least this time, the same unwanted relational pattern did not occur. Among other important effects, the new or corrective response from the therapist creates *greater interpersonal safety* for the client. This is not the general safety that comes from being taken seriously and treated with respect, but a more significant safety that results when the client is not hurt again by receiving certain specific interpersonal responses that are familiar, expected, and often dreaded. As we have noted already, this therapeutic moment is not the end point in treatment, but a window of opportunity for a range of important new behaviors to emerge. Clients may be empowered to:

- Feel better about themselves or forgive themselves for behavior they have unrealistically felt shameful or guilty about
- Have the security to feel or experience more fully the pain of how much this has hurt them in other relationships
- Remember more fully when, how, and with whom this particular relational conflict was learned or originally occurred
- Make connections between their current behavior and historical relations that are useful to them in resolving their current problems with others
- Feel bolder and risk addressing—or making overt—conflicts, misperceptions, or reenactments that have been occurring with the therapist
- Make decisions or attempt new ways of responding with others that have been too threatening or unattainable before

Thus, repeated corrective emotional experiences with the therapist begin to break old relational templates and repetitive relational patterns, and a wider interpersonal range becomes available to the client. We will see later how the reparative response from the counselor can also change clients' internal or subjective experience of themselves.

Corrective emotional experience has been the cornerstone of many dynamic and interpersonally oriented therapies. For example, Franz Alexander and Thomas French (1980) originally developed this conceptualization in 1946. Although misunderstood at that time by the traditional psychoanalytic community and disdained as merely superficial role-playing, their more active, direct approach provided clients with a corrective emotional experience in the here-and-now relationship with the therapist.

Weiss and Sampson (1986) elucidate how the therapist must "pass transference tests" and respond to conflicted developmental needs in a more helpful way than parents and other caregivers did originally. If therapists do this consistently and programmatically throughout treatment, they will gradually disconfirm the client's "grim, unconscious pathogenic beliefs" that developed in problematic parent-child interactions (for example, "If I get stronger and do what I want, I will hurt my parent," or "I must always be left because my needs are too demanding"). Focusing more specifically on the process dimension, Kell and Mueller (1966) also illuminated how the client's central conflicts are metaphorically reenacted with the therapist. These conflicts can then be either resolved or recapitulated, depending on the nature of the relationship that the therapist provides. These and other clinical theories we will be exploring (especially the contemporary, short-term dynamic therapies of Strupp, Sifneos, Luborsky, Davanloo, and others) use different terms to describe the same direct, immediate experience of change in the real-life relationship between the therapist and client. Perhaps Frieda Fromm-Reichmann (1960) originally captured this central tenet best by saying that the therapist must provide the client with an *experience* rather than an *explanation*.

In order to provide this corrective emotional experience, however, the therapist must be able to work with the process dimension. Let's return to our example of the client who complains that his wife tells him what to do. When the client asked his therapist what he should do, one option would have been for the therapist to offer a *process comment* that described their current interaction and made it overt as a topic for discussion. For example, the therapist might have said, "It seems to me that you're asking me to tell you what to do, which will bring up the same problem for us in therapy that you are having at home with your wife. Let's see if you and I can do something different in our relationship. Rather than having me tell you what to do, let's try and work together to understand what is going on for you when you are feeling indecisive. Where do you think is the best place to begin?"

In this instance, the therapist is being supportive by offering the client a new and different *collaborative* relationship in which they can work together on the client's problems. If this collaborative effort continues to develop over the course of treatment, their working relationship will provide the facilitative context necessary for change to occur. Of course, the therapist's initial attempt to offer the client a corrective response will need to be repeated in many different ways for change to occur. If the therapist can establish and maintain this type of collaborative relationship, however, it will facilitate other types of therapeutic interventions as well. Educational inputs, behavioral alternatives, dynamic interpretations, and interpersonal feedback will all be more effective when they occur in this type of interpersonal context.

As we have seen, the guiding principle in the interpersonal process approach is to provide a corrective emotional experience in which clients experience a new and more satisfying response to their long-standing conflicts. This often demands significant personal risk and involvement on the part of the therapist, who must be personally engaged and emotionally available to the client. The relationship must hold real personal meaning for both participants in order to have the emotional impact necessary to effect change. In this simple human way, it is the relationship that heals. If the therapist is merely an objective technician, psychologically removed and safely distant from the client, the relationship will be too insignificant to be utilized as a vehicle for change.

However, therapy will also falter if the therapist becomes inappropriately close. If therapists overidentify with the client or become too invested in the client's changing in order to shore up their own feelings of adequacy, therapy will not progress. In this situation, therapists lose sight of the process they are enacting with the client and typically begin to respond in problematic ways that recapitulate the client's conflict.

The most common problem of beginning therapists is to experience the client's situation or concerns as identical to their own. The solution is a supervisory relationship that helps therapists see the ways in which their own experience differs from the client's. Once able to differentiate their own issues from those of the client, therapists will often be able to see how the therapeutic process has been reenacting aspects of the client's conflict. This understanding usually enables therapists to return to the therapeutic relationship and provide the client with the new type of interpersonal experience that will facilitate change.

Beginning therapists should expect to make many "mistakes" with their clients—to become overinvolved or underinvolved with certain kinds of clients, to recapitulate the client's conflict at times, and so on. Few clients are fragile, however, and therapeutic relationships are often remarkably resilient. Unfortunately, most beginning therapists do not know this, and concern about making mistakes remains one of their biggest anxieties. Beginning therapists will do better therapy—and enjoy it more—once they discover that *mistakes can be undone.* In fact, mistakes provide important therapeutic opportunities when therapists are willing to work with the process dimension—to talk with clients about problems, misunderstandings, and impasses that may be occurring right then in the therapeutic relationship. (For example, the therapist may say, "I think I might have misunderstood what you just said, and you seem to have become more distant from me in the last few minutes. What did you see happening between us right there?") By clarifying the miscommunication or responding to whatever is problematic for the client (who might say, for example, "When you are being quiet like that and don't respond very much, I feel judged by you. Would you talk more?"), the therapist helps to make this relationship different from past scenarios.

As their performance anxiety declines, beginning therapists will find that they are better able to identify and work with the process dimension and, in so doing, to undo mistakes and realign the therapeutic relationship when it has gone awry.

In closing, remember the maxim that clients will not hold it against you for making a mistake, only for not owning it. Sometimes we need to apologize.

Client Diversity and Response Specificity

We have already seen that beginning therapists need a theoretical framework to guide their therapeutic interventions. What must such a theory provide if it is to be of help to therapists and their clients?

One feature of an effective clinical theory is that it must have the flexibility and breadth to encompass the diversity of clients who seek treatment. Although unifying patterns in personality certainly exist, every client is different. Each client has been genetically endowed with a unique set of features, and each has been raised differently in his or her family. Socialization is different for women and men; members of different cultures have different experiences; and economic class shapes opportunity and expectations. To be helpful, a clinical theory must be able to help therapists work effectively with all of these highly diverse clients.

Moreover, each therapist is a different person. Like clients, therapists differ in age, gender, ethnicity, sexual orientation, and developmental background. Therapists also bring diverse training, theoretical orientations, and personal styles to their clinical work. How can any theory help such a diversity of therapists respond to the extraordinary range of human experience that clients present?

Client response specificity will be one of our best tools. This approach requires therapists to be *flexible*, so that they can respond sensitively on the basis of each client's personal history, experiences, and ways of viewing the world. While it is helpful to understand or be familiar with the experiences of particular groups (for example, to know that African Americans' history includes slavery and racial discrimination), this approach attempts to understand and respond to each client as a unique individual whose experiences are embedded within a particular familial and cultural milieu. Thus, we will not seek rules or intervention guidelines that apply to all clients, or even to specific diagnostic categories or groups, such as men, Latinos, or Christians; rather, we will be trying to find case-specific recommendations. Let's examine further what this means.

Interpersonal process psychotherapy is a highly idiographic approach; it emphasizes the personal experience or subjective worldview of each individual client. Diagnostic categories and personality typologies are certainly useful, but they are only the map and not the territory. They highlight *potential* issues and directions that therapy might take, but the actual therapeutic process and issues covered will be client-specific. In other words, throughout each session, therapists are encouraged to try to find the subjective meaning that this particular experience holds for this particular client at this point in their lives. To respect and empathically enter into the client's *subjective worldview* is a challenging enterprise, because the meaning that this particular experience holds for the client often differs greatly from the meaning the same experience holds for the therapist or for other clients that the therapist might see. Of course, counselors are not expected to know or grasp all the nuances of a particular developmental or cultural experience. However, awareness of the potential importance of developmental and cultural experiences will lead therapists to encourage clients to explain their own subjective experiences more fully and to reveal their own particular worldviews. Therapists must be careful not to make assumptions. Rather, therapists would be well advised to listen to clients' language and choice of words, which will clarify their concerns as they experience them; and to seek to collaboratively explore clients' own metaphors, which can bring clarity to the clients' narratives. According to the interpersonal process approach, each individual client's developmental history, cultural context, and current life circumstances can serve to guide the therapist's treatment plans and intervention strategy. Thus, rather than advocating a particular therapeutic stance—directive, nondirective, neutral, or active—toward clients in general, the interpersonal process approach asks therapists to respond to the unique circumstances of each client and to provide the specific interpersonal experiences that this client needs in order to change.

To illustrate, consider a much debated question: Should therapists self-disclose to their clients? Client response specificity emphasizes that self-disclosure (or any other response) will hold very different meanings for different clients. In fact, the same response will often have exactly the opposite effect on two different clients with contrasting developmental histories and cultural contexts. For example, if a client's parent was distant or aloof, the therapist's judicious, well-timed self-disclosure may be very enabling for the client. In contrast, the same type of response early in treatment may be anxiety-arousing and counterproductive for a client who had been the confidant of a depressed parent or had an emotionally seductive caregiver.

Greater intimacy and sharing with the therapist may help the first client learn that he or she does matter and is of interest to other people. For the second client, greater intimacy may imply the imposition of the unwanted, threatening, or burdensome needs of others. Either response may work fine with a particular client. However, *therapists will be most effective if they first consider how the client's developmental history and relational templates may shape the impact of their interventions.*

To illustrate further, let's examine another question: Should therapists give opinions when clients ask for advice? As before, the therapist's response is informed by the client's current life circumstances. For example, offering advice may be counterproductive for a compliant client who is trying to become more assertive with a dominating, dependency-fostering spouse. For such a client, it may be better for the therapist to ask, "What do you think would be best to do?" In contrast, giving advice and directives may work well with clients whose parents were incapable of, or uninterested in, preparing them for aspects of adult life. To withhold information, advice, and opinions from these clients may only reenact their developmental conflict and impede progress. Once again, the same intervention or response from the therapist will have very different effects on different clients. From this metaperspective, techniques or interventions from any theoretical approach may help or hinder the client; it depends on whether the particular response serves to reenact or resolve core relational patterns for this particular client. Thus, in order to conceptualize client dynamics and guide treatment interventions, counselors will be given specific guidelines:

1. To clarify the specific relational patterns that originally created problems for the client
2. To anticipate how these problematic patterns and responses could be expressed or played out in the therapeutic relationship
3. On the basis of this awareness, to track the process dimension and provide a different and more satisfying relationship than the client has found in the past

Another aspect of client response specificity is that therapists can assess the effectiveness of their interventions by observing *how the client utilizes, or responds to, what they have just done.* For example, if the therapist observes that the client gets anxious and distant in response to the therapist's self-disclosure, the therapist is learning how to respond most effectively to this particular client. When the therapist provides a corrective response, many clients will make progress right away by feeling safer, acting more boldly, bringing forth relevant new material, becoming more engaged and available, and so forth. In contrast, when the therapist responds in a way that recapitulates problematic old patterns, clients will behaviorally inform the therapist of this by acting "weaker" in the next few minutes—for example, by being more distant, confused, or compliant (Weiss, 1993). In this way, we will see that, if the therapist tracks clients' immediate responses to various interventions (directing, challenging, self-disclosing, making process comments, advising, supporting, interpreting), and the therapist has the flexibility to adjust to clients' verbal and nonverbal responses, clients will behaviorally teach the therapist what they need and how best to respond.

To sum up, working in this highly idiographic way adds complexity to the therapeutic process and places more demands on the therapist. However, such an approach gives therapists the flexibility to respond to the specific needs and unique experiences of the diversity of clients that seek help. Client response specificity is a core concept in the interpersonal process approach, and we will develop it further in subsequent chapters. In particular, we will begin utilizing guidelines for keeping process notes (Appendix A) and writing case conceptualizations (Appendix B) to help therapists formulate more systematically the specific relational experiences that their clients need.

Model of Therapy

Drawing on some of the concepts introduced in this chapter, we now provide a brief overview of the therapeutic model outlined in detail in Chapters 2–10. For most clients with Axis I and Axis II disorders, certain basic developmental tasks were left unfinished in their family of origin. Although clients' problems are often complex and multiply determined, in many cases this unfinished business is reflected in the interpersonal coping strategies and characterological accommodations that clients have had to make in order to cope with insecure attachment histories. The purpose of therapy is to complete these developmental tasks by providing a corrective emotional experience within the therapeutic relationship. The therapist provides a relationship that enacts a resolution of the client's conflict rather than a repetition of it. Again, providing such corrective experiences is much easier in theory than in practice. As we will see in subsequent chapters, clients are often adept at engaging the therapist in the same conflict or maladaptive relational patterns that characterize their other relationships.

The client simultaneously seeks to avoid reexperiencing painful conflicts and to find a new and more satisfying response to them. To this end, *the client tries to assess safety and/or danger* in the therapeutic relationship. The client's relational problems with others are activated with the therapist and are especially likely to be played out along the process dimension. Throughout treatment, the client continues to assess, with different degrees of awareness, whether the therapist responds in a new and safer way than others have done or whether the therapist responds in the same problematic way. For example, if the client risks becoming stronger and begins successfully pursuing his or her own interests and goals, will the therapist take pleasure in the client's strength and success or feel threatened and competitive, as the client's parent used to feel? If the client reveals vulnerabilities and has emotional needs of the therapist, will the therapist respond by offering comfort, understanding, and acceptance or will the therapist instead feel overwhelmed and responsible for this sadness, as the client's parent used to feel?

Clients resolve their conflicts when the therapist's response repeatedly disconfirms their pathogenic developmental experiences. Through this corrective emotional experience, clients find that it is safe to act in new and more adaptive ways—at least with some people, sometimes. Clients can then begin exploring new aspects of themselves and the social world and begin the next phase of treatment: generalizing their experience of change to other relationships outside of therapy. We will see that clients can do this, in part, by figuring out how they can change their own responses to others in current relationships—even when significant others continue to respond in the old, problematic ways.

Therapists cannot simply tell clients that relationships different from those in the past are possible; they must show the client that this is so. Deeply ingrained relational templates that have fundamentally organized the client's relational world for decades will change only if the therapist and client confront the client's conflict, struggle with it, and jointly work through to its resolution in their real-life relationship. *Finding this interpersonal solution commonly entails conflict and anxiety for the therapist as well as for the client.* Thus, the power to effect enduring change does not come from reassurances, explanations, or directives but rather from the client's having lived out with the therapist the actual conflict and discovering that relationships can exist in another way. This experiential relearning is the basis of interpersonal process psychotherapy, which places real demands on the personhood of the therapist but holds the great reward of helping people change.

Limitations and Aims

Limitations   In this text, beginning therapists are introduced to the practice of interpersonally oriented psychotherapy with individual adult clients. This therapeutic approach is best suited to clients with long-standing problems in close interpersonal relationships, including issues of dependency, trust, control, and commitment. It is also directed toward resolving developmental ego conflicts that may include identity and self-esteem problems; concerns over initiative and adequacy; and accompanying symptoms of anxiety, depression, guilt, and shame.

Other types of therapeutic approaches are better suited for some clients, although the principles presented here can inform treatment guidelines for therapists working within any therapeutic modality. For example, many clients who function well need only supportive measures; these clients need help coping with situational crises that do not reflect broader problems in living or with problems that do not tap into other enduring personality conflicts. Behavioral approaches are the treatment of choice for clients who have circumscribed problems, such as an isolated phobia, or who are seeking only symptomatic relief from their problems. Still other clients will not be suited for the more intensive, interpersonal process approach because they are too impulsive or are psychotically disorganized. These clients may not have the *observing ego* necessary to explore their relational patterns and current interaction with the therapist, or may not be able to manage effectively the anxiety aroused by this approach.

Finally, some clients require only educational and skill development approaches; sex education, time management, assertion training, or parent training are among the treatments of choice for these clients.

In this text, we focus on the process dimension in psychotherapy and on how the therapist-client relationship can be used as a vehicle for change. This applied focus prevents us from addressing other important aspects of clinical training. Basic information about ethics, confidentiality, report writing, record keeping, informed consent, legal reporting, and other practical concerns is beyond the scope of this text, although guidelines for keeping process notes and writing case conceptualizations are provided in Appendixes A and B. Similarly, we do not address clinical interventions in many important situations, such as responding to client emergencies, managing suicidal crises, mandated reporting of abuse, and treatment complications when clients are required to attend therapy by a court or employer. However, the information presented here will be relevant to most of these treatment concerns.

It is essential for beginning therapists to learn how to work in different treatment modalities and in varying treatment lengths. Although many will ultimately choose to specialize in one modality, clinical trainees must develop the skills necessary for child, marital, group, and family therapy. While the treatment model presented here emphasizes individual adult treatment, the process orientation provides basic skills that are especially applicable to group and family treatment. (For instance, the first journal published in the new field of family therapy was titled *Family Process.*)

The therapeutic model presented here is designed for an intermediate treatment length of 6–15 months. This treatment length is not necessary for many clients and is not possible in many clinic and managed care settings. However, the treatment fundamentals presented here comprise the basic clinical skills necessary for briefer modalities, such as initial intakes, crisis intervention, and short-term therapy. Guidelines for working in these essential treatment lengths, and in other treatment modalities, will be provided at various points throughout the text.

In particular, the interpersonal process approach can readily be adapted to short-term and time-limited treatment modalities of 10–12 sessions. In recent years, interpersonal dynamic theorists have developed effective, time-limited treatment approaches that are grounded in a relational or process approach (Davanloo, 1980; Levenson, 1995; Luborsky & Marks, 1991; Malan, 1976; Mann & Goldman, 1982; Sifneos, 1987). Before they can succeed in these short-term models, however, beginning therapists must first acquire certain basic clinical skills and gain more experience working with the interpersonal model. For example, short-term work with this model requires therapists to "bring the conflict into the relationship" and to make overt how the therapeutic process may be reenacting aspects of the client's conflicts with others. Most beginning therapists are not ready to be so forthright and to address these relational reenactments directly with the client as quickly as time-limited approaches demand. Also, short-term treatment approaches require therapists to conceptualize clients' dynamics, and establish a treatment focus, by identifying or assessing *core conflictual relational patterns* more quickly and more accurately than most beginning therapists are able to do. Furthermore, most beginning therapists will be unable to manage effectively the intense emotions and negative transference reactions that are commonly evoked in clients in the closing sessions of time-limited treatments. However, some relevant guidelines to working in this modality will be provided where appropriate. Interested readers are also encouraged to examine other excellent short-term, interpersonal treatment approaches to depression (Klerman, Weissman, Rounsaville, & Chevron, 1984), personality disorders (Benjamin, 1995; Horowitz et al., 1984), and psychotherapy (Strupp & Binder, 1984).

Finally, medical interventions will be an important aspect of treatment with some clients. Temperamental and biological factors do play a causative role in some disorders, and medications will be a necessary and highly facilitative accompaniment to treatment at times. Once again, such interventions fall outside the scope of the present book, whose purpose is to present only the essential elements of interpersonal process psychotherapy to clinical trainees.

Aims The interpersonal process approach presented here emphasizes three broad aims that are essential to helping clients change. First, therapists must try to establish a significant emotional relationship with their clients. In addition to the relationship issues discussed earlier, this means that therapists must maintain their own internal commitment to helping clients change. That commitment is especially important during periods of intense client resistance, ambivalence, and negative transference toward the therapist. Conceptions of therapeutic change often vary on a continuum. At one end, discrete, problem-solving techniques are applied. At the other, there are vague allusions to "the relationship," which is somehow supposed to produce change, although how or why is not clear. The interpersonal process approach presented here offers a middle ground between these two extremes. Change *is* predicated on the nature of the therapist-client relationship, and specific guidelines are provided to show how the therapeutic relationship can be utilized to effect change.

Second, in order to effect change, therapists must be able to respond effectively to the client's emotions and inner life. Therapy is a private and intimate sharing in which therapists respond to clients' pain and help them understand their conflicted emotions. Making contact with clients in these very personal ways is a critical aspect of therapy, yet it is the dimension along which therapists are most likely to stumble. Most therapists, like other people reared in our culture, have been socialized to avoid rather than approach strong feelings such as anger, despair, fear, and especially shame. In addition, it is often difficult for therapists to respond to clients' emotions because they arouse the therapist's own conflicted feelings as well. The most effective way for therapists to facilitate change, however, is to help clients integrate emotional reactions that they have not been able to resolve on their own. Thus, specific guidelines will also be provided to help beginning therapists understand and respond to their clients' conflicted emotions.

The third component of effective therapy is to conceptualize the client's personality and problems and to formulate what experiences in therapy the client needs in order to change. This conceptualization should enable the therapist to identify the client's central conflict, to recognize the maladaptive coping strategies that have kept the client from changing, and to provide direction for the ongoing course of treatment. Typically, beginning therapists' least developed skill is their ability to formulate a useful conceptualization of the client's personality and problems. (Specific guidelines for writing case conceptualizations and treatment plans within an interpersonal process framework are provided in Appendixes A and B.) However, the most effective therapists are as adept in the conceptual domain as in the affective and relational domains. Thus, this text aims to facilitate the therapist's effectiveness in all three areas.

Readers of this text will also gain an overview of the nature and course of a therapeutic relationship. Therapeutic relationships usually have a coherent life course and follow a predictable developmental pattern from beginning to end. Although variations and exceptions occur with every client, therapists can often identify an ordered sequence of stages. At each successive stage of therapy, the therapist must negotiate certain therapeutic tasks with the client. If the therapist achieves this, the next set of issues and concerns will often emerge from the client. An awareness of this developmental structure will help the therapist to formulate intervention goals and strategies for each successive stage. Chapters 2–10 are organized to reflect this developmental schema and to parallel the course of therapy from beginning to end.

Finally, readers of this book must have realistic expectations. Beginning therapists who are seeing their first clients will find much to help them in their initial work with clients. However, more information is presented here than a novice clinician can fully integrate and apply. The concepts embodied in the interpersonal process approach are complex and challenging. Beginning therapists are advised to incorporate these concepts into their practice gradually, at their own pace. Some may begin to feel that the more they learn, the less they know. With more experience, however, second-year students will be able to employ many of these concepts successfully. Typically, it takes about three years before clinical trainees have the confidence to employ these concepts routinely and to respond to the powerful effects the interventions generate. Since it will take some time to learn how to work in this way, the best approach is to be patient and enjoy the learning.

With these general considerations in mind, we can now examine more specifically the first stage of therapy: establishing a collaborative relationship.

Suggestions for Further Reading

1. One of the most useful books for trainees interested in the interpersonal process approach is Michael Kahn's *Between Therapist and Client* (New York: Freeman, 1997), which provides an informative historical overview of the major concepts and theorists of the therapist-client relationship. In particular, Kahn eloquently articulates the cardinal issue of *nondefensiveness* in the face of clients' attempts to pull the therapist into their own relational reenactments. Repeatedly, Kahn succinctly expresses complex relational issues. This illuminating book deserves to be read and reread by beginning and experienced therapists alike.

2. Chapters 3 and 5 of Salvadore Minuchin's *Families and Family Therapy* (Cambridge, MA: Harvard University Press, 1974) provide a succinct presentation of structural family relations and family developmental processes. This discussion will help therapists who work with individual clients to understand the genesis and familial context of their clients' problems.

3. A highly readable application of object relations and attachment theory to psychotherapy can be found in Chapters 6–9 of John Bowlby's book *A Secure Base* (New York: Basic Books, 1988). An excellent conceptual overview of object relations theory may be found in *Object Relations in Psychoanalysis* by J. Greenberg and S. Mitchell (New York: Basic Books, 1984).

4. Readers interested in the large body of empirical research on interpersonal theory and a history of seminal ideas in the field should examine *Handbook of Interpersonal Psychotherapy*, edited by J. Anchin and D. Kiesler (New York: Pergamon, 1982). See especially Chapters 1 (pp. 14–20), 2, and 15. A summary of research findings on process variables in psychotherapy and how they are related to treatment outcome is provided in Chapters 4–6 of F. Walborn's *Process Variables* (Pacific Grove, CA: Brooks/Cole, 1996).

5. The great novelists best bring to life the profound impact of childhood and familial experience on adult personality. See, in particular, John Steinbeck's *East of Eden* (New York: Viking, 1952) and Franz Kafka's *Letter to His Father* (New York: Shocken, 1966).

PART TWO

Responding to Clients

CHAPTER TWO

Establishing a Collaborative Relationship

Conceptual Overview

Psychotherapy is a profession based on trust. Clients enter therapy with a need: They are in pain and asking for help with something they have not been able to resolve on their own. A trustworthy response is to honor the need in clients' requests for help and to respond compassionately to their pain. Yet to be most effective, the therapist must also respond in a way that helps clients achieve a greater sense of their own personal *efficacy*; the goal is not only to resolve specific situational problems, but to do so in a way that leaves clients more aware of their own competence and mastery. In this way, the interpersonal process approach strives to foster personal growth as well as to provide symptom relief.

Clients cannot resolve problems and achieve a greater sense of their own personal power in a hierarchical or one-up/one-down therapeutic relationship. Clients need to share ownership of the change process and must be active, egalitarian participants when working with the therapist, rather than being passively "cured" or told what to do. Thus, this chapter presents a model for a *collaborative relationship* that accepts the client's need for comfort, understanding, and guidance, but equally encourages the client's own initiative and autonomy.

Chapter Organization

One of the most important concerns for beginning counselors is whether their clients will return after the first session and remain in treatment. Many clients do indeed drop out after the first session; such experiences painfully exacerbate beginning counselors' concerns about their own adequacy and performance. This chapter, and the one to follow, provide basic guidelines for conducting a successful initial session.

This chapter is divided into two sections. The first introduces the concept of a collaborative relationship and then elaborates this cardinal concept in terms of a balance between directiveness and nondirectiveness in the therapeutic relationship. Sample therapist-client dialogues will be used to illustrate how the therapist can begin to establish a collaborative working alliance in the initial session.

The second section presents ways to foster the therapeutic alliance and further engage the client in treatment. Principally, these goals are achieved by listening with presence, empathically entering the client's subjective worldview, and understanding or grasping the central meanings in what the client is saying. The therapist can better *understand* the client's experience by (1) clarifying the *repetitive relational themes* that recur throughout the different stories and vignettes the client presents; (2) identifying the *pathogenic beliefs, automatic thoughts,* and *problematic expectations* that provide unifying links for the client's varying problems and concerns; and (3) approaching the client's *core conflicted feelings* in a direct but caring way. Following the central theme of this book, both sections will focus on interpersonal process characteristics of the therapist-client relationship, including how these three themes are replayed in the therapist-client relationship and how they can be used to facilitate change.

A Collaborative Relationship

At each successive stage of treatment, therapists have a different overarching goal to guide their interventions. In the first stage, the therapist's principal goal is to establish a *collaborative relationship* or *working alliance* with the client. Ralph Greenson (1967) wrote lucidly about the therapeutic relationship and identified three interrelated components in all counseling relationships: the real relationship; the transference configuration; and, most importantly, the working alliance. A collaborative alliance is established when the client perceives the therapist as a powerful, trustworthy, and committed ally in his or her personal struggles. In order to "join" with clients and establish such a working alliance, the therapist must be able to *be* somebody to clients. The therapist becomes somebody for clients when he or she can successfully communicate that the therapist:

- Sees their predicament and recognizes their distress
- Feels with them and is moved or touched by their pain
- Is on their side and has their best interests at heart
- Has an abiding commitment to help them through this predicament

Although the concept of the collaborative alliance was originally developed by psychoanalytic theorists, it can also be described in many other theoretical contexts. For example, it is akin to Bowlby's *holding environment* in object relations or attachment terms, where the client's need or distress is emotionally "held" or contained in the safe relational envelope that the therapist's compassionate understanding provides. In existential therapy, Rollo May (1977) elucidates the therapist's *presence* with the client; he emphasizes that the therapist must be fully present and intensely involved with the client in an authentic manner throughout each session. Most importantly, the collaborative alliance is closely linked to Carl Rogers's core conditions of *empathy, genuineness,* and *warmth* (Rogers, 1981).

To sum up, the collaborative alliance entails a generous and affirming responsiveness from the therapist and a clear invitation to join together and work collaboratively as partners on the client's problems. Researchers find that the therapist's ability to establish a working alliance in the initial sessions strongly predicts successful outcomes in both short-term and longer-term treatment (Gelso & Carter, 1985). Thus, the therapist's primary aim in the initial sessions is to join with the client in these ways; to articulate clear expectations for working in a collaborative manner and, most importantly, to behaviorally enact these spoken expectations by giving the client the *experience* of working together on his or her problems. If a collaborative relationship is established and maintained throughout treatment, this interpersonal process will go a long way toward helping clients resolve their presenting problems and will meet our broader goal of helping clients achieve a greater sense of their own mastery and competence as well.

Balancing Directive and Nondirective Initiatives

At the outset, the therapist should explore clients' expectations regarding their relationship with the therapist and how change will occur. Perhaps the most widely held misconception among clients is that the therapist is a doctor who will prescribe their route to mental health. The therapist is seen as the sole agent responsible for change—via advice, explanations, interpretations, and simply telling the client what to do. Therapists need to explain to clients that the therapeutic relationship is similar to the medical model, but also differs from it. One similarity is that the therapist is trained to work with problems such as those the client is presenting. An important difference is that the clients must be active participants in their treatment planning—for example, by describing symptoms, identifying what's been helpful and what has not, and agreeing with the therapist on a particular course of treatment. Therapists may also explain that clients' participation enhances the outcome of treatment and the clients' sense of efficacy and personal power. Unless the therapist attends to the process dimension and takes corrective action as necessary, a hierarchical doctor-patient or teacher-student relationship is likely to develop under pressure of clients' expectations.

Such a one-up/one-down relationship brings with it a number of problems. As we would expect from client response specificity, a doctor-patient relationship will work well with some clients in the short run. Clients who expect to be compliant in close relationships will be content to follow the therapist's lead at the beginning of therapy. However, this *compliance* will ultimately evoke shame and anger, which, although often unrecognized by the clients, will prevent them from utilizing the therapist's help and making progress later in therapy. Moreover, clients remain dependent on the therapist in this hierarchical model. They will not be able to gain a greater sense of their own personal power as long as they hold the dysfunctional belief that the source of potency resides in the therapist, rather than in themselves. Let's examine this further.

Is the interpersonal process approach a nondirective approach to treatment? Not at all. Some clients bring relational templates to the treatment setting that do not encompass collaborative relationships or mutuality, and some of these clients will insist that the therapist assume a directive or leadership role. If clients bring to therapy an authoritarian upbringing and hierarchical internal working models, it would most likely be fruitless to insist that they go along with a more egalitarian relational process, which they have not experienced in the past and cannot yet encompass. Instead, the therapist can accept the client's request and simply provide a more directive or advice-giving stance. At the same time, however, the therapist must start the change process by giving the client permission to be more initiating whenever he or she would like, and by tracking the process that results from this hierarchical interaction—that is, watching for opportunities when the client seems to be rejecting the directives they have elicited from the therapist (by saying, for example, "Yes, I'll try that, but . . . ").

Whereas a directive stance may work in the short run, a purely nondirective approach will often sputter right from the start. Typically, clients feel frustrated if their requests for help or direction are continuously reflected back upon them. A negative cycle may ensue: The client becomes increasingly angry and demands direction from the therapist, who further eschews this role and talks about inner direction and finding one's own answers. This, in turn, further frustrates the client, who may not feel that he or she has the answer to anything at that moment, and only sees the therapist's attempts to be nondirective as manipulative and evasive. Unfortunately, many clients do indeed drop out of therapy prematurely because the therapist was too quiet, not engaging, or unwilling to accept their realistic requests to be more actively involved, responsive, and informative. Therapists need to have a genuine presence in the relationship and to provide feedback about the relational and cognitive patterns they observe in the client's life, empathic responsiveness to the client's feelings, and process comments or observations about what is occurring in their current interaction. Beginning therapists who lose clients through lack of responsiveness often report feeling afraid of making a mistake or not knowing what to do. In addition, therapists' inactivity is often prompted by concerns about acting more strongly or expressing their own thoughts and observations and by apprehension at the responsibility implied in becoming important to the client and having an impact on his or her life.

Thus, clients do need advice and direction from an expert, and the therapist should provide guidance for how to proceed. However, the therapist should try to provide this structure and direction without falling back into a counterproductive authoritarian role. How can such a structured but nonauthoritarian relationship be achieved? In the initial interview, the therapist must make an overt bid to establish a collaborative relationship in which the therapist and client work together to resolve the client's problems. For example, this attitude may be conveyed by the following overture*:

Therapist:
Let's work together to figure out what's been wrong. Tell me your ideas about what the problem has been. I'd also like to know where you think is the best place to begin and what you'd most like to change or get help with.

The therapist is trying to communicate in words and actions that he or she will be an active and responsive ally. It is clear from this invitation, however, that the client is not going to be a passive recipient, but an active participant with the therapist. The therapist must follow through on this invitation behaviorally, by enacting the therapeutic process as a collaborative enterprise, in which the client contributes valuable information, abilities, and resources. This establishes a new, middle ground of shared control and responsibility, between the extremes of directively taking charge or merely following the client's lead nondirectively. For many clients, this will provide a corrective relational process that they have not experienced in other close relationships, and will be an essential aspect of helping them change. *Whether the therapist and client can continue this process of mutual collaboration will be one of the most important determinants of the eventual outcome of therapy.* Let us look now at some specific ways to establish a collaborative relationship in the initial therapy session.

Beginning the Initial Interview

Therapists must begin to establish a collaborative relationship in their initial contact with the client. The therapist structures the session by providing the client with guidelines and direction for what is going to occur in the interview. However,

* Throughout this book, sample dialogues are provided to illustrate the concepts presented. Beginning therapists should not use these exact words (or try to say what a supervisor might say in a similar situation) but should find their own ways to express these concepts. Beginning therapists will lose their creativity with their clients and their own sense of personal efficacy when they are merely trying to mimic someone else.

it is essential that the treatment focus reflect the client's own wishes and that the treatment goals be experienced by the client as his or her own. In other words, the therapist is trying to help clients articulate their own goals and agenda, and then collaboratively join clients in addressing their concerns. As suggested before, this can be done simply and briefly by offering clients an open-ended invitation to talk about what brings them to therapy.

Therapist:

> I'm looking forward to working together with you on the problems that have brought you in. First, I'd like to learn more about the concerns you have. What do you see as the difficulty?

OR

Therapist:

> Why don't you help me understand what's wrong in your life or hurting you right now. It might be difficult to describe all the issues at first, but I think we can work through this together. Tell me what you think we should start with—or what concerns you feel are most pressing—and what you think needs to change.

This type of inquiry communicates several important things to the client. At the simplest level, it ends the opening phase of social interaction that occurs as the therapist and client are introduced and walk to the interview room. More importantly, it tells clients that the therapist is someone who is willing to talk directly about their personal problems—as they experience or perceive them— and is ready to respond to their need for help by listening. It does so, however, in a way that still gives clients the freedom to choose where they want to start and leaves them in charge of how much they want to disclose. From the outset, clients are sharing control of the interview by choosing what *they* want to talk about, but the therapist is an active participant who has offered some direction for where they are heading. This type of collaborative alliance does not occur if the therapist gives the client a more specific cue.

Therapist:

> When we talked on the telephone you said you were having trouble with your boss. What is the problem there?

OR

Therapist:

> You've been having anxiety symptoms. How long has this been going on, and how severe have they been?

The difference between these two openings may seem insignificant. However, it is important to communicate that clients should talk about what they want to talk about and not feel that they have to follow the therapist's interest or agenda. From the start, we want the client to take an active role in directing the course of therapy, while still feeling that the therapist is participating as a supportive ally. *Enacting this process dimension in the initial session is more important than the content of what is discussed.*

In most cases, clients will readily accept the therapist's open-ended invitation and begin to share their concerns. The therapist can then follow the client's lead and begin to learn more about this person and his or her problems. Therapy is underway when this occurs. Before we go on to the next step, however, we must examine two exceptions in which the client does not accept the therapist's offer to begin: when another therapist has conducted an initial screening interview and when the client has conflicts over initiating.

Previous screening interview   If another therapist has previously conducted an intake interview, the client may not be so ready to begin.

Client: *(impatiently)*
> I've already been through all of this in the intake with Dr. Smith. Do I have to go over it all again just for you?

OR

Client:
> I don't know how much you already know about me. What has Dr. Smith told you about me?

Our initial goal is to develop a working alliance between the therapist and the client. In these two examples, the previous intake therapist is a third party who is psychologically still in the room with them. This is especially problematic for those clients who have come from families in which a third person was *triangulated* into every two-person relationship (Bowen, 1966); that is, whenever two people were close, or in conflict, a third family member would be drawn in and would disrupt the dyad (Haley, 1967). To keep therapy from recapitulating this dysfunctional family interaction, the therapist and client must begin their own relationship as a stable dyad. It is always important to maintain a dyadic therapeutic relationship that does not allow others to disrupt the therapeutic alliance.

Therapist:
> I know that you have already spoken with Dr. Smith, and I have learned a little bit about you from his intake notes. But just you and I are going to work together from now on, and I'd like to hear about you in your own words. It may be a little repetitious for you, but this way we can begin together at the same point.

Most clients appreciate this offer for a dyadic relationship and will begin to share their concerns with the therapist. However, triangulation may also occur in the initial therapy sessions because of the unseen presence of the beginning therapist's clinical supervisor. Through insecurity, compliance issues, or other factors, beginning therapists may overtly triangulate their supervisors into the therapeutic dyad; for example, the therapist might tell the client, "I'll have to ask my supervisor about that." More commonly, beginning therapists may silently invoke their supervisors—by wondering what they would say or do at a particular point in the session or how they would evaluate the therapist at that moment. This worry or self-critical monitoring does not facilitate therapy; it diminishes the therapist's engagement and presence with the client. It also disempowers therapists by preventing them from being themselves in the session, from finding their own words and acting on their own perceptions, and from developing their own personal styles. Beginning therapists are encouraged to trust themselves enough to simply be a person sharing the story of another person. Supervisory input and evaluation, while necessary, will be utilized more productively in most cases when it is processed outside of the therapy hour. The therapeutic process is awry when the dyadic relationship between the therapist and the client is broken by third-party interference.

Conflicts over initiating   There is another common circumstance in which the client does not respond to the therapist's initial request to begin. The central conflict or problem that some clients bring to treatment involves issues of leading, initiating, or accepting responsibility. This type of client cannot begin at the therapist's request. By asking the client to begin, the therapist has inadvertently presented the client with his or her central conflict. It is difficult for this type of client to decide what to talk about or to assume any responsibility for the course of therapy. Thus, a therapist who nondirectively sits and waits for the client to lead places an impossible demand on this client.

On the other hand, a directive therapist who begins the session by telling the client what to talk about often merely reenacts the problematic relational scenario. Thus, therapy stalls right from the start if the client has problems with initiative and the therapist responds in either a directive or nondirective manner. What can the therapist do in this case? One way to find the more effective middle ground between directive and nondirective approaches is to make a *process comment*, as introduced in Chapter 1. An effective intervention is simply to describe, ask about, or make overt what is occurring between the therapist and client at that moment.

Therapist:
It seems to be hard for you to get started. Maybe we can begin right there. Is it often hard for you to begin, or is there something about this situation in particular that is difficult for you?

By first identifying this as a problem and then encouraging the client to explore it, the therapist has offered the client a focus and helped the client move forward. However, the therapist has provided this focus without taking over and telling the client what to do, which would only recapitulate the client's conflict. The therapist's open-ended inquiry is supportive, in that it responds to the client's immediate concern. Yet the client can take this issue of initiating wherever he or she wants, and can share responsibility for the course of treatment. This type of response provides a new opportunity for clients to explore their problems in a supportive environment. Our first goal is met as the client *experiences* a collaborative interaction with the therapist, rather than merely having a conversation about "the need to work together." Such a collaborative experience sets important expectations for the future course of therapy.

Finally, age, class, gender, and cultural issues are involved in initiating therapy. For example, the beginning therapist needs to be sensitive to the fact that, in some cultures, it may be seen as disrespectful to presume to lead with one's elders, an educated person, or an authority figure such as a therapist. A person of color working with a white therapist has probably experienced racism and may be suspicious or hesitant to begin this relationship. Therapists can best respond to such concerns by acknowledging differences and inviting clients to express or explore the concerns together. In the next chapter, we provide more specific guidelines to help beginning therapists respond to cultural issues that emerge in the initial stages of the therapeutic relationship.

Understanding the Client

The therapist has begun the session by giving the client an open-ended invitation to talk about whatever feels most important. Now imagine that the client has begun to share his or her concerns with the therapist and is clarifying the background and context of his or her problems. The therapist must find the subjective meaning that each successive story or vignette holds for the client. It is critical for the therapist to grasp what is most relevant or central to the client *from the client's point of view* (Kelly, 1963). In other words, therapists must have the cognitive flexibility to decenter, enter into the client's subjective experience or worldview, and try to recognize what is most significant to the client and why this particular issue holds the meaning it does for this particular client (McClure & Teyber, 1996). The client will feel *understood* by the therapist when the therapist can find a common meaning that links together the client's concerns and/or distills the central feeling in his or her experience. When this occurs, it facilitates the working alliance: Clients become more interested in exploring their problems and more engaged with the therapist, and they are encouraged to reveal more about themselves and their problems. Let's examine this more closely.

As used here, *understanding* connotes warmth and a feeling of concern for clients. It signals to clients that their situation is seen or grasped and their distress really matters to the therapist. As we will see, when therapists can capture and express this understanding to the client, they are providing Bowlby's secure base or holding environment to *contain* the client's distress. Although this is not sufficient to resolve clients' problems, such understanding will often ease clients' initial distress and, for some, their presenting symptoms may somewhat abate. A secure attachment configuration is established, and the entire family of attachment affects may be temporarily relieved by the therapist's *attuned responsiveness:*

- Anxiety may diminish from this reassuring emotional contact.
- Depression may lessen; the experience of being seen and accepted, rather than dismissed or critically judged, allows the client to feel hope.
- Anger from frustration over not being seen or feeling invalidated by others may be assuaged by the therapist's affirmation.
- Shame over having needs revealed by entering therapy and asking for help may be lessened by the therapist's caring respect.

The sections that follow examine further what it means to *understand* the client's experience. We will see why it is important to acknowledge and affirm the client's subjective experiences; explore what therapists can do to better understand the client; examine how process comments can be used to clarify misunderstandings and miscommunications; and see how beginning therapists' own performance anxieties may make it difficult for them to listen to clients with presence.

Clients Do Not Feel Understood or Affirmed

Most clients are concerned that others do not really listen to them, take them seriously, or understand what they are saying. Clients often describe themselves as feeling invisible, alone, strange, or unimportant. *Many clients feel this way because their subjective experience was not validated or acknowledged in their family of origin.* While growing up, most clients repeatedly received messages that denied their feelings and invalidated their experience.

Why would a silly thing like that make you mad?
You can't possibly be hungry now.
I'm cold. Put your sweater on.
How can you be tired? You've hardly done anything.
You shouldn't be upset at your mother. She loves you very much.
You don't really want to do that.
You shouldn't feel that way.
You musn't be like that.

Further, family members often changed the topic or simply did not respond when the client expressed a feeling, concern, or interest. One of the most effective ways therapists can help their clients change is to validate and affirm their subjective experience. R. D. Laing goes so far as to suggest that people stop feeling "crazy" when their subjective experience is validated (Laing & Esterson, 1970).

Consistent invalidation, disconfirmation, or mystification in the client's family of origin has profound, long-lasting consequences. In its extreme forms, some authors describe it as "soul murder" because clients lose themselves when they lose the validity of their own experience (Schatzman, 1973). Whereas invalidation has occurred to some extent in most clients' families, victims of emotional, physical, and sexual abuse have experienced extreme levels of invalidation. Disempowerment is one of the most serious consequences of such systematic invalidation. When feelings and perceptions are denied repeatedly throughout development, clients are incapable of setting firm limits with others, saying no, and refusing to go along with what does not feel right to them.

Consistent denial of their experience leaves clients unsure of what has actually happened to them and of the subjective meaning that events hold. They no longer trust their own perceptions of what may be making them uncomfortable. Denying the validity of their own experience, these clients characteristically say to themselves, "Oh, that didn't really happen" or, "I'm just exaggerating this; it wasn't that bad," when it actually was significant. Further, they often cannot find words to communicate their experience and, even if they could, do not expect others to respect or understand what they say.

When their subjective experience has been denied repeatedly, as commonly occurs in alcoholic, abusive, or highly authoritarian families, clients do not know what they are feeling, what they like or value, or what they want to do. In place of clear feelings and confident perceptions, a vague, painful feeling of internal dissonance results. Fortunately, this undifferentiated, dissonant feeling state can be replaced with emotional clarity and an accompanying sense of greater personal power if the therapist consistently listens to, accepts, and validates clients' experience. To affirm clients and support their efficacy, therapists listen intently, take seriously whatever matters to the clients, and communicate understanding and acceptance of what clients say. This type of empathy and validation may sound like a simple and common human response. However, as we have seen, many clients have not had their feelings and perceptions acknowledged in current and past relationships. Thus, *therapists must validate the client's experience by identifying and articulating the central meaning that this particular experience seems to hold for the client.* For example, in the following questions and comments, the therapist is trying to capture the central meaning or issue in what the client has just said.

It didn't seem fair to you.
I'm wondering if you were frightened when he did that?
It's been too much for you, more than you can stand.
It felt great to be so effective and in charge!
It was disappointing; you wanted more than that.
Here again, are you having to take care of everyone else?

The therapist does not have to be an exceptionally insightful or perceptive person to understand the client's experience, and therapists need to know they will not be and cannot be accurate all the time. Usually they will be right often enough to form an empathic bond with their client; Winnicott (1965) referred to this as "good enough mothering." When the therapist does not understand what clients are saying, however, the therapist should not feign understanding by saying "uh-huh," "yes," or "OK." Instead, the therapist should acknowledge his or her confusion and/or ask the client to say something in a different way, so that the therapist can understand and they can get back on track. Further, if therapists feel they have been inaccurate, they can simply check out their perceptions and invite the client to clarify them. For example, the therapist replies: "As I listen to you, it seems as if you feel so hopeless that you just want to give up. Is that how it is for you?" This invitation for a dialogue takes the performance pressure of having to be right off the therapist, diminishes unwanted concerns about mistakes, and furthers the collaborative alliance. Even though the therapist's perceptions will be inaccurate at times, the therapist's sincere efforts to understand clients' experience and ask for true communication will often make clients feel cared for.

In most instances, a primary working goal for the therapist is to provide validation throughout each session by grasping the client's core messages and affirming the central meaning in what the client relays. To appreciate how important validation is, beginning therapists may reflect back on what others have done to help them during their own crisis periods. Almost universally, helpful responses include an empathic understanding and validation of one's feelings and perceptions.

Providing validation is particularly important when working with people of color, gay men and lesbians, economically disadvantaged clients, and others who feel "different." These clients will bring issues of oppression, prejudice, self-hate, and injustice into the therapeutic process, and their personal experiences have often been invalidated by the dominant culture. These clients, in particular, will not expect to be heard or understood by the therapist. The first step in working with all people is to listen empathically and hear what is important to them. Therapists respect the personal meaning that experiences hold for different people when they enter into and affirm the client's subjective experience. The most effective therapists, of any theoretical orientation, offer these basic human responses profoundly well. Pioneering clinicians as varied as Beck, Bowlby, Kohut, Rogers, Satir, Wolpe, and Yalom all share this genuine concern and respect for the client's subjective experience, and emphasize the need for therapists to empathically and affirmingly enter the client's worldview.

Demonstrating Understanding

To engage a client in a working relationship, the therapist must listen to the client's experience, find the feeling and meaning that the stories hold for the client, and accurately reflect back what is most significant in the client's experience. An accurately empathic reflection that captures the meaning of the client's experience is a complex and significant intervention. It is never just a rote parroting of what the client has said. An effective reflection is more akin to an interpretation; it goes beyond what the client has said and communicates that the therapist understands the core message, emotional meaning, or what is most important in the client's experience. Rogers (1951) believes that communicating such understanding helps provide a deep acceptance that is a prerequisite for any meaningful change. Beginning therapists will find that they achieve credibility with their clients and become very important to them when they *demonstrate* their understanding in this tangible way.

According to one common stereotype, a therapist is someone who says, "I hear you. I know just what you mean. I understand completely." However, this type of global, undifferentiated response is not effective and, paradoxically, often furthers the client's sense of never being seen, heard, or understood. The therapist does not simply say, "I understand" but demonstrates that understanding by *articulating* the central meaning of what the client has said. Therapists show the client that they can be helpful when they capture and express the specific meaning in the client's experience, rather than offering well-intended but vague reassurances.

How can the beginning therapist put these ideas into practice? As an illustration, we now examine a brief case study of a client who initially had the experience of not being heard by her therapist but later felt understood with a second therapist.*

While growing up, Marsha did not feel heard or understood by her parents. Her father was distant. He was not comfortable talking with his adolescent daughter and believed his wife should handle the children. Her mother was critical, demanding, and intrusive. Whenever her mother felt or believed something, she demanded that her children see it the same way. For example, if Marsha felt something that her mother did not feel, her mother would angrily respond, "That's ridiculous. What's wrong with you?" As an adolescent, Marsha felt sad, lonely, and afraid. She had no idea why she felt this way, however, and often thought, "There's just something wrong with me." Marsha frequently cried alone in her room.

Marsha had always thought that everything would get better when she went away to college. To her great dismay, though, she found herself depressed during her first semester. She could not stop crying and started gaining weight. More confused about herself than ever, Marsha began seeing a counselor at the Student Counseling Center. Although she did not really have words for what was wrong, Marsha tried to help her counselor understand her problems:

---

*The clinical examples in this text are based on actual cases, but identifying information, including the gender of the client or therapist, has been altered to ensure confidentiality.

Marsha:

I feel empty—kind of lonely.

Therapist:

Do you have any friends in your life? What are your peers like?

Marsha:

I guess I have friends. I have a roommate in the dorm.

Therapist:

What do you do with your friends—go to the movies, shopping?

Marsha:

Yeah, I do those things. I belong to a foreign language club, too.

Therapist:

Do you like your friends? You're new to the university; maybe you need some new friends here at school.

Marsha:

Well, I've had friends, but I just feel empty.

Therapist:

But you've just left your family and come to college. You must miss your family and feel lonely. It's natural to feel lonely when you move away from home. Most of the other kids in the dorm feel that way, too. I know I sure did when I moved away to college.

Marsha:

Oh.

Therapist:

This is not unusual at all. You're going to be just fine.

Marsha:

I hope so. Maybe my family is different, though. In high school, I always thought that my family had more problems than my friends' did.

Therapist:

Yeah, but, like I said, it's real natural for you to be kind of emotional at this time in your life.

Marsha:

It is? But I still feel different from everybody else.

Therapist:

Sure, it's real natural for you to feel different from everybody else. Late adolescence, moving away to college—it's a tough time in life. You're going to be just fine. Do you have a boyfriend?

Marsha:

Yes. I try to tell him what's wrong, but it doesn't work. And then he gets angry at me because nothing helps. I just don't know why I get depressed a lot. My mother used to yell at me a lot, though.

Therapist:

Do you have trouble eating or sleeping?

Marsha:

No trouble eating; I've gained ten pounds. And I wake up at night crying sometimes.

Therapist:

Are you eating alone? Maybe you should be eating with friends.

Marsha: (*with resignation*)

Maybe that would help.

Marsha could not say exactly why, but she did not like seeing the counselor and she did not go back. She became even more depressed as the semester went on, wondered what was wrong with her, and struggled in her classes. During advising for second semester, a concerned professor saw her distress and talked her into going back and trying another counselor. Reluctantly, Marsha agreed.

Therapy began more slowly this time. Marsha found herself angry at the therapist and reluctant to share much. She alternately acted blasé and then distressed, but would never allow the therapist to stay with her experience for very long. The therapist was not frustrated, however, and was effective at communicating his continuing interest and concern for her. After a few weeks, Marsha sensed that this relationship might be safe enough to risk disclosing her feelings again.

Marsha:

I feel empty.

Therapist:

What's it like to feel empty?

Marsha:

I just feel empty.

Therapist:

Show me in some way what that emptiness feels like.

Marsha: (*long pause; nothing is said*)

Therapist:

Maybe you can use just adjectives, or make a sound or a gesture that will describe the emptiness.

Marsha:

There's just an emptiness inside. The wind blows right through me. I'm always hungry. It never gets filled up.

Therapist:

That's right. You just feel empty inside. Sex, alcohol, food—nothing fills it up. The hole just stays there.

Marsha: (*doesn't speak; nods and becomes teary; looks at therapist*)

Therapist: (*holding her gaze kindly*)

I feel for you. I can see how much you are hurting. It's very sad that things have happened to make you feel this empty and alone.

Marsha: (*cries and nods again*)

There's something wrong with me.

Therapist:

Tell me about it; help me understand what's wrong inside.

This time, Marsha was heard and felt understood: she was not crying alone in her room anymore. Consistently, in subsequent sessions, the therapist continued to listen and respond to what was most important for Marsha. Although there were ups and downs, a collaborative working relationship had begun, and Marsha ultimately went on to resolve her depression and successfully complete therapy.

The concept of understanding the client sounds simple enough. All the therapist has to do is listen carefully to what clients say, grasp the underlying feelings, and find a way to communicate to clients that their feelings or experiences are shared and appreciated. However, this is not so easy for most beginning therapists. Most of us have been strongly socialized to "hear" in a limited, superficial way that denies the emotional or conflicted messages embedded in clients' remarks. This was why Marsha's first counselor could not hear what she was truly saying and avoided the difficult emotions involved. Although the overt content of what the counselor said was apt enough, the process he enacted with her was problematic. By reassuring her and trying to talk her out of her feelings, he metaphorically reenacted her developmental conflict of not being seen or heard.

Most beginning therapists are aware that they possess a "third ear"—that they already have a highly developed ability to hear the basic meaning in what people say. However, they often feel they must avoid acknowledging the true content of these underlying, and often nonverbal, messages because of their strong emotional content. Therapists want to avoid or quickly close these emotional responses for many reasons: feeling responsible for causing a client's emotional wound; feeling unprepared or inadequate to respond to a strong emotional expression; identifying with or sharing the client's difficult affect; feeling reluctant to violate cultural proscriptions; or, perhaps most common for those entering the helping professions, feeling an unspoken need to take care of others or protect them from experiencing their own pain or discomfort. As a result, many beginning therapists have a habit of switching automatically to a more superficial level to avoid the painful material that emerges from empathic listening.

Marsha's second counselor broke the social rules, however, and responded to the central affective meaning in her message, "I feel empty." As a result, Marsha felt understood for the first time. In a small but significant way, this was a corrective emotional experience. The therapist joined her in her experience. He did not move away from her feelings by trying to talk her out of them, as had happened in the past. He made her aware that he was interested in knowing how *she* was feeling. His acknowledgement was experientially important for Marsha and gave her permission to identify and express her true feelings and needs. For many beginning therapists, stepping out of the social norms they grew up with is an exciting but stressful component of clinical training. Therapists can then start to respond in a more direct, empathic way, use the ability they already possess to hear what is most important to the client, and take the risk of saying what they see.

To summarize, therapy offers clients an opportunity to be understood more fully than they have been in other relationships. When this understanding is evident in the initial session, clients begin to feel that they have been seen and are no longer invisible, alone, strange, or unimportant. At that moment, the client begins to perceive the therapist as someone who is different from most other people and possibly as someone who can help. In other words, hope is engendered when the therapist understands and articulates the personal meaning that each successive vignette holds for the client. With this in mind, let us look more specifically at how the therapist can find the central meaning in what the client presents.

Identifying Recurrent Themes

As we have seen, the therapist's goal at the beginning of the session is to:

- Encourage the client's initiative
- Actively join the client in exploring his or her concerns
- Identify the core messages or central meaning in what the client says
- Articulate this empathic understanding to the client

Although this may sound simple to do, it is not. It requires the therapist to *give up a great deal of control over the direction of therapy and over the timing with which issues are brought up for discussion. The therapist is placed in the far more demanding position of responding to the diverse and unpredictable material that the client produces, rather than simply directing what the client will cover.* This approach achieves our overarching goal of enacting an interpersonal process that fosters clients' self-efficacy by giving them more control and responsibility over resolving their problems. However, it also puts significant personal demands on the therapist, who must be able to relinquish control over what will occur next in the therapeutic relationship, tolerate the *ambiguity* of not knowing what clients will produce or where the current topic will lead, and understand or make sense of the varied material that clients present.

What assistance is available to help therapists meet the demands of working in this way? If therapists can identify *recurrent themes* in what the client relays, it will help them make sense of the client's subjective experience and better understand what is most central or important out of the wide-ranging material the client presents. This empathic understanding, in turn, provides therapists with the only legitimate control they can have in therapeutic relationships—over their own responding—as opposed to ineffective attempts to direct or control the material that clients produce. Additionally, the appropriate inner controls that come from this understanding will enable therapists to tolerate the ambiguity inherent in this work.

Suppose the client tells the therapist about his or her reasons for coming to therapy. As the therapist follows the different recollections and descriptions the client chooses to relate, the therapist needs to find an *integrating focus* for the wide diversity of material that the client is presenting. The therapist can do this by identifying three types of common factors that recur throughout the different material the client presents:

1. Relational themes or interpersonal scenarios
2. Pathogenic beliefs, automatic thoughts, or faulty expectations
3. A core affect or central feelings

As we will see, identifying these superordinate patterns provides unifying themes that help counselors conceptualize their clients' dynamics and formulate treatment plans. To beginning therapists, the disparate material that clients present often seems unrelated and disconnected. As they progress in their training, however, beginning therapists will increasingly recognize that synthesizing themes are present and can be found in all three domains if they listen for them. In order to help therapists develop this essential conceptualization skill, guidelines for writing process notes within an interpersonal process framework are provided in Appendix A. Keeping these process notes after each session is one of the best ways for counselors to begin recognizing the organizing *patterns* that exist in their clients' functioning. Let's examine each of the three types of common factors in turn.

Repetitive relational patterns   Although terminology varies and different facets are emphasized, the bedrock of all interpersonal dynamic treatment approaches is identifying the *repetitive relational patterns* that are central to the client's problems and distress. The therapist listens for the relationship themes that are most pervasive or characteristic across the different narratives the client relates. For example, "Anyone I trust betrays me by leaving"; "They expect or need so much from me, and I feel overwhelmed by their demands"; "I have to give up what I want in order to be close. If I ever try to meet my own needs or say what I want, my partner will leave me." As beginning therapists review the process notes they have been writing for their clients, they usually find that no more than two or three central themes, conflicts, or relational patterns recur throughout the varied problems their clients present.

Interpersonally oriented psychotherapy has always been focused on clients' repetitive, self-defeating relationship patterns. Since the mid-1970s, however, this approach has been developed more systematically as an effective focus in short-term or time-limited treatments (Davanloo, 1980; Levenson, 1995; Luborsky &

DeRubeis, 1984; Sifneos, 1987; Strupp & Binder, 1984). These effective, programmatic treatments all focus on the client's maladaptive relational patterns, which are variously labeled as the core conflictual relationship theme, cyclical maladaptive pattern, maladaptive transaction cycle, or repetitive transactional pattern. In each approach, the therapist attempts to assess the most common or central relational patterns that occur in the different interpersonal conflicts the client is experiencing, and tries to help the client work them through. To do this, the therapist identifies the relational patterns and clarifies them to clients, so that they learn to recognize these repetitive patterns themselves. The client's pattern, for example, might be to feel controlled by others, or else let down and abandoned; perhaps the client always feels criticized, or idealized, or disdained. It is compelling for clients when the therapist can highlight the same themes across three different spheres: in current interactions with significant others; in formative relationships with family members; and, most importantly, in the current, here-and-now interaction with the therapist. Let's restate and extend this basic approach.

The therapist highlights repetitive relational patterns as clients are relaying vignettes from current or historical relationships. More importantly, the therapist is actively looking for opportunities to address instances in which these relational patterns are occurring in the current interaction or process between the therapist and the client. In order for clients to change, *the therapist must have the nondefensiveness, flexibility, and willingness to work through this reenactment with the client and find a different and more satisfying solution in their real-life relationship than the client has found with others.* Throughout this text, we will focus on how therapists can help clients rework this repetitive relational pattern; we will elaborate further treatment guidelines for this essential process.

Pathogenic beliefs   Clients' repetitive relational patterns involve problematic beliefs about themselves and others and expectations for what is going to occur in relationships and in the future. Weiss and Sampson (1986) and Weiss (1993) have clarified the pathogenic beliefs that help to create and sustain such maladaptive relational scenarios. For example, regarding guilt/compliance issues, the client's pathogenic belief might be: "I am being selfish when I do what I want or pursue my own interests; others are hurt or wounded by my success or independence; I must sabotage or not enjoy my own success or I will be disloyal." A pathogenic belief regarding rejection/shame dynamics might be: "I am unimportant and do not matter; others will see me as childish and needy; they would ignore or reject me if they knew what I felt or needed." Beck (1976) and Beck and Freeman (1990), in particular, have best illuminated the automatic thoughts and dysfunctional interpretations that are central to clients' problems. They write compellingly about the schemas people develop—that is, ways of viewing themselves, others, the world, and the future. In particular, these authors have identified a *cognitive triad* in which people come to view themselves as defective, inadequate, worthless, or unlovable; the world—including others and their environmental experiences—as unmanageable, uncontrollable, or overwhelming; and the future as bleak and hopeless. Thus, as a result of formative developmental experiences, clients begin to exercise a *selective bias* in processing information. As a result of this cognitive filter or selective attention process, clients synthesize the same repetitive themes from diverse interpersonal experiences. They continuously construe events to arrive at the same conclusions or themes, such as loss, defeat, or unfairness, which come to *characterize* their life experience.

These dysfunctional beliefs and expectations are often learned in repetitive interactions with caregivers and may operate either consciously or without much awareness. Once again, therapists try to identify these core beliefs that provide integrating threads throughout the client's relational problems; highlight or make them more overt to clients—especially as they are occurring in the therapist-client relationship; and challenge or question the truthfulness or utility of these beliefs in current relationships. As Beck emphasizes, automatic thoughts and dysfunctional beliefs are not so evident when clients are functioning well. It is *when clients are upset or distressed* (as Beck would say, in their *hot cognitions*) that pathogenic beliefs and relational templates will be revealed. At these crisis points, they are expressed as the clients' reality and experienced as the only way that relationships have ever been and ever will be! In other words, when clients are not upset, they can imagine a wide range of interpersonal scenarios and outcomes, entertain a more realistic set of beliefs about themselves and others, and have flexible expectations about what may occur in relationships. In contrast, clients beliefs and relational expectations become rigid or *unidimensional* when life circumstances have activated their central conflicts. Thus, it is when clients are in distress or conflict that therapists can best assess their orienting cognitions and relational templates.

Emotional patterns Therapists can also identify common themes in clients' emotional reactions. Recognizing and responding to these *recurrent affects* is a very meaningful gift therapists offer their clients. (The therapist might say, for example, "This is so hard for you. Here again, when _____ happens, you are always left feeling _____.") Often, a single, core affect comes up again and again for the client (Kell & Mueller, 1966). As the therapist listens to the client, an overriding feeling such as sorrow, bitterness, or shame may pervade the client's mood or characterize the different experiences the client relates. (The therapist might then say, for instance, "As I listen to you, it sounds as if you have felt hopelessly burdened all of your life.") When the therapist can identify and accurately name this core affect—the feeling that the client experiences as the central or defining aspect of his or her existence—it has a profound impact. Clients often feel that the therapist sees or understands who they really are and what their existence is like in a way that friends and others have not been able to do. Perhaps nothing goes further toward establishing a collaborative alliance than the therapist's ability to identify a core affect and to communicate the profound meaning it holds for the client.

In most cases, responding to the recurrent affects or repetitive emotional themes that result from clients' conflicted interpersonal relations is one of the most important interventions therapists can provide. Unfortunately, beginning therapists often feel insecure about responding to clients' feelings, and, to complicate matters, this topic often receives too little attention in clinical training. We will explore these ideas at length in Chapter 5.

To sum up, if the therapist can identify the repetitive interpersonal themes, pathogenic beliefs, and primary affects that link together the clients' experience and problems, clients will feel that the therapist understands them in a way that others do not. This feeling of being understood helps them progress in treatment. The process notes provided in Appendix A are designed to help therapists identify the recurrent patterns in all three arenas.

The basic model described here also applies to short-term therapies. Although therapists must be more focused and limited in their goals, their principal aim in both modalities is to establish and maintain a focus on maladaptive relational patterns. The primary challenge of successful time-limited therapy is that, far more quickly and accurately than in longer-term treatment, clinicians must be able to establish a collaborative alliance, identify and respond to the recurrent themes in all three arenas, and make these patterns overt by highlighting them *in the current interaction* between the therapist and client. Once they can enact these processes with clients in longer-term counseling, trainees will be successful in adapting these skills to short-term modalities.

Facilitating the Collaborative Alliance by Means of Process Comments

Carl Rogers legitimized concern about the quality of the therapist-client relationship. His contribution to the field of counseling cannot be overstated: He "changed the game" (Kahn, 1997). His conviction was that the therapist does not cure clients but, rather, offers them understanding and caring (agape). Rogers's core conditions for the success of the therapeutic enterprise are empathy, genuineness, and warmth (Rogers, 1951). In practice, however, the therapist may find it difficult to remain empathic when clients insistently distance themselves by discussing only irrelevant or superficial issues; to remain genuine when they repeatedly diminish or compete with the therapist; and to remain warm when they are consistently critical, demanding, or controlling toward the therapist. In a series of studies of psychotherapy process and outcome known as Vanderbilt I,

Strupp and Hadley (1979) found that *clients' maladaptive relational patterns often successfully elicited hostility and control from seasoned therapists and led to poor therapeutic outcomes.* Clients with angry, distrustful, and rigid relational styles tended to evoke countertherapeutic hostility and control—even in a carefully selected sample of highly trained and experienced therapists! Rogers's core conditions are invaluable and provide a good working definition of a collaborative alliance, but something more is needed. *Process comments* provide therapists with a method of intervening to help resolve, rather than merely reenact, the client's problematic relational patterns.

Process comments make the interaction between the therapist and client overt and put the relationship "on the table" as a topic for discussion. Process comments are not confrontations, accusations, intrusive demands, directives, or judgments. They are simply observations about what may be occurring between the therapist and client at that moment. They are offered tentatively, as an invitation for further dialogue and mutual sharing of perceptions. Often they lead to greater closeness. (For example, the therapist might say, "As you're talking about this, I'm wondering if you might be trying to tell me how you are feeling about me and our relationship as well?") Process comments enliven the relationship and are especially effective when the interaction has become repetitive or distant or has lost its direction. They bring *immediacy* to the therapeutic interchange (Turock, 1980; Carkhuff, 1969). They also provide therapists with a vehicle to alter the client's relational patterns and to change problematic reenactments when they occur.

Therapist:

> I think that something important might be going on right now. Can we talk together about what just happened between us? I know people don't usually talk together this way, but I think it could help us understand what's been going wrong with your wife and others as well.

In addition to activating stalled relationships and altering maladaptive relational cycles, process comments, when offered respectfully and in an egalitarian manner, also do the following:

- Invite a fuller or more intimate sharing between therapists and clients
- Promote further self-exploration and help clients see discrepancies between what they are saying and doing
- Provide interpersonal feedback and help clients see themselves from others' eyes and learn about the impact they have on others

Other terms besides *immediacy* have been used to describe interventions similar to process comments. For example, the interpersonal theorist Kiesler (see especially Kiesler & Van Denberg, 1993) discusses *therapeutic impact disclosure* as interpersonal feedback from the therapist to make overt how the client's maladaptive relational patterns are affecting the therapist at that moment and are being played out in their current interaction. Similarly, object relations theorists such as Cashdan (1988) provide *metacommunicative feedback* to register the unspoken emotional quality of a relationship (or, in object relations terminology, to make the client's *projective identifications* overt).

Therapist:
> Although you've never actually said anything like this, I'm afraid that, if I disagree with you or see something differently, you'll be angry and leave our relationship. What do you think about this feeling I'm having? Is there anything to it?

Closely akin to process comments are *self-involving statements,* which may be especially useful for beginning therapists. Whereas *self-disclosing statements* refer to the therapist's own past or personal experiences, self-involving statements express the counselor's current feelings or reactions to what the client has just said or done (McCarthy, 1982; McCarthy & Betz, 1978). (For example, the therapist might say, "Right now, I'm feeling _____ as you're telling me this.") Contrast these two responses:

Therapist:
> I have a temper too, sometimes.

VERSUS

Therapist:
> As I listen to you speak about this, I find myself feeling angry too.

As we would expect on the basis of client response specificity, self-disclosing comments may be useful at times. Too often, however, when therapists reveal personal information, clients' focus is shifted away from themselves. More productively, self-involving statements keep the focus on the client and *reveal information about what is happening in the relationship* and/or how it makes the counselor feel.

Therapist:
> I can feel my stomach tighten and almost turn over right now as you are reading that letter to me.

Sharing personal reactions to what clients have just said or done conveys personal involvement with clients and acknowledges emotional resonance. It also gives clients feedback about the impact they are having on the therapist. When sincere, this type of self-involving statement or process comment often goes a long way toward facilitating a collaborative alliance.

Depending on the therapist's sensitivity and tact, of course, any of these process comments or related interventions that focus on the current, here-and-now interaction can be used ineffectively or effectively.

Therapist:

I'm feeling a little bit bored by this. Maybe we should talk about something else.

VERSUS

Therapist:

I think I'm missing you right now, and not grasping the real meaning this holds for you. Can you help me get closer to what is most important for you here?

In particular, there are two situations when process comments will fail or be counterproductive. First, therapists should not jump in with process comments without previously considering the possibility that their observation or reaction reflects more about their own personal or *countertransference* issues than it does about the client. As a rule of thumb, therapists may want to wait until they have seen an interaction occur two or three times and had time to reflect upon it—or discuss it with a supervisor—before venturing this observation with the client. Differentiating the client's concerns from the therapist's own issues is an important topic; we will return to it in subsequent chapters.

Second, process comments create problems for clients when therapists promise something they cannot provide. The process comments and related concepts introduced here all encourage open, honest, and direct communication, which is also an invitation for further closeness. However, if the therapist then responds in impersonal, distancing, or judgmental ways that do not honor the openness and honesty invited, clients are set back by this mixed message. This double-binding situation will be especially problematic for clients whose caretakers gave promises of genuine emotional contact or involvement but repeatedly failed to follow through.

To sum up, process comments are powerful interventions that facilitate a collaborative alliance, reveal important new issues for further exploration, and provide the most effective way to alter problematic relational patterns that are being reenacted with the therapist. However, beginning therapists should not utilize process comments until they feel comfortable communicating in this forthright way and are ready to meet the client in the genuine and emotionally accessible manner that process comments invite. For masterful illustrations of how process comments facilitate group interaction, see the classic text by Yalom (1975).

Performance Anxieties

Therapists, like other people, often do not really hear what they are told. They are not accurately empathic and often do not capture the feeling and meaning in the client's experience. This section examines *performance anxieties* that make therapists less empathic and cause therapists to push clients to change prematurely.

Therapists will not be as effective with clients when they are burdened by their own excessive performance demands. Commonly, beginning therapists are trying too hard (to be helpful, to win approval from a supervisor, to prove their own adequacy to themselves, to be liked by the client). When therapists are trying too hard, it is almost impossible to decenter, to empathically enter into the client's subjective worldview, and to be emotionally available to the client.

Too often, novice therapists are thinking about where the interview should be going next, rehearsing or wondering how best to phrase what they are going to say next, preparing advice or reassurances, or worrying about what their supervisors would expect them to do at this point in the interview. Such self-critical monitoring often immobilizes student therapists; blocks their own creativity; and, sadly, keeps them from enjoying this meaningful and rewarding work. Further, it prevents therapists from really listening to what the client is saying and keeps them from understanding and being as empathic as they could be. Thus, beginning therapists would be well-advised to slow down their own inner monitoring process and to focus more on the moment, on their current interaction with the client, and on the meaning that this particular experience seems to hold for this particular client right now. Outside the therapy session is usually a better situation for assessing therapeutic interventions and conceptualizing client dynamics.

Identifying recurrent themes, as described in this chapter, can itself become a disabling performance demand. If therapists try too hard to find the broader meaning of the stories that the client tells, they may lose the social context or the whole picture of the client's life—the integrating picture that will clarify the conflicts and patterns therapists are trying so hard to find. It is better for therapists to relax, to be more patient, and simply to help clients unfold their story as they experience it. Undistracted by performance demands, therapists will be better able to see the repetitive themes and core messages in clients' experiences.

When they feel anxiety about their helping abilities, beginning therapists may develop a need to *do* something to make the client change. Novice therapists often believe they are supposed to make the client think, feel, or act differently within the initial therapy session. As a result, beginning counselors often change the topic abruptly (at least from the client's point of view) and pull clients away from their own concerns into what the therapist thinks is important. Although therapists will certainly want to help clients make bridges to new topics or issues they haven't considered, these attempts will usually fail when they are driven by the therapist's own internal pressure to fix the problem. Paradoxically, when the therapist acts on these unrealistic expectations, clients are more likely to drop out of therapy because they feel that the therapist is not attuned to them, does not understand their experience, or is demanding or controlling.

To sum up, therapists with excessive performance anxieties are less effective because they respond more to their own internal needs (to be helpful, competent, or liked) than to the client's need to be understood. Beginning therapists may be alerted to their own excessive performance demands if clients express that they are not being heard, that they are being told what to do, or that they are being rushed. When this occurs, therapists may have become inappropriately invested in making the client change in order to manage their own performance anxieties. The most effective way for therapists to manage this concern is to discuss it with a supportive supervisor or colleague. We will elaborate on this point later.

Care and Understanding as Preconditions of Change

The therapist does need to help the client change, of course, but change is most likely to occur if the client first experiences the therapist as someone who understands and cares. In the initial sessions, the therapist's primary goal is to establish an emotional connection with the client and to begin a working alliance. This is more important than obtaining any specific information about the client or effecting any change. In this regard, one of the best ways to evaluate the success of an initial session is for therapists to ask themselves, "Do I feel like I made contact with the client and have a genuine feeling for who this person is?" Therapists are building an affective bridge to the client and creating the inter-personal context necessary for change, which will not begin until clients feel that the therapist is available, concerned, and sincerely trying to understand. If therapists provide this type of responsiveness throughout the initial session, they are already being helpful to the client. As noted earlier, anxiety and depressive symptoms may even abate somewhat as the therapist's understanding and respon-siveness provides a secure holding environment where the client's pain or distress can be contained in the therapeutic relationship (Bowlby, 1988).

Finally, we must address the most important component of what it means to understand the client. Therapists must actively extend themselves to the client and directly express their feeling and concern for the client. That is, *therapists must articulate their understanding of the client's experience in a way that also communicates their compassion and care for the client.*

This ability to articulate the client's experience in an accurate and caring way is illustrated beautifully in a classic article, "Ghosts in the Nursery," by Selma Fraiberg and her colleagues (Fraiberg, Adelson, & Shapiro, 1975). The therapist in this case study is working with a very depressed young mother whom social services has judged to be at high risk for physically abusing her infant daughter. Early in treatment, the therapist is disconcerted as she observes the mother holding her crying baby in her arms for five minutes without trying to soothe it. The mother does not murmur comforting things in the baby's ear or rock it; she just looks away absently from the crying baby. The therapist asks herself the question, "Why can't this mother *hear* her baby's cries?"

As the young mother's own abusive history began to come out in treatment, the therapist realized that no one had ever heard or responded to the mother's own profound cries as a child. The therapist hypothesized that the mother "had closed the door on the weeping child within herself as surely as she had closed the door upon her own crying baby" (p. 392). This conceptual understanding led the therapist to a clinical hypothesis and basic treatment plan: When this mother's own cries are heard, she will hear her child's cries. .

The therapist set about trying to hear and articulate compassionately the mother's own childhood experience. When the mother was 5 years old, her own mother had died; when she was 11, her custodial aunt "went away." Responding to these profound losses and the mother's resultant feeling that "nobody wanted me," the therapist listened and put into words the feelings of the mother as a child:

> How hard this must have been. . . . This must have hurt deeply. . . . Of course you needed your mother. There was no one to turn to. . . . Yes. Sometimes grown-ups don't understand what all this means to a child. You must have needed to cry. . . . There was no one to hear you. (p. 396)

At different well-timed points in treatment, the therapist accurately captured the mother's experience in a way that gave her permission to feel and remember her feelings. As a result, the mother's grief and anguish for herself as a cast-off and abused child began to emerge. The mother sobbed; the therapist understood and comforted. In just a few more sessions, something remarkable happened. When the baby cried, the mother, for the first time, gathered the baby in her arms, held it close, and crooned in its ear. The therapist's hypothesis had been correct: When the mother's own cries were heard, she could hear her baby's cries. The risk for abuse ended as this beginning attachment flourished.

This poignant case study illustrates how the therapist can use the therapeutic relationship to resolve the client's conflict. A corrective emotional experience occurred when the therapist responded to the mother's pain in a caring manner. This case study illustrates how powerful it is when the therapist articulates an understanding of the client's experience in a way that also communicates the therapist's genuine feeling for the client.

Using role models such as this, beginning therapists need to explore and develop their own personal ways of communicating that they are moved by the client's pain and are concerned about the client's life and well-being. Clients' inability to care about themselves is central to many of their conflicts, and most clients cannot care about themselves until they feel someone's caring for them (Gilligan, 1982). Therapists provide this care when they recognize what is important to the client, express their genuine concern about the client's distress, and communicate that the client is someone of worth and will be treated with dignity in this relationship. These are the therapist's goals in the initial stage of therapy.

Closing

The interpersonal process approach tries to resolve problems in a way that leaves clients with a greater sense of their own self-efficacy. This independence-fostering approach to psychotherapy can be achieved only through a collaborative therapeutic alliance. The client must be an active participant throughout each phase of treatment—not a good patient who waits to be cured and told what to do by the doctor. This process dimension of how the therapist and client work together is more important than the content of what they discuss or the theoretical orientation of the therapist. In this chapter we have also seen the profound therapeutic impact of listening with presence and understanding the client's core messages. Offering the deep understanding described here is a gift to clients—one of the most important interventions that therapists of any theoretical orientation can offer. Perhaps because they are so simple, these basic human responses are too easily overlooked. They are the foundation of every helping relationship, however, and the basis for establishing a collaborative therapeutic alliance.

Suggestions for Further Reading

1. For a marvelous explication of the therapist's basic values and attitude toward the client, see Chapter 3, "The Therapist's Stance," in *Psychotherapy in a New Key*, by H. Strupp and J. Binder (New York: Basic Books, 1984).

2. *Presence* is an important but elusive therapeutic dimension. For a powerful illustration of a therapist's presence with an adolescent sexual abuse survivor, see F. McClure's "Case Study of Sheila: A 15-Year-Old African American Female," which is Chapter 4 of *Child and Adolescent Therapy: A Multicultural-Relational Approach*, edited by F. McClure and E. Teyber (Fort Worth: Harcourt Brace, 1996).

3. For some, client-centered therapy connotes passivity on the part of the therapist and the lack of an effective treatment focus. However, Carl Rogers's original discussion of the therapeutic reflection—(Chapter 4 of his book *Client-Centered Therapy* (Boston: Houghton Mifflin, 1951)—remains an enormous contribution to psychotherapy.

4. Chapter 1 of Ralph Greenson's *The Technique and Practice of Psychoanalysis*, Volume 1 (New York: International Universities Press, 1967), lucidly describes a collaborative therapeutic relationship. The reader does not have to be interested

in psychoanalysis in order to glean valuable insights about this basic therapeutic dimension.

5. Illuminating ideas about the effects of multigenerational family relations on individual personality are found in two articles on triangular coalitions: J. Haley's "Toward a Theory of Pathological Systems," in *Family Therapy and Disturbed Families*, edited by G. H. Zuk and I. Boszormenyi-Nagy (Palo Alto, CA: Science & Behavior Books, 1967'), pp. 11–27; and especially M. Bowen's, "The Use of Family Theory in Clinical Practice," *Comprehensive Psychiatry, 7* (1966): 345–376. Both of these pioneering articles will help therapists apply family systems concepts to the practice of individual psychotherapy.

CHAPTER THREE

Honoring the Client's Resistance

Joan was nervous about seeing her first client, but the initial session actually went very well. The client talked at length about his concerns and, to her relief, she found it was easy to work with him. The client expressed some difficult feelings, and Joan felt that she understood what was important to him. She was thinking they had begun a good working relationship; the client seemed so friendly and appreciative when he left. One week later, however, Joan received a telephone message from the client saying that he was "unable to continue therapy at this time." Confused and dismayed, Joan sat alone in her office wondering what had gone wrong.

Conceptual Overview

Just when the therapist feels that something important is getting started, some clients put their foot on the brakes. The client cancels the second appointment, shows up 25 minutes late, or asks to reschedule for Sunday at 7:00 A.M. This *resistance* is puzzling and frustrating for the novice therapist: "Why didn't that client return? We had a great first session!" Although most clients will not be resistant in this particular way, other forms of resistance will occur regularly throughout therapy. The purpose of this chapter is to help beginning therapists recognize, understand, and respond to client resistance.

All clients have both positive and negative feelings about entering therapy, although the positive feelings are usually more apparent at first. Clients seek therapy in order to gain relief from their suffering. We must look further into the complexity of the client's feelings, however. Often, as clients seek help and genuinely try to change, they simultaneously resist or work against the very change they are trying to attain. How can we understand this paradox? For many clients, painful feelings of *shame* are associated with having an emotional problem they cannot solve on their own. In some cultures, in particular, revealing problems to people outside one's family meets with disapproval or is seen as disloyal. Other clients feel *guilt* about asking for help, meeting their own needs, or having someone else care for or respond to them. Still other clients, because of their relational templates, have *anxiety* evoked by the expectation that significant others will respond to them in hurtful ways. Thus, if clients have a problem, need to ask for help, or reveal problems to individuals outside the family, then shame, guilt, anxiety, and other painful emotions may be evoked. Further, we will also see that this may occur when clients feel better, improve in therapy, and resolve their problems. As a part of formulating case conceptualizations (see Appendix B), therapists need to identify the specific issues or conflicts that make it threatening or difficult for each particular client to approach his or her problems. Although specific concerns must be clarified for each individual client, the following are some common themes:

> If I let myself depend on the therapist, he or she might leave me, or take advantage of me, or try to control me, as others have done when I needed them.
>
> I cannot ask for help or need anything from others because I must be perfect and in control all the time. Besides, I don't really deserve to be helped anyway.
>
> I cannot ask for help or need anything from others because I must be perfect and in control all the time or I will bring shame to my family.
>
> Asking for help is admitting that there really is a problem, and that proves there really is something wrong with me.
>
> I am afraid of what I will see or what a perceptive therapist will learn about me if I stop and look inside myself.
>
> If I cannot handle this by myself it means that I really am shamefully weak and inadequate, just as they always said.
>
> If I don't see a therapist, my spouse might leave me.

Many clients are struggling with feelings of *debasement* at the prospect of losing control over their emotions or lives or of being controlled by another. As we would expect from the principle of client response specificity, there will be many different reasons for clients' resistance and defense. Most commonly, perhaps, resistance and defense are driven by shame.

It is easy to respond to the approach side of clients' feelings—the pain or need that motivates them to seek help and enter therapy. If you gently scratch the surface, it is usually plain to see, and the client wants you to respond to it. In contrast, the other side of the client's feelings—the reactive emotions and the interpersonal consequences of having a problem and needing help—may not be so accessible. These concerns over what it means to enter therapy or ask for help, even though the client may not be aware of them, act as a countervailing force to the client's motivation to enter therapy. If unaddressed, these ambivalent feelings will draw some clients out of therapy prematurely, as in our opening vignette.

Chapter Organization

Resistance is not a welcome concept to most beginning therapists. In the first section, we examine why it is difficult for both therapists and clients to address this important dimension of therapy. In the second section, we look at a working definition of resistance, distinguish between psychological resistance and reality-based constraints on the client's ability to participate in treatment, and observe that both may be operating simultaneously for the client. We also see how therapists can formulate *working hypotheses* to better understand their client's resistance, which, in turn, enables therapists to respond more compassionately and effectively. The third section presents intervention guidelines for responding to resistance and further information to better understand and conceptualize clients' resistance and defense throughout treatment. Common expressions of client resistance are illustrated at three different stages: during the initial telephone contact; at the end of the first session; and during subsequent sessions. Sample therapist-client dialogues illustrate effective and ineffective responses at each stage.

Most of this chapter outlines practical guidelines to help manage resistance effectively in the early stages of therapy. These guidelines provide trainees with tools to use in keeping their clients from dropping out of treatment prematurely. The clients most likely to drop out of treatment after the initial telephone contact or first session are people of color, men, and clients who have been vague or evasive about their problems. Resistance and defense will occur at every stage of treatment, however, and the final section includes guidelines for conceptualizing and responding to resistance later in treatment.

Reluctance to Address Resistance

Imagine your client has missed, come very late for, or twice rescheduled the second appointment. Perhaps this has no significant psychological meaning at all. Cars do break down; traffic jams occur; children get sick; and employees get called for work at the last minute. When these *reality-based* problems occur, most clients do not respond well if the therapist questions their motivation to be in treatment. On the other hand, if this behavior reflects the client's *ambivalence* about some aspect of being in therapy, the therapist must be willing to address and help resolve it. Therapists usually do not know whether the client's behavior is reality-based, psychologically motivated, or both. As we will see, however, by trying to explore this behavior with the client in a respectful and accepting manner, therapists can greatly increase the chances of the client's continued participation in treatment. In these explorations, therapists' manner is as important as what they say: Unless they are warm and compassionate, the client will hear their observations as blame or criticisms, which never help and only serve to exacerbate client resistance.

More specifically, the model proposed here will suggest that therapists first enter supportively into clients' frame of reference, affirm or validate the reality-based constraints that they perceive, and do whatever possible to flexibly accommodate or help resolve the problem. Only after taking seriously clients' concerns as they see them can therapists begin to inquire about other conflicted feelings or meanings that entering treatment may hold. The therapist can begin this joint exploration by wondering aloud, in a tentative manner, whether entering treatment may be evoking other more psychological concerns as well. If therapists approach resistance without following this two-step sequence, most clients will feel misunderstood or blamed, and their ability to engage successfully in treatment will be impeded.

The Therapist's Reluctance

To a greater or lesser degree, every client will be ambivalent, defensive, or resistant. This push-pull occurs at the beginning of therapy and will continue to wax and wane throughout treatment. Therapists need to address signs of potential client resistance, but many beginning—and experienced—therapists find this difficult. Let us examine three reasons why therapists may be reluctant to approach their clients' resistance.

Many beginning therapists are unaware of the multiple meanings and conflicts associated with the decision to enter therapy. These therapists are surprised to find that, in the initial sessions, clients actually resist the help they are overtly seeking. Other therapists are familiar with the concept of resistance, but they view it pejoratively as a way to keep the therapist in a superior position and to deny the validity of the client's own experience. This occurs, for example, when a therapist says to a working, single mother of three, "Hmm, I notice that you are five minutes late for our appointment today." Resistance does not have to be associated with this type of hierarchical therapist-client relationship. Therapists can still be sensitive to the social context of the client's life. Poor clients who use public transportation will not be able to arrive on time consistently. Clients from other cultural contexts may not think that being five minutes late is of any significance. For some, it may even be impolite to come on time; arriving late allows the host extra time to prepare for the visit. Therapists can educate clients about the bounds and procedures of therapy and, as we will see, respond to resistance in other ways that strengthen the client's commitment to therapy, enhance the therapeutic alliance, and empower rather than invalidate the client.

Another reason why therapists are reluctant to address their clients' resistance is more personal. Most novice therapists have strong needs for their clients to like them, to find them helpful, and to keep coming to therapy. If the client does not show up or comes late, therapists may feel that they have failed. When this occurs, therapists become increasingly concerned about their ability to help others and overly invested in the next client's satisfaction with therapy.

In order to ward off unwanted and/or feared criticism, therapists often do not inquire about signs of potential resistance. Like their clients, most therapists are not eager to approach issues that arouse their anxiety. Thus, the beginning therapist may be hesitant to invite clients to express their negative reactions to the therapist's own behavior or their conflicted feelings about having a problem, asking for help, or needing to be in treatment. Although it is difficult to invite critical feedback and approach conflicts, it is necessary to do so, especially if the client expresses conflict or suggests that there may be some dissatisfaction with the therapist. The paradox is that, if therapists allow their own anxiety to keep them from dealing with clients' resistance, their clients will be far more likely to act on these unexpressed concerns and drop out of treatment prematurely.

An effective approach is to make a process comment that acknowledges the resistance and suggests addressing it through the therapist-client interaction. It is possible to do this in a way that is respectful and will not violate cultural or social norms.

Therapist:

You seem more discouraged than before, and you've had more trouble getting here. I'm wondering if something isn't going right between us. Any ideas?

Client:

What do you mean?

Therapist:

I'm wondering if something about our relationship or being in treatment isn't working for you right now. We seem to be missing each other in some way. Any thoughts about what could be going wrong between us, or what I could do to make this work better for you?

Client:

Well, maybe you could talk a little more or give me some more feedback. You're pretty quiet and don't say very much, so I don't really know what you're thinking.

Therapist:

I'm so glad you're telling me this. Sure, I'll be happy to be more active and responsive to you. Let's check in on this together every so often and make sure it's changed. Can we talk some more about my being quiet, too? What's it like for you when I am quiet or you don't know what I am thinking?

The Client's Reluctance

Unfortunately, the client usually shares the therapist's reluctance to address resistance. Clients are often unaware of their resistance and externalize it to others or outside events. For example, the client may say, "I'm always late for our sessions. The traffic always seems so bad." While there may be truth to this statement, the consistency or frequency of the behavior suggests the need for further exploration. In order to acknowledge both the possible reality-based constraints and the probable psychologically based resistance, the therapist can ask if leaving earlier for the appointment is possible or if an alternative session time needs to be negotiated. The therapist also needs to ask if it is possible that the client has some concerns about attending therapy or other feelings about the therapy process. Even if the client is not ready to address whatever psychologically based resistance may be operating, the therapist has alerted the client to the possibility of underlying psychological factors and opened the door to future discussions of this issue.

In contrast, when clients see themselves resisting, they often feel confused, frustrated, and embarrassed by their own contradictory behavior. For example, clients often exclaim with dismay:

Why do I go to all the trouble and expense of coming here to see you, when I can't think of anything to say as soon as I walk in the door?

Why would I forget our next appointment after we had such a great session last week? It doesn't make sense!

Why do I keep asking you for advice, and then say, "Yes, but . . . ," to whatever you suggest? What's wrong with me?

When it is clear to clients that they are sabotaging their own efforts, they typically evaluate themselves and their own resistance harshly—as bad in some way—and assume that the therapist shares this critical attitude. Thus, when the therapist begins to inquire about resistance, most clients want to avoid the topic because they are afraid the therapist is going to blame them—for not really trying, failing, being manipulative or unmotivated, and so on. Because it is never therapeutic for the client to feel blamed or criticized, the therapist must make this self-critical or blaming attitude overt and correct it. The best way to do this is to help clients reframe their critical attitude toward their resistance. Both the therapist and client must *honor* the client's resistance, because it originally served a self-preservative and adaptive function: It was the best possible response to an unsolvable conflict that the client had available at particular stages in his or her development. The therapist then helps the client to realize that he or she no longer needs this coping strategy but can develop more flexible response styles. Let's examine what this means.

Clients' resistance and even their symptoms once served as a survival mechanism that was necessary, adaptive, and often creative. For example, anxiety signals danger—perhaps not a current threat but a historical danger that really did exist. Depressive symptoms may reflect a client's attempt to cope with painfully unfulfilled wishes for protection or love from a caregiver and/or a way to cope with the responses of rejection, exploitation, or derision that they came to expect from significant others. In other words, past experiences with family members and others have given clients very good reasons for not wanting to ask for help, not sharing a painful feeling, or not risking a disclosing relationship. If the therapist and client explore how significant others have responded in the past, the client's resistance will make sense and become understandable. The feelings that underlie the client's resistance always make sense historically, although they may no longer be necessary or adaptive in current relationships.

Suppose that a client is having trouble entering therapy. The therapist might ask a question such as the following:

What might I do to hurt you or make things worse if you seek help from me?
How have others responded to you in the past when you have asked for help or needed someone?
If you and I begin to work together productively on your problems, what could go wrong between us?
How could our ethnic or cultural differences be a problem for us in therapy?

Among other things, such questions will help the therapist and the client identify the aversive consequences when the client asked for help in the past. For example, clients may report that, when they asked for help in their family, they were ignored and felt powerless, were made fun of and felt ashamed, or were told that they were selfish and too demanding. They may have been told that they didn't really need or want what they asked for. They may have felt that everything they received brought burdensome obligations. In this way, therapists and clients can clarify both why and how clients originally learned to defend themselves.

Thus, the first thing the therapist can do is help clients recognize problematic relational patterns, understand why they originally needed to defend themselves, and clarify how they are continuing to do so in therapy. Second, the therapist must confirm or validate the necessary protection this resistance once provided. The maladaptive resistance in the therapeutic relationship was once an adaptive response: The original defensive behavior (withdrawing, diminishing, provoking, avoiding, arguing, pleasing, intellectualizing, and so on) minimized the hurt and allowed the client to cope with the painful or unwanted interpersonal response. Third, therapists must also differentiate their own responses in the current relationship from the aversive ways that significant others have responded in the past.

Therapist:

>   In the past, you learned that if you allowed your parents to help or take
>   care of you, they wanted you to remain a child and continue to need them.
>   I would like to respond to your need, but I'm different from your parents.
>   I don't need you to remain dependent on me. I would like to help you
>   achieve your own independence, so that you will not need me any longer,
>   and will be able to go on with your own adult life.

Thus, therapists must follow this three-step sequence:

1. Identify the original transactional pattern.
2. Validate the client's unmet need and defensive adaptation to it.
3. Differentiate the therapist's current response from the past dynamic.

This sequence will need to be repeated with each manifestation of resistance in the course of treatment. In this way, the therapist will be providing the client with the experience of no longer needing to use the old problematic relational patterns.

We have seen, then, that clients are far more likely to become stalled in therapy, or to drop out prematurely, if their resistance is not addressed. One of the most important ways to keep clients in treatment, and to identify conflict areas in clients' lives, is to invite clients to discuss any negative or problematic reactions they may have toward being in therapy or to anything the therapist has done or might do. The guiding principle is that if clients talk about their ambivalence or discontent, they will be less likely to act on it and drop out. Before going on to illustrate specific ways of implementing this general principle, we must learn how to identify clients' resistance and differentiate it from reality-based concerns that must be affirmed.

Identifying and Conceptualizing Resistance

Early in their training, most therapists will have clients that drop out of therapy within the first few sessions. The therapist might conclude that the client wasn't really motivated or wasn't ready to look at his or her problems, or that the therapist unwittingly made some irrevocable mistake or error in the previous session. The main reason clients prematurely terminate, however, is that they are *acting on*, rather than *talking about*, their conflicted feelings about entering therapy. In order to help clients remain in treatment, the therapist must be prepared to identify when resistance is occurring; to approach this issue in a noncritical and accepting manner that enables clients to express their concerns more fully; and to help them resolve these concerns by responding affirmingly and accommodating flexibly whenever possible.

Identifying Resistance

How does the therapist know when a client is resisting treatment? Therapists never know what any behavior means for a particular client, but resistance may be operating when clients consistently have difficulty participating in treatment. For example, after reality-based constraints have been considered, psychologically based resistance is probably occurring when any of the following occur repeatedly or in combination:

> The client misses the appointment or comes late.
> The client needs to reschedule appointments frequently.
> The client has very limited hours available for therapy.
> The client cannot make a firm commitment to attend the next session.

It is important to stress that the same behavior often means very different things to different clients. However, when clients have trouble attending sessions, some form of resistance is probably at work. In such cases, the therapist should generate working hypotheses about the possible meaning this resistance holds for the client.

Formulating Working Hypotheses

Therapists must begin to formulate tentative working hypotheses about the client's repetitive relational conflicts, faulty expectations, pathogenic beliefs, core conflicted feelings, and interpersonal coping styles right from the initial contact. One systematic way to do this is to formulate answers to three questions. (Further guidelines are provided in Appendixes A and B.)

1. What does the client elicit from others?   Even in the initial telephone contact, the client's interpersonal style and what it tends to elicit from others is an important source of information. For example, if the client sounds helpless and confused on the telephone, might he or she adopt a victim stance and invite others' rescuing behavior? If the client presents in an angry and demanding way, might others often withdraw and leave him or her alone? If so, perhaps this client is testing whether the therapist can tolerate the angry challenges and still see the underlying need and hurt. As therapy progresses and the therapist learns more about each client, many of these initial hypotheses will prove inaccurate and will need to be discarded. As therapists learn to trust themselves and attend to the feelings and reactions evoked in them by the client, however, they will find that many hypotheses do indeed fit the client and can be further developed and refined. Thus, in order to begin conceptualizing the client's dynamics and developing initial treatment plans, therapists must formulate working hypotheses about the potential meaning of the client's behavior, evaluate these tentative initial hypotheses as they interact further and learn more about the client, and revise these hypotheses over time to more accurately reflect each client's personality and problems.

2. What is the threat?   On the basis of information the therapist has started to gather about the client's current and past relationships, the therapist can begin speculating about the different ways that therapy could be aversive for the client. Therapists should formulate working hypotheses such as the following about the feelings, issues, or concerns that are likely to be difficult for each client.

> Is it guilt-inducing for this highly responsible mother, who grew up taking care of a depressed parent, to ask for something for herself?
>
> Is it incongruent for this older Latino male to seek help from a younger female therapist?
>
> Is it a shameful failure experience for this blue-collar worker to have an emotional problem that he cannot solve by himself?
>
> Does this Christian client believe that her inability to resolve her problems and need to ask for help are evidence that she has failed in her faith?
>
> Is it disloyal for this Asian client to discuss family problems with someone outside of the family?

Therapists may also keep in mind that resistance and defense are reflections of the client's fear or felt lack of safety, and are attempts to manage unwanted feelings of shame, guilt, and anxiety. Although clients may be unaware of it, they often struggle with the worry that, if they continue in treatment, the therapist will hurt them in the same way their caregivers did when they were children. Clients who were more traumatized may worry that they will hurt the therapist and others as they were once hurt. Over time, the therapist can help clients clarify such concerns and invite clients to take a risk and see whether the therapist can offer a potentially more satisfying relationship than they have come to expect.

3. How will the client express resistance?   Finally, therapists should formulate a third set of hypotheses that anticipate how clients will enact these concerns with the therapist. For example, suppose that a therapist has observed that her depressed client feels guilty and believes he is selfish whenever he does something for himself or whenever someone does something for him. The therapist then hypothesizes that this client may withdraw from treatment as soon as he starts to feel better or emotionally disengage from the therapist when he realizes that she genuinely cares for him.

To illustrate further how the therapist generates working hypotheses on the basis of these three questions, let's consider another example. Suppose that a young male client whose problem is substance abuse telephones the therapist to schedule an initial appointment. During this telephone call, the client tells the therapist that his alcohol and drug usage is more problematic than it has ever been and that his mother believes that he really needs to be in therapy. On the basis of the helplessness and confusion expressed in his telephone call and the allusion to how much more problematic his drug usage now is, the therapist first hypothesizes that an important part of the client's interpersonal style is to let others know he is hurting, to elicit help from them, but then to avoid taking responsibility for his own needs and behavior.

Second, the therapist hypothesizes that this client may become resistant to treatment when the therapist addresses the issue of how he meets his dependency needs, or when the client is given reality-based confrontations about how he eschews responsibility for his own decisions (for example, to enter therapy or to continue drinking). Third, the therapist hypothesizes that this client is impulsive and may act out his resistance by bolting out of therapy as soon as these anxiety-arousing issues are addressed (just as, perhaps, he has used alcohol and drugs to avoid dealing with certain feelings, internal psychological issues, or external stressors).

The therapist knows that these initial hypotheses may not be accurate, and is ready to assess, revise, or discard them as more is learned about the client. However, *the therapist must try to anticipate the concerns that might cause each client to drop out of treatment and the way each client is likely to express these concerns.* Prepared with working hypotheses such as these, the therapist will be able to respond more effectively once clients have begun to act on their resistance and will be able to help clients anticipate their own resistance before they act on it.

As therapy proceeds, the therapist will learn much more about the client, and many of these initial hypotheses will need to be discarded. Some will be accurate, however, and can be elaborated to better understand this particular client. In this way, the therapist identifies the enduring issues that arise for the client. The therapist will then be better prepared to center treatment on these repeated themes, which provide structure and focus for the ongoing course of treatment. As we will see in Chapter 7, this is part of the continuing process of formulating and refining conceptualizations of client dynamics and treatment plans.

Responding to Resistance

In this section, we see how therapists can respond to the common types of client resistance that occur at the beginning of therapy. Sample therapist-client dialogues illustrate effective and ineffective responses at three critical points: during a telephone conversation in which the client has difficulty scheduling the initial appointment; at the end of the initial session; and during later sessions with clients who have difficulty keeping appointments. As before, student therapists should find their own words to express the principles embodied in these dialogues.

Resistance during the Initial Telephone Contact

In trying to schedule the initial appointment, the therapist wishes to obtain a firm commitment from the client. The client, however, may be ambivalent about entering therapy and express this during the initial telephone contact. The therapist should be prepared to respond directly to this issue.

An uncertain commitment   In the following example, the client is only somewhat uncertain about attending the first session.

Therapist:
> It's been good talking with you, and I'll see you on Tuesday at 4 P.M.

Client:
> OK, I guess I'll see you then.

Therapist:
> You sound a bit uncertain; let's talk a little more about coming in.

Client:
> Well, maybe a bit uncertain, but I do want to come.

Therapist:
> Good, I'm looking forward to meeting with you. If you would like, we can talk together then about therapy and any questions or concerns you may have about me or treatment. See you next Tuesday at 4 P.M.

An effective response to the client's ambiguity ("I guess . . . ") is modeled in this dialogue. The therapist hears the indecision in the client's commitment and addresses it directly. An ineffective response would be for the therapist to let it pass by unnoticed, on the assumption that it probably doesn't mean much. Perhaps it doesn't, and chances are that the client will still arrive at the appointed hour. However, the chances of the client not arriving are far greater if the therapist does not directly address the signs of uncertain commitment.

Further, the therapist should mentally note this indecision as a possible sign of resistance to treatment. This is an instance in which the therapist can begin to generate working hypotheses—derived from the first of our three orienting questions—about the possible meaning of relevant client behavior.

A more ambivalent client   Now imagine the same telephone conversation, but with a more difficult client.

Client:
> OK, I guess I'll see you then.

Therapist:
> You sound a bit uncertain; let's talk a little more about coming in.

Client:
> Well, in my family, we don't really talk to others about family problems.

Therapist:

> I appreciate and respect your concern about being loyal to your family. Maybe one of the things we can talk about is how you can ask for help from others and get your own needs met without feeling that you are being disloyal.

Client:

> Let's do that. My family is really important to me and some people don't understand that.

Therapist:

> Yes, we can work together to try and find ways to sort out your own needs in conjunction with your feelings about your family. Also, we can agree to meet only one time on Tuesday at 4 P.M. After talking together a little more, you can better decide for yourself whether you'd like to continue.

An ineffective response would be to accept the client's uncertain commitment ("OK, I guess I'll see you then") and deny that an important issue is being played out. In this dialogue, instead, the therapist has tried to understand the client's concern first, and only then asked for a firmer commitment—even if that is only to attend the first session. By listening for, and responding to, the client's ambivalent feelings, the therapist has done much to enable this client to take the first step and attend the initial session.

Why must the therapist take such a firm stance? Wouldn't it be more supportive to let clients leave the appointment a little bit tentative, if that is what they need to do? No. Little or no therapeutic change will occur until clients take responsibility for the decision to enter therapy and work on their problems. Therapists must be flexible, and clients can make a very limited commitment—for one session or just a few sessions—to get a sense of whether therapy is for them. However, without this commitment, the therapist has little to work with. In fact, it is better for the client to remain out of therapy than to enter without commitment and have an unsuccessful therapeutic experience. As Yalom (1981) emphasizes, the therapist must be concerned about preventing failed hope in the client. If clients have one or more unsuccessful therapy experiences, it may discourage them from trying to seek help with their problems again.

Clients who try to make the therapist responsible   Let's return to the telephone conversation between the therapist and a prospective client. Some clients will try to make the therapist take responsibility for their decisions.

Client:

> I'm not sure if I should start therapy or not. What do you think I should do?

Although it is important to reach out to clients and welcome them into treatment, the therapist does not want to resolve clients' ambivalence by assuming responsibility for their decision or by trying to provide the necessary motivation for therapy. With certain clients, a helpful response might be as follows:

Therapist:

You're really hurting right now, and I would like to try and help. Would you like to make a commitment to come in for one appointment? We could discuss your concerns further at that time?

OR

Therapist:

I can't offer you any guarantees, of course, but yes, I do believe that therapy may be of help. I'd like to try and work with you. Let's agree to one session. At that time, we can assess further your concerns about therapy being helpful.

As we would expect on the basis of client response specificity, such direct invitations may be useful at times. When clients have felt unwanted while growing up or believe that others are uninterested in them, it is especially necessary for therapists to express genuine interest in working with them. Even with clients such as these, however, therapists must still respond in a way that ensures clients take responsibility for their own decision to enter treatment; the therapist does not want to cajole, coerce, or win the client into therapy. At the same time, the therapist does not want to be cool, aloof, or indifferent. The therapist needs to reach out emotionally and work supportively with clients on their ambivalence, without taking the responsibility or providing an external solution for the client. Thus, with many clients, an effective response to the client's question may be as follows:

Therapist:

It seems as if one part of you wants to be in therapy, but another part of you doesn't. Tell me about both sides of your feelings.

OR

Therapist:

It sounds like a part of you wants to try this and a part of you doesn't. Why don't we agree to meet for just one or two sessions. After we've worked together for a little while, you will be able to decide for yourself whether I can be of help.

Especially in the first response, the therapist is inviting the client to express all concerns about entering therapy. In both responses, the therapist is working with the client to reach a decision, but leaves the client responsible for his or her own decision. Here again, we are working with the process dimension in psychotherapy. Although this process distinction is subtle, it will make all the difference in the eventual outcome of therapy.

The client in this dialogue is providing the therapist with potentially important information. As before, the therapist should begin generating working hypotheses about what this behavior may mean. For example, uncertain clients may not have been supported when they have tried to take responsibility for their own actions. These clients may now try to avoid the aversive consequences they have experienced in the past by getting others to take responsibility for them. If so, this could make it difficult for clients to initiate new activities (such as applying for college) or to follow through on their own wishes and interests (such as selecting the major or career they are most interested in). Being held back in this way may leave clients feeling resentful and frustrated.

Having generated such hypotheses, therapists can be alert for evidence of these themes in other areas of the client's life. If these prove to be enduring and pervasive issues, the therapist is better prepared to focus on them throughout treatment. The therapist must not become wed to these early hypotheses, however, and must be ready to discard them if further supporting data is not forthcoming. As therapists become more experienced, their initial hypotheses will become more fruitful. This gradually developing skill will become an especially important aid to counselors as they begin working in short-term and crisis-intervention modalities that require quicker, more accurate assessment during the initial phase of treatment.

As noted in Chapter 1, the therapist does not want to recapitulate the client's conflict in the therapeutic relationship. This can easily occur in a metaphorical or encapsulated way in the initial negotiations between therapist and client. This is another reason why the therapist should not take over and tell the client seeking advice what to do. For example, the following response is not recommended:

Therapist:
On the basis of what you've told me, I think that therapy can help you, and I believe that you should give it a try now.

Assuming such a directive stance often reenacts a problematic relational scenario for clients and confirms an underlying pathogenic belief. Clients often feel that they cannot act on what they want, or that they need to be dependent and let others assume responsibility for their decisions. Although such clients feel they need to ask the therapist to tell them what to do, it also makes clients feel that their own autonomy is being suffocated. Paradoxically, clients resist the control they have just elicited by avoiding or resisting the therapist. As this scenario continues, clients come to feel badly about themselves—confused as to why they are rejecting the help they have just asked for; and guilty for rejecting the therapist who was trying to help. Therapeutic responses that give the client the responsibility of deciding whether to follow through prevent this reenactment from occurring and may provide clients with a new and different response to an old conflict.

Resistance at the End of the First Session

At the end of the first session, no matter how well it seems to have gone, the therapist should ask clients how the session felt to them and whether they feel any difficulty over being in therapy or any dissatisfaction with the therapist. Unless the therapist inquires about the clients' ambivalence, it is likely to remain unspoken. A client who is beginning to feel significant conflict will be far less likely to drop out prematurely if the therapist has given the client permission to express his or her concerns, has taken them seriously, and has tried to resolve them to the extent possible.

Addressing interpersonal conflict   Ten minutes before the end of the first session (to allow clients enough time to talk about their reactions to the therapist and the session, even if it went very well), the therapist should check in with the client:

Therapist:
> How was it to be here today?

Most clients will answer, "Fine." Thinking that the therapist may need some reassurance, or trying to win approval, clients will often go on to say how helpful the session has been. If so, the therapist might respond as follows:

Therapist:
> Good, I'm glad you've found our first session helpful. You've told me a lot about yourself today, and I feel we have gotten off to a good start too. It may sound like I'm just asking for approval here, but I want us to be able to talk about our relationship more directly than others usually do. Was there anything about coming today, or anything about you and me or our work together, that did not feel good to you? If so, it's important that we talk about that.

Client:
> Oh no. I was eager to begin, and you have been very nice and understanding.

Therapist:
> That's fine. If, in the future, you ever have any uncomfortable feelings about our relationship or being in therapy, I would like for you to talk with me about them. Do you think you would be able to do that?

In this vignette, the therapist is trying to establish an important set of rules and expectations for what is going to occur in therapy. The therapist is telling the client that, in contrast to others with whom the client may have been in relationship, the therapist can tolerate the client's angry or dissatisfied feelings. The therapist is also saying that he or she is not afraid of conflict and wants to deal with problems between them straightforwardly. Although many clients will not be able to accept this offer initially, the therapist is laying the groundwork for an honest and direct relationship. Therapists want to establish these important expectations with their actions, and not just their words, early in treatment.

A more assertive client   Now imagine that the therapist raises the same type of query, but this time the client is critical of the therapist.

Therapist:
How has it been for you to talk with me today?
Client:
Well, just fine for the most part, but I did feel you were trying to hurry me up or move me along a few times.

The therapist must learn to approach the conflict, rather than avoid it by moving on to another topic, ending it with a punitive response, or expressing hurt and communicating that the client's criticism is too much or unacceptable. The best way to approach the conflict is to encourage the client to express fully his or her critical reaction.

Therapist:
I'm sorry that happened. Let's look at that together and try to understand what happened for each of us. Tell me more about your feeling of being pushed or hurried up.

Of course, it is often difficult for beginning therapists to do this. Such confrontations may arouse the therapist's anxiety, especially when the therapist has strong needs for the client's approval. If they are to approach such anxiety-arousing confrontations, therapists must first become aware of their own characteristic responses to interpersonal conflict. For example, many therapists will want to avoid or ignore the client's confrontation or will readily agree with, and apologize to, the client in order to abate the criticism. Other therapists will automatically begin to defend themselves and offer long justifications and explanations to the client. A few therapists will counter with their own hostility. Each therapist must become aware of how she or he typically or initially reacts to challenging confrontations.

One of the best ways for you to do this is to recall how your own family of origin dealt with interpersonal conflict. Regardless of age, your initial reaction to conflict usually follows closely how your parents dealt with conflict in their marriage and how each parent dealt with conflict with you as a child. Rather than reflexively following their initial propensity, therapists must learn to respectfully approach conflicts with clients.

After inviting clients to express fully their concerns, therapists must accept the validity of the clients' perception. Perhaps the client is right and the therapist was rushing the client. The therapist must honestly examine his or her own behavior and see if the client's perception is accurate. If so, the therapist should not be afraid to admit the mistake to the client.

Therapist:

> Yes, I think you're right. I was aware of the time going by, and I wanted to touch on a few more issues before we had to stop. I probably was pushing you there, and I can see why that bothered you. I'm glad you made me aware of that; I'll try not to do it again.

It is liberating for beginning therapists to learn that they do not need to be afraid of making mistakes, although they may need to acknowledge them. In this example, the therapist is responding to the client in a reality-based and egalitarian way. The therapist validates the client's perception, which tells the client that the therapist is willing to have a genuine dialogue. It also tells the client that the therapist will respond to the client's concerns with respect. Too often, therapists and clients enact a relationship in which the client is the sick and needy one and the therapist is the healthy one who does not make mistakes or have problems. The response suggested here discourages that illusion. Overall, these messages can have a powerful effect on clients' expectations of therapy and accelerate the level of disclosure and exploration that clients undertake in subsequent sessions.

In contrast, therapists can handle the confrontation in a way that hides their own humanity, keeps the relationship safely distant, and puts them in a superior position.

Therapist:

> So you thought I was hurrying you up. How was that for you?

OR

Therapist:

> That's interesting. Do you sometimes find that you feel this way in other relationships, too?

These responses are a profound misuse of the client-centered reflection or psychoanalytic neutrality. With this type of *deflection*, the message to the client is, "It's always your problem. I will not look at my own contribution here. This will not be a genuine interpersonal encounter." Such a response limits the relationship to a superficial encounter and sets up a power battle between client and therapist. Little therapeutic gain will be realized as long as the interpersonal process continues in this problematic mode.

But what if the therapist does not agree with the client's complaint? The therapist should still accept the validity of the client's perception, but without agreeing to the comment.

Therapist:

> I'm sorry you saw me as being impatient with you; that wouldn't feel very good to me either. I wasn't aware of being in a hurry or trying to push you,

but I'll be alerted to that in the future. If you ever feel that happens again, stop me right then and we'll look at it together.

The therapist should never agree to something he or she doesn't believe is true. In this example, the therapist is telling the client:

I will take your concerns seriously even if I do not see it the same way.
We can have differences between us and still work together.
Your feelings about our relationship are important to me.

This type of response also tells the client that problems between them can be resolved, which lends hope that the therapist and client together will be able to resolve the client's problems. By addressing conflicts between the therapist and the client directly, the therapist provides the client with the opportunity to have a new, more adaptive means of conflict resolution modeled in vivo. Especially for clients who have learned to dread interpersonal conflict, such experiential re-learning is empowering indeed.

Finally, it is also possible that the client is systematically misperceiving the therapist's behavior. For example, this client may readily experience the therapist and most other authority figures as demanding more than the client would like to produce and then being dissatisfied with whatever the client does produce. If so, this client may have grown up with a parent who repeatedly demanded that the client do everything on the parent's timetable and/or ignored the child's own wishes and accomplishments. Even if the therapist has gained further evidence to support such a hypothesis, this type of historical or transference interpretation should not be presented until later in therapy or after the therapist and client have resolved the dispute in their current relationship; therapists stand to lose a lot if clients see them as sidestepping a reality-based confrontation. However, the therapist can use this information to generate working hypotheses that can be utilized later.

Clients who test the therapist's adequacy   Because it is usually difficult for beginning therapists to respond to client confrontations, we will examine another example. In the following situation, the client challenges the therapist's compe-tence. Although this taps into an anxiety-arousing area for the novice therapist, the therapist is still able to approach the issue directly and invite the client to express the concern more fully.

Therapist:
    I wonder if there is something about our relationship, or being in therapy, that just isn't feeling quite right to you.
Client:
    Are you just a student here? Isn't that what you said when we first spoke on the phone?
Therapist:
    Yes, I'm a second-year graduate student, working on my master's degree.

Client:

Well, I don't really feel like a guinea pig or anything, but you really are just practicing on me, aren't you? I don't want to sound unkind or anything, but maybe you haven't had enough experience to help me?

Therapist:

Are you worried that if you go to all of the trouble of coming here and talking about difficult problems with me that I just won't know enough to be able to help you?

Client:

Yeah. After all, you really are just a beginner, and I must be 15 years older than you. What do you think? Are you certain that you can help me?

Therapist:

I certainly cannot offer you any guarantees, but I will do my best to try to help you. And as we continue to meet together a few times, I think you will be able to decide for yourself if I can help you or not. But for now, let's see if there are any questions I can answer for you about my training status and how our age difference might get in the way.

In this vignette, the therapist was able to approach the client's concern directly and respond effectively because she remained nondefensive, even though this was anxiety-arousing for her and difficult to do. The therapist did not act on her initial impulse, which was to try to assure the client of her competence and ability to help. Instead of responding to her own personal need, the therapist was able to respond to the client, by inviting the client to express reservations directly and fully. The therapist tolerated her own personal discomfort well enough to be able to discuss the client's concerns and, in so doing, behaviorally demonstrated her competence. This is always more effective than offering verbal reassurances, which would only sound hollow to the client and still leave the burden of proof on the therapist.

Clients' concerns about entering therapy   In the previous two examples, the therapist inquired about resistance and the client expressed concerns that personally challenged the therapist. In the following example, the therapist asks about the client's potential resistance, and the client expresses concerns about entering therapy. Such concerns are far more common than challenges to the therapist and are usually much easier for the therapist to address.

Client:

It's hard for me to ask for help. I guess I'm not used to talking about myself and my problems.

OR

Client:

I feel a little awkward talking to a stranger. In our family, we always kept problems to ourselves.

First and foremost, it is crucial to let clients know that you have heard their concerns, take them seriously, and are willing to be flexible and do something about their concern if possible. Again, the best way to begin demonstrating this respect and responsiveness is to invite clients to discuss or elaborate their concerns more fully.

Therapist:

That really does sound like an important concern for us. Let's talk some more about it, so that we can both understand it better. What does it mean for you to talk to others about your problems or to ask for help?

OR

Therapist:

I'm so glad you can tell me about this awkward feeling. I think you're right: It is hard to talk to someone you don't really know about personal things. Let's be sure to go at your pace in here. You decide when and how much you wish to share with me. I'd be very comfortable with that way of working. In fact, I'd even prefer it.

Listening and responding to the client's concerns is even more important when ethnic or cultural differences exist between the therapist and client (Pinderhughes, 1989; McClure & Teyber, 1996). Cultural differences may be a particularly sensitive issue; neither client nor therapist may feel comfortable addressing them. It sometimes seems as if there is an unwritten societal rule against recognizing differences. As before, however, the best way to work with these important differences is to acknowledge them in an open-ended and nondefensive way that gives clients permission to express any concerns they may have. For example, suppose that the client is a person of color.

Therapist:

Was there anything about coming today or about our work together that concerns you?

Client:

I don't know if you can help me because you're white.

Therapist:

I'm glad we're talking about this, because our cultural backgrounds really are very different. I think there probably will be important things about you that I won't understand as well as I want to without your help. Tell me some of the things you are thinking about now that could keep us from working together well.

The therapist responds by accepting the client's concern, validating his or her point of view, and being willing to explore it further. *Different* is an important word as therapists work with people of color, gay men and lesbians, and other clients who feel on the outside. For example, many people of color will feel different in a predominantly white school or workplace—especially if few attempts are made to reach out and include them. In some settings, people who are different may be regarded pejoratively as deficient. The therapist's willingness to acknowledge differences and invite clients to express any concerns or misgivings they may have will go a long way toward establishing a collaborative relationship and keeping clients who may feel different from dropping out prematurely.

In all three of the examples here, the therapist is able to approach the client's conflict. As the therapist and client discuss the conflict, some of the client's concerns will abate with a more complete understanding of the process of therapy. Other concerns will be resolved as the client's inaccurate assumptions about the therapist are corrected. However, some of the concerns that clients express cannot be assuaged so easily. The therapist can offer to be sensitive to these concerns and express a willingness to work with the client on them over the course of therapy. One of the best ways to do this is by enlisting clients' help in better understanding their background and experience. In other words, the therapist can invite clients to tell the therapist when they are not feeling understood. The therapist can then ask clients to share their experience more fully and begin a mutual dialogue that will bring the therapist closer to clients' actual experience.

As we have seen, when obvious ethnic, religious, or other cultural differences exist, this approach of enlisting the client's help is even more apt. Ethnic, class, gender, and religious differences are complex, and it is counterproductive for therapists to labor under the misconception that it is their responsibility to understand everything about a client from a different culture. However, therapists do have a responsibility to educate themselves about clients from different cultures or social contexts, for example, by reading and consulting with informed colleagues. The therapist can also invite the client to educate the therapist whenever the client feels that the therapist is misinformed or does not understand. By welcoming this clarification instead of feeling threatened by it, the therapist signals acceptance of their very real differences, openness to a true dialogue, and invitation for a real relationship. In so doing, the therapist will gain the client's appreciation.

The therapist also needs to explore more fully with clients how they believe their cultural or social context shapes or influences their current problems and subjective experience.

Therapist:

Would you help me understand this problem in terms of the expectations from your culture?

Even when obvious cultural differences (of race or religion, for example) are not noted, therapists will be more effective and understand more fully the clients' worldview if they recognize that "all counseling is cross-cultural or multicultural because all humans differ in terms of cultural background, values, or life-style" (Speight, Myers, Cox, & Highlen, 1991, p. 29).

To sum up, the therapist must give all clients an opportunity to express their concerns about treatment or the therapist and must sincerely try to accommodate those concerns. This will go a long way toward diminishing those concerns and will allow many more clients to enter treatment and receive help.

Resistance during Subsequent Sessions

Addressing resistance directly   When a client cancels, does not show up, or comes late to appointments later in treatment, the therapist must consider the possible meaning that this behavior holds for the client. In most cases, clients are not fully aware of their conflicted feelings about therapy. When their feelings are consistently invalidated and their perceptions are disconfirmed, people eventually lose awareness of many aspects of their experience. Thus, resistant clients are not lying or deceiving the therapist; they are simply unaware of the multiple contradictory feelings they have about seeking help and being in treatment. For example:

> Getting help is relieving, but may also arouse detested feelings of humiliation.
> Feeling cared about is comforting, but can also evoke sadness about the many times this need went unfulfilled.
> Being listened to and heard is reassuring, but may trigger guilt over not having a right to one's own feelings.

Unless addressed and clarified, these conflicted emotional reactions will cause some clients to leave therapy. More commonly, when these concerns are present later in treatment and go unaddressed, therapy will simply become stalled—repetitive or intellectualized—and clients will be unable to move forward and resolve their problems. At the beginning of their clinical training, however, many therapists do not fully appreciate the ambivalent, push-pull nature of emotional conflicts. Let's examine this further by recalling Marsha from Chapter 2.

It was profoundly reassuring for Marsha when her second therapist approached her feelings directly and validated her experience. At the same time, however, this very positive experience also aroused other contradictory feelings that she hated. Marsha now felt heard and understood, but she also felt:

- Sad, as years of unacknowledged loneliness welled up
- Angry at not being heard so many times
- Guilty and disloyal to her parents for feeling angry at them
- Anxious because her ties to her internalized parents felt threatened

Marsha did not want to experience any of these feelings. She did not understand them, felt threatened by them, and could not make them go away. Fortunately, her therapist was comfortable with the push-pull nature of conflict. He continued to help her understand the different, contradictory aspects of her experience that emerged, and responded in an accepting way that also helped integrate these conflicted feelings and gradually resolve them. An important aspect of this process was that the second therapist provided Marsha with containment or a holding environment that increased her sense of safety so she could address—enter, explore, experience—these internal psychological conflicts with a significant other.

Clients are usually unaware of, and threatened by, their conflicted feelings about continuing in treatment and changing. As a result, they often feel blamed when the therapist inquires about potential signs of resistance. Clients often feel that the therapist's questions imply that they are doing something wrong or are not trying hard enough. Concerned that the therapist is angry or disappointed with them, clients may try to justify their good intentions. (For example, they may say, "Oh no, you don't understand. I really do want to see you and get here on time. It's just that. . . .") As we have already noted, therapists do not want clients to feel blamed; rather, the client is invited to join the therapist in trying to understand the danger or threat that some aspect of treatment has aroused.

To achieve this, the therapist must be direct but accepting when approaching resistance. This may sound contradictory to some beginning therapists. Without warmth and affirmation, however, approaching resistance directly can be experienced as an aggressive stripping away of the client's defenses. In actuality, few beginning therapists respond in this aggressive or intrusive way. Instead, many beginning therapists are too worried about hurting their clients or evoking their disapproval; as a result, they do not respond as forthrightly or clearly as they could. If beginning therapists are worried that their explorations may be too intrusive, aggressive, or exposing for the client, they should simply stop and find another way to respond. Better yet, they can check out their concern with the client by posing the question directly.

Therapist:

Is it okay for us to be talking about this so directly? I think it's important for our work together that we talk about this, but I don't want to be intrusive or make you feel blamed in any way. Are you feeling uncomfortable about anything like that?

The therapeutic goal is to address signs of client resistance directly but in a collaborative, nonthreatening, warm, and respectful manner. Thus, therapists can wonder aloud with clients about the possible meanings certain behavior may hold rather than insist on exploring a client's resistance. In other words, therapists must address resistance in a way that leaves clients' dignity intact. One of the best approaches is through a series of gradual steps that are progressively more direct. A three-step sequence for approaching resistance is often effective.

In the first step, the therapist offers a *permission-giving and educative response* to encourage clients to express their conflicted feelings about therapy; that is, the therapist tells the client that resistance and ambivalence are understandable and expectable feelings. If this does not work, the therapist can take a second step and encourage the client to explore the potential danger these feelings hold; this focuses on the *defense*. (For example, the therapist says, "What could the threat or danger be for you if . . . ?") If the client continues to show signs of resistance but cannot talk about it, the therapist can draw on previous working hypotheses and try to interpret the *content*. (For example, the therapist says, "I wonder if it is difficult for you to become involved in therapy because . . . ?") Let's consider examples of these three steps.

Therapist:
I'd like to talk about how it feels for you to see me in therapy. I noticed that you were 20 minutes late today and had to reschedule your appointment last week. I was wondering if there was something about coming here that didn't feel good to you. If so, it would be helpful for us to talk about that.

Client:
Oh no, I really do want to be here. And you've been very helpful already. It's just that my boss called.

Therapist:
OK, but if in the future you ever find yourself having any difficulty about being in therapy, I want you to tell me about it. In therapy it is very important that we talk about any problems that come up between us. How would it be for you to tell me about something I did that you didn't like or something about being in therapy that didn't feel good to you?

Client:
Well, that might be kind of hard for me to do.

Therapist:
What about that would be hard for you?

At this point the therapist and client are off and running together. The therapist can help clients work through the issues that would prevent them from expressing dissatisfaction or disagreement (for example, clients' concern that the therapist will not like them or others will leave them unless they are always nice). Not only will this help keep clients in therapy; it also opens up important conflict areas that need to be addressed in therapy.

In this example, the therapist offered a permission-giving and educative response in the hope that the client would then feel free to talk about ambivalent feelings toward therapy. If the client continues to show signs of resistance without being able to talk about it (missing, canceling, rescheduling, coming late, discussing only superficial issues), the therapist moves to the second step by addressing the defense.

Therapist:

It's important that you feel free to tell me about critical or troubled feelings you may have because they are bound to occur. Would you be able to tell me if something about our relationship or being in therapy troubled you?

Client:

I think so.

Therapist:

Well, maybe so, but I guess I'm not convinced of that yet. I'm wondering if it might be pretty difficult for you to tell me if I did something that you didn't like or if there was something about being in therapy that was uncomfortable for you. Sometimes people believe good clients or nice people never have irritated, frustrated, or disappointed feelings, or at least that they don't ever express them.

Client:

Well, yes, that's probably true.

Therapist:

Let's look at that. You are having difficulty being in treatment, but don't feel comfortable talking with me about that. What danger or threat might arise for you if you were less worried about being nice and took the risk of telling me what doesn't feel good about coming here? What might happen between us or go wrong?

In this example the therapist is trying to identify the threat that keeps the client from expressing conflicted feelings about therapy. *The therapist is not trying to find out why the client is not showing up for therapy but why the client is having trouble talking about it.* In other words, it is more effective to address the defense than to probe for the conflict that is being defended against. In most cases, this approach will prove effective. If focusing on the defense does not work, however, it's time to move to the third step. The therapist can try to interpret the content of the client's resistance, as a last resort. For example, if the therapist thinks the client is about to pull out of treatment and the client cannot talk about the conflict, the therapist can draw on working hypotheses and interpret the resistance:

Therapist:

It's been hard for you to get here; you've missed the last two sessions. Maybe you just forgot, but I'm wondering if there is something about being in treatment that just doesn't feel safe to you. On the basis of some other things we have talked about, I wonder if . . . ?

Why does the therapist take this more direct interpretive stance only as a last resort? The interpretation, accurate or not, is the therapist's issue and puts the ball in the therapist's court. Whenever possible, it is better to try to follow the client's lead or focus on the therapeutic process rather than to pull the client along in the therapist's direction. In other words, it is usually more effective to explore why the client is resisting or to make a process comment and address the current interaction than to make interpretations or tell the client what to do. The goal is to leave as much responsibility as possible with the client.

Conceptualizing resistance to change later in treatment   Resistance, defense, and ambivalence will occur throughout treatment. As we would expect from client response specificity, the reasons for clients' resistance to beginning treatment, and then to change later in treatment, are complex and varied. Later in treatment, when therapy has stalled or clients have stopped making progress on a problem, therapists should consider three possible sources of resistance: relational reenactments between the therapist and client; shame; and guilt. Although there are many exceptions, we will see that these three factors sometimes emerge for clients in a temporal sequence.

First, therapists should consider the possibility that some aspect of the client's core relational conflicts is being played out between the therapist and the client; that is, the client's resistance is occurring because the interpersonal process between the therapist and client is reenacting rather than resolving some aspect of the client's problem. Using process comments or finding other ways to change the maladaptive relational patterns in the therapeutic interaction will often allow the client to resume productive work. Early in their training, most beginning therapists will not possess the objectivity necessary to identify these predictable relational reenactments and will need assistance from their supervisor. All therapists can also benefit from personal exploration and therapy, so that they are aware of their own relational propensities and how these may be interacting with the client's interpersonal style to facilitate reenactments rather than resolutions of the client's conflicts.

When such relational reenactments are occurring between therapist and client along the process dimension, clients do not have the *interpersonal safety* necessary to disclose or enter more fully into their conflicts, and therapy remains stalled on a more superficial or intellectualized level. Once the ship has been righted and the therapeutic relationship is enacting a solution to clients' relational patterns, rather than a recapitulation of them, they experience the interpersonal safety they need to begin entering more fully into their conflicts.

The second mode of resistance is shame, which begins to appear as clients enter more experientially into their conflicts. Shame is more apt to generate resistance and defense than any other issue. As counselors develop the ability to hear shame and begin to recognize its many faces, they will detect it in clients' concerns about entering treatment as well as in exploring more closely clients' central conflicts or issues later in treatment. With shame, clients feel worthless, inadequate, or defective in the core and totality of their being. With guilt, in contrast, clients feel only that they have done something wrong or bad to hurt another. Thus, shame is a more primary and pervasive affect than guilt. Clients often will be without words to communicate their experience when they are suffering a profound shame reaction.

Clients employ a wide variety of characterological or interpersonal coping styles to keep their shame concealed from others and to defend against the agonizing experience of a full-blown shame reaction. For example, clients may do this by acting arrogantly and self-righteously toward others; by intimidating, controlling, or inducing shame in others; by adopting perfectionistic standards around their own work, cleanliness, or religious practices; and by developing addictions. As the affect of shame is evoked, therapists may see *shame-anxiety*— anxiety signaling a threat to clients' defenses against their shame that might be revealed to themselves and others. Perhaps more disturbing to observe is a client's *shame-rage cycle* of anger and outrage, an attempt to restore a sense of power and control after the experience of being shamed. These and other emotional expressions commonly occur to protect the self from being wounded or feeling annihilated by intense shame experiences.

Once clients stop these defenses and begin to actually experience the shame that accompanies a conflict, therapists seem to have as much difficulty in witnessing and responding to the shame as clients have in bearing it. Guidelines for responding to shame and helping clients contain their painful self-loathing or self-hatred will be provided in Chapter 5. For the time being, therapists need to begin listening for the emergence of shame and to allow themselves to consider the possibility that resistance often serves to protect clients against painful feelings of shame.

After clients have been able to tolerate, contain, and ultimately integrate their shame and other conflicted feelings that are central to their problems, they begin to make healthy new changes. At this point, some clients will exhibit the third mode of resistance: *binding feelings of guilt or disloyalty over success in treatment.* Therapists will observe that some clients retreat from progress in treatment or undo successful changes they have just made. Often, clients cannot sustain positive changes they have just achieved; because the healthy new behavior is inconsistent with internal working models and threatens their attachment ties to internalized caregivers. *Becoming stronger or improving in therapy makes some clients feel cut off from parental approval and affection, disloyal to caregivers, or guilty about hurting, leaving, or surpassing the parent in some way.* These clients may report feeling either: (1) alone, empty, or disconnected following meaningful change; or (2) guilty, selfish, or bad. When these reactions occur, clients need to be reassured of the therapist's continued presence and support for this stronger self, or else they will have to restore ties by sabotaging their own progress or success. These fascinating and complex sources of resistance to change will be elaborated throughout the chapters that follow.

Closing

Listening to the client with concern and respect is the most powerful intervention that the therapist can make in the beginning of therapy. Such listening is the basic tool the therapist uses to begin a working alliance with the client. The next step is to address ambivalence or resistance to therapy in order to maintain the relationship that has just begun. Resistance will be more of an issue with some clients than with others, but it will occur to some extent for every client. As we have seen, if the client expresses these ambivalent or resistant feelings, and the therapist responds to them, the feelings will be far less likely to pull the client out of therapy. Resistance will remain a concern throughout therapy and will emerge repeatedly as clients address difficult personal issues. Responding effectively to the many types of resistance that occur for clients is an important way to keep clients engaged and progressing in treatment. If therapists formulate working hypotheses, it will help them to anticipate and recognize client resistance when it occurs and to respond more effectively; better prepared for what may occur, they won't have to think on their feet so quickly. At the same time, therapists must be careful not to steer their clients toward confirming or refuting these working hypotheses. Rather, therapists continue to follow the client's lead but are ready to address any issues related to working hypotheses that may appear.

Resistance, ambivalence, and defense provide a window to certain inner aspects of the self. Each provides an opportunity to observe the fascinating workings of internal conflict. People do not possess a unified self; humans are so complex and multifaceted that it often feels as if one part of ourselves is working against another part. By resolving an internal conflict, a person becomes a little more integrated or whole. Integrating disparate parts of the self is often an important part of resolving problems and is the avenue to personal growth and change. When this process centers on issues of resistance, the therapist and client stride forward together. This early success can show clients that they possess the internal resources, and have the help they need, to resolve the bigger issues that lie ahead.

Suggestions for Further Reading

1. Beginning therapists are encouraged to read further about resistance. One classic reference in this area is Ralph Greenson's *The Technique and Practice of Psychoanalysis*, Volume 1 (New York: International Universities Press, 1967).

Chapter 2 provides a lucid introduction to the topic of resistance and offers informative case studies to illustrate effective therapeutic responses. Again, the reader does not have to be psychoanalytically oriented to be informed by this discussion.

2. Beginning therapists will find it useful to examine other approaches to working with resistance. For example, the family therapy literature has highlighted therapeutic responses to resistance. Even students who are interested only in individual psychotherapy may find it useful to examine such basic texts in this area as *Mastering Resistance: A Practical Guide to Family Therapy*, by C. M. Anderson and S. Steward (New York: Guilford Press, 1983).

3. A process-oriented approach to multicultural issues in counseling can be found in Chapter 1 of *Child and Adolescent Therapy: A Multicultural-Relational Approach*, edited by F. McClure and E. Teyber (Fort Worth: Harcourt Brace, 1996). An excellent discussion of ethnographic and cross-cultural issues in counseling can be found in H. Pinderhughes's, *Understanding Race, Ethnicity, and Power* (New York: Free Press, 1989).

CHAPTER FOUR

An Internal Focus for Change

Conceptual Overview

The first stage of therapy is complete when therapist and client have established a collaborative relationship and begun to work together on the client's problems. This collaborative alliance is a necessary prerequisite to the second stage of therapy: the client's journey inward. In order to change, clients must become less preoccupied with the problematic behavior of others and begin to explore their own internal and interpersonal responses. Clients need to stop focusing exclusively on historical events, past relationships, and the problematic behavior of others in their lives. Instead, clients must begin to clarify their own thoughts, feelings, and reactions to the troubling events they are experiencing. Why? *Clients will usually fail in their attempts to change or control others in their lives, whereas they can often resolve problems by understanding and changing themselves and their own role in problems.* In this process, clients can begin redefining how they want to construct future relationships and can begin to identify internal responses based on old problematic templates and to determine their usefulness in their current lives. This process gives clients greater control and the freedom to act purposefully in ways they choose.

The therapist's task is thus to help clients make the transition from seeing the source and resolution of problems in others to adopting an internal focus for change. This is a twofold process. First, the therapist must help clients to look within; that is, the therapist must focus clients away from complaining about or trying to change others, and toward identifying, understanding, and changing their own reactions. Second, therapists must help clients become active agents in their own therapeutic work. This will occur as clients become aware of their old templates and usual styles of responding. With greater awareness of their own internal and interpersonal responses, clients begin to recognize that they do not have to respond in their old maladaptive ways and to receive the same hurtful or disappointing responses from others. This new awareness, coupled with the therapist's support for the clients' own self-direction and initiative within the therapeutic relationship, will empower them to find new and more effective responses to old problems.

Self-efficacy develops out of this collaborative process between therapist and client; when clients feel ownership of the change process, enduring change results. However, it will be difficult at times for clients to shift their focus from others and begin looking at their own role. To take this rewarding journey inward, clients need the support provided by a secure and caring relationship with the therapist. In this chapter, we will see how therapists can provide that safety.

Chapter Organization

This chapter has four sections. In the first section, we examine the need for clients to shift from an external to an internal focus for change. We describe how clients often externalize their problems onto others, which perpetuates their feelings of powerlessness, frustration, helplessness, "stuckness," or depression. Thus, we discuss how therapists can help clients achieve this inward focus and gain a greater sense of legitimate or internal control; and we examine how some clients can misunderstand the therapist's shift to an internal focus at first. Some clients initially misperceive the internal focus as implying that they have to accept unwanted blame for problems with others, or as denying the reality of others' problematic behavior. We will see how therapists can resolve this misunderstanding if it occurs.

In the second section, we examine ways in which clients can become active agents in resolving their own problems; we discuss how therapists can use the therapeutic relationship to foster the client's own initiative and sense of control; and we explore specific ways of intervening that facilitate the client's ownership of the change process. Sample therapist-client dialogues illustrate effective and ineffective ways to achieve these goals.

In the third section, we see how therapists can actively enlist clients in understanding their own problems by extending the collaborative alliance into the problem-solving phase of treatment as well. A case study illustrates how therapists can help clients share responsibility and remain contributing agents throughout each phase of treatment.

In the final section, we will examine how therapists can help the client adopt an internal focus by tracking the client's anxiety. Anxiety is a *signal* that can guide therapists closer to the heart of the client's concerns and indicate when it is most productive to focus the client inward.

Shifting to an Internal Focus
Clients' Externalizing Their Problems

In the initial stage of therapy, many clients see the source of their problems in others. Clients often want to spend more time describing others' problematic behavior than discussing their own experience of and response to the problems. For example, many clients begin the first few therapy sessions by announcing that the problem is really with another person:

My wife is always on my back.
My husband won't pay any attention to me.
My children are impossible; they won't do anything I say.
My boss is a demanding tyrant. If he doesn't back off, I'm going to have a heart attack.
My mother won't stop criticizing me; I can't do anything right according to her.
I'm 27 years old, and my father treats me like a child.
My boyfriend has a drinking problem.

In most cases, *it is essential for the therapist to acknowledge these complaints as valid and legitimate concerns that are genuinely troubling.* Usually, these complaints are not merely transference distortions to be challenged or dismissed but contain much truth. As emphasized in Chapter 2, the therapist must validate or affirm, whenever possible, that the behavior of others is troublesome; when that is not possible, the therapist should at least affirm the subjective reality of the client's perceptions of what the problematic behavior of others is doing. If the therapist does not provide this affirmation first, many clients—especially those who have suffered much invalidation in their lives—will feel that the therapist doesn't understand or care. This is counterproductive; it will often reenact the client's developmental conflict. Thus, to keep this potential invalidation from occurring, the therapist's first responsibility is to hear and validate the client's concerns. Only after making this genuine empathic connection, which often requires the therapist to decenter and see issues from the client's subjective viewpoint, can the therapist take the next step.

The second step is to begin focusing clients back on their own thoughts, feelings, and reactions to the problematic behavior of others, rather than joining the client in focusing exclusively on the external problem. Often, it will be easy for therapists to pair the first two steps together.

Therapist:
I'm sorry your father does that to you. I think it would be upsetting, too. What do you say and do when he . . . ?

OR

Therapist:
It sounds so frustrating to have your boss respond that way over and over again, no matter what you do. How do you feel and what are you thinking inside when she . . . ?

The same principle applies when working with people of color, women, gay men and lesbians, and others who accurately explain the social inequities underlying their problems. Once again, the therapist must join such clients in the reality of their situation, as they experience it, before focusing inward on their reactions to, and ways of coping with, these inequities. Why is therapy more productive when therapists focus clients inward? In many cases, clients who feel depressed, anxious, distrustful, enraged, or helpless because of the behavior of another person will try to enlist the therapist in blaming, criticizing, or trying to change the other person. However, therapy will not progress very far if the therapist merely joins clients in focusing on the other person's behavior, no matter how problematic this behavior may be. Why? As the existential therapists inform us, clients' attempts to change the other person will usually fail; clients are much more likely to resolve the problem by changing their own way of responding (Kaiser, 1965; Wheelis, 1974). Thus, the therapist's task is to expand clients' focus beyond the other person and include themselves as well.

Note that the therapist should be flexible in adopting an internal focus, however. For example, some clients will benefit from understanding the behavior or potential motivations of problematic others. If a client has grown up befuddled by a parent with a borderline or narcissistic personality disorder, it may be deeply validating for the client to read the descriptions of these disorders in the *Diagnostic and Statistical Manual* (DSM-IV) and to realize that all of this was not their fault and did not happen solely to them.

How can therapists increase clients' awareness of their own thoughts, feelings, and reactions in problematic situations? Recall the externalizing comments cited at the beginning of this section. Each of the following questions can be used to focus those clients back on themselves.

How do you feel when your wife is nagging you?
What do you find yourself thinking when your husband is ignoring you?
What do you do when your children disobey you?
What is it like for you when your boss is so demanding?
What thoughts were you having as your mother was criticizing you?
How do you react to your father when he treats you that way?
What is most troublesome for you about your boyfriend's drinking?

Simple inquiries of this type serve two important functions. First, they tell clients that the therapist is listening to their concerns and is taking them seriously. The therapist has not changed the topic or brought up something that was discussed 10 minutes ago but is responding directly to their concerns. Second, while inviting clients to say more about their concerns, the therapist is also shifting clients' focus away from others and encouraging them to look more closely at themselves. Inviting clients to become more aware of their own reactions is a critical step toward understanding and resolving their problems. As we will see, the simple response of focusing clients on their own experience is a powerful intervention that will elicit strong feelings, memories, and associations.

In many cases, clients will welcome the therapist's offer to talk more about themselves. When clients begin to focus inward and learn more about themselves, they begin to change. Clients' active exploration of their own lives and problems will lead to increased awareness of their own responses and of the options available to them. This results in new feelings of mastery and control, which, in turn, lead to growth and change. However, not all clients will respond so positively to the therapist's initial invitation to look more closely at themselves. Clients may avoid or actively reject an internal focus and continue to talk about the problem out there in others. In that case, the therapist continues to inquire about the personal meaning that this particular situation holds for the client.

Therapist:
Where would you like to begin this morning?

Client:
My wife is impossible to live with. She complains constantly; nothing ever pleases her.

Therapist:
It sounds like you've had a difficult week with her. What's been the hardest thing for you?

Client:
Do you know how hard it is to live with an angry, demanding wife who keeps trying to tell you what to do all the time?

Therapist:
You're very angry at her. And I can see why; it would be hard to have someone after you like that. Tell me, how do you respond to her when she is doing this?

Client:
I don't know; I just hate it. I guess I yell back sometimes, or just try to get away from them. It's not just her, you know; her whole family is like that.

Therapist:
They really are very critical of you, and I can imagine how hard that would be to live with. But it seems like criticism, in particular, really gets under your skin. What's it like for you to be criticized so much?

Client:
I hate it. I just hate it. They make me feel like I can't do anything right—that I'm doing it wrong and failing all the time.

Therapist:
Doing it wrong and failing—ouch! Sounds like her family's constant criticism taps into your own painful feelings about yourself of not measuring up or not being enough. And having those shameful feelings of inadequacy aroused all the time would be infuriating.

Client:

> Yeah, I hate them for making me feel this way. If I could just make them stop, everything would be okay.

Therapist:

> This is very important for you, and we need to work together on it. I would like to understand better what you do when they criticize, so that I can help you learn some more assertive, limit-setting responses. But your own feeling of not measuring up is also a part of this problem that we need to work on, too, so that you can stop charging at the red flag they are waving. If we can change the internal part of you that overreacts as if you believe what they are saying is true, it would be much easier for you to handle this than it has been in the past.

Client:

> What do you mean, "overreacts"?

Therapist:

> Sounds like their criticism makes you feel you can *never* do anything right. Is that accurate?

Client:

> Yes, but that's not true. They just expect too much and I can't do it all. I can never do it good enough for them.

Therapist:

> No matter what you do they will never be pleased. But their expectations still upset you. It's the way you end up feeling inside, about yourself, that becomes the problem?

Client:

> Yes.

Therapist:

> So being able to work on your feelings would be helpful?

Client:

> All right, it probably would be. What should I do?

Therapist:

> Tell me more about your feeling of not measuring up.

Client:

> Well, I guess I've always sort of felt like I'm not really good enough.

In this dialogue, the therapist validated the client's experience while at the same time encouraging the client to look at his own reactions and role in the conflict. Although the client kept trying to focus on the problematic behavior of his wife and her family, the therapist's repeated attempts to have the client look at himself as well eventually served to mitigate his defensive, externalizing stance. As a result, the client moved closer to his own problematic feelings and beliefs about himself that contributed to the marital conflict. By gaining a better understanding of his own feelings of inadequacy, the client will become less *shame-*

prone and overreactive to others' criticisms. This, in turn, will be an important part of learning to respond more assertively and set limits more effectively, rather than merely yelling back or withdrawing.

Many clients have not been encouraged to focus on themselves before. As in this example, some clients may resist this internal focus initially, perhaps because it asks them to come face to face with difficult feelings that they have not been able to understand, contain, or resolve alone. The paradox is that, as long as clients avoid the internal or personal aspects of their relational conflicts by externalizing their problems onto others, they will feel powerless, dependent, frustrated, and out of control.

Many clients who are struggling with conflicts are inappropriately invested in changing others as a means of managing their own problems or insecurities. In attempting to shape or direct others, the client is seeking an external solution to an internal problem. The reality is that, beyond clearly and directly expressing our preferences and personal limits ("I would like . . . "; "I will not . . . "), we cannot readily influence how others think, feel, and act. For example, clients routinely fail in their attempts to make their spouse stop drinking, smoking, or overeating. Similarly, some clients try for decades to win the approval or recognition from others that they never received from a parent. Others try unsuccessfully for years to have their grown offspring choose a different mate, religion, or career. As a result of these failed attempts to change others, many clients enter therapy with feelings of helplessness and depression.

A therapist can offer clients the more productive alternative of decreasing their attempts to change others and gaining more understanding and control over their own reactions instead. This internal focus is a necessary prerequisite before clients can adopt new, more effective responses to old problems. Only when clients begin to focus on understanding and changing themselves will they begin to feel in charge of their lives and capable of change. Clients experience these feelings as *empowerment*. Thus, as we have seen, a consistent therapeutic response should be to gently focus clients back onto themselves. The following types of questions will help clients explore their own responses:

What is the main feeling you are left with when . . . ?
What were the thoughts you were having when . . . ?
What was the most difficult thing for you when . . . ?
How would you like to be able to respond when . . . ?

Focusing Clients Inward

Focusing clients on their own behavior often reveals how they are contributing to, or participating in, their own conflicts. Often, clients will not have realized their own role in these conflicts. As we will see, clients who can focus on themselves and see their own participation in a conflict are usually motivated to change their own part in it. This, in turn, may allow the other person to respond differently as well. To illustrate, let's return to our example of the husband who complained about his critical wife and in-laws.

Although the client did begin to explore his own feelings of inadequacy and the ineffective ways he responded to his wife, he kept complaining about his "obnoxious" wife and "superior" in-laws. As before, the therapist was empathic, validated his experience, and supported his anger. At the same time, the therapist did not join the client in focusing exclusively on his wife or blaming her as the sole source of his problems. Instead, the therapist continued to focus the client away from his preoccupation with his wife's behavior and toward his own reactions to her.

What would have happened if the therapist had not done this? If the therapist had responded to the client's eliciting pull to blame his wife, the client would have remained an angry but helpless victim. In contrast, if the therapist had emphasized just the client's contribution to his marital conflict, without first validating his experience, it would have recapitulated for him a prototypic developmental conflict. Exploration of the client's feelings of inadequacy revealed that these shameful feelings stemmed in part from rarely being supported by his parents and consistently being blamed for whatever went wrong. Thus, by validating the client's experience first, and not prematurely focusing him inward, the therapist helped the client find a new and more satisfying response to the old conflict of constant critical intrusions. By coupling this validation with a consistent invitation to adopt an internal focus, the therapist allowed the client to become more aware of how his own current reactions to his wife contributed to the ongoing conflict in his marriage.

For example, the client learned that he was quiet and unresponsive to his wife when he arrived home from work. The therapist helped him see that his silence escalated his wife's attempts to gain his attention and, in response, he became even less communicative. Their problematic interaction worsened as the wife became more insistent and demanding of his attention and he withdrew even further—by taking a nap, for example. By focusing on his own emotional reactions at that moment, the client realized that his wife's increasing requests for attention also made him feel overwhelmed by the prospect of trying to meet all of her needs; he had held the unrealistic belief that if he did not always respond to her, he was failing as a husband and again being inadequate.

By encouraging the client to become more aware of his own internal and interpersonal responses to his wife, the therapist gave the client an opportunity to change his part in the marital conflict. Specifically, the husband began to express more directly to his wife when he would like to be close to her and when he would prefer to be alone. He explained to his wife that he wanted 30 minutes alone to unwind when he returned home from work, after which they could sit down and share their day together.

More important, as the husband gained a better understanding of his tendency to feel overwhelmed by others' demands, he became less reactive to his wife's requests. He learned that he felt shamefully ruled by her, in part because he could neither say no to her requests nor express his own wishes. If his wife wanted something from him, he believed that he was failing if he didn't respond immediately. And because he couldn't set limits or ask for what he wanted without feeling selfish, he resented the "unfairness" in their relationship. The client also learned that he expressed this resentment to his wife indirectly—in his own critical and withholding ways. Over the next few months, his new awareness allowed him to respond to his wife's requests in the manner that felt most appropriate to him; he no longer felt he must comply with them, and his resentment abated. As the client set these limits, he began to feel more control over himself in their relationship and was able to make some tentative steps toward greater closeness with his wife.

Thus, as clients begin to focus on their own behavior, they often start to feel less stuck, powerless, and depressed. As they start to see more alternatives available to them, clients begin to feel more hopeful and in charge of their lives. Thus, the therapist's task is to discourage clients from passively complaining about others or from trying to manipulate and control others' behavior. However, therapists must do this gradually, after affirming clients' legitimate concerns about the problematic behavior of others.

Reluctance to Adopt an Internal Focus

Therapists' reluctance   Most beginning therapists feel comfortable in diminishing or assuaging the client's fears or concerns. However, focusing the client inward means that the therapist is *drawing out* the client's fears and concerns. There are many reasons why therapists may have difficulty in approaching client conflicts directly. For some, such directness goes against unspoken familial rules or cultural norms. Other therapists may equate being forthright with being aggressive, intrusive, or exposing, because they have only seen forthrightness utilized in these hurtful ways. Some therapists are reluctant to address clients' internal conflicts directly because they believe that good people don't make others hurt or feel bad. These therapists try to make clients feel better by reassuring them about their insecurities and by emphasizing only their strengths and successes. However, moving away from client conflicts is an ineffective response that, although well-intended, keeps clients from fully experiencing, expressing, and exploring their problems. In this way, therapists rob clients of the opportunity to participate in finding solutions to problems and to develop the sense of efficacy and empowerment necessary for enduring change. Still other therapists may avoid an internal focus in order to please clients and maintain their approval;

these therapists fear that clients will become angry as the therapist invites them to explore the uncomfortable feelings or difficult conflicts they have been avoiding. Therapists are also afraid at times that they may not be able to remain present and responsive to the client when strong emotions are expressed. This issue will be addressed more fully in subsequent chapters.

For these and other reasons, therapists often join with clients in looking away from the internal aspect of the conflict or the client's participation in the problem. Therapists collude with the client in externalizing problems when they repeatedly:

- *Give advice* and tell the client what to do or how to respond to others
- *Interpret or explain* what the other person's motives or behavior may mean
- *Reassure* clients that their problems will go away or are not something to be concerned about
- *Disclose* what the therapist has done to cope with a similar person or problem

As we would expect from client response specificity, each of these responses will be effective at times. However, if they come to characterize the ongoing course of treatment for most clients, clients will be prevented from exploring their own experience and, ultimately, from being able to resolve their own conflicts. Thus, throughout the course of treatment, therapists must repeatedly help clients expand and elaborate their awareness of their internal experiences and interpersonal responses. As we will see, however, this is difficult to do. A consistent focus on self and internal experiences often arouses clients' anxiety as dreaded feelings, pathogenic beliefs, or expectations of hurtful responses from the therapist and others may all be evoked.

Clients' reluctance   Why do clients tend to avoid an internal focus? As we have seen, a shift away from their focus on others can make some clients feel that the therapist does not really understand them or is not sympathetic to their concerns. Other clients fear that, if they give up their attempts to change others, they will either have to accept the blame for the problem or remain forever resigned to "defeat." Still other clients may not yet feel held or contained by the therapist's compassion and understanding; they do not have the secure base they need to approach the painful feelings or difficult choices that looking inside entails. Clients will also feel unable to share certain concerns with the therapist if the therapist and client are reenacting some aspect of the client's conflict in their interpersonal process. Both, however, are usually unaware that this is occurring.

Cultural factors may also impede an internal focus. For example, Native American, Asian, and other clients from communally centered cultures may initially react negatively to looking within because it sounds self-centered. If clients perceive a difference in values and goals, they may question the therapist's credibility. Therapists can help with this by educating clients about the treatment process and, more importantly, by responding in a manner that is congruent with the clients' worldview.

Therapist:

> How does your response to this affect everyone else?

OR

Therapist:

> Let's look at their role in this—and your role, too—and see if we can make more sense of what's going on for you.

Thus, the therapist should work with clients to explore their reasons for not wanting to look within. This is another aspect of working with resistance. However, when clients are not ready to look at their own issues, the therapist should not be insistent. How can the therapist best respond when clients resist an internal focus? One thing the therapist can do is to follow clients along for a while, while offering clients repeated invitations to reveal a little more about themselves. In most cases, clients will gradually accept the therapist's invitation and soon come to welcome the opportunity. If the clients' externalizing stance does not begin to change over the course of several sessions, however, the therapist can make a process comment that simply describes their interaction.

Therapist:

> I've noticed that you talk very easily about your husband and your daughter, but you don't say very much about yourself. Are you aware that that happens?

OR

Therapist:

> I think we've been missing each other the last two sessions. I keep asking what you were thinking about or trying to do in a particular situation, and you keep responding by telling me more about the other person. What do you see going on between us?

These process comments will help clients to become aware of their externalizing style with the therapist and to learn that they probably respond in the same way to other people. For example, clients who cannot share themselves in a personal way with others will often be perceived as boring or aloof. In that case, part of the clients' presenting problem—for example, clients' sense of loneliness, isolation, or lack of meaningful contact with others—may also be enacted in their relationship with the therapist. Although it is often frustrating and discouraging for clients to realize that they are recapitulating their conflict in the therapy session, this reenactment provides them with the opportunity to resolve the problem by changing the distancing pattern *within* the therapeutic relationship. The therapist helps clients become aware of how they interact with others and how others experience them, and then offers a new and more effective way of relating in the therapeutic relationship. For example, the therapist can describe the current interaction and offer to develop a more meaningful and sharing relationship:

Therapist:

I feel that I am being held away from you when you talk about others rather than yourself. I feel like I am missing you, and I don't want that. I would like to learn more about you, or your concern in talking about your-self. Can we work together on this?

When the therapist uses questions or process comments to focus clients inward, clients will often reveal a variety of new and important concerns that they have not talked about before.

Well, I guess there really isn't very much about me that people would want to know.

I'm not used to telling people what I'm thinking or feeling.

Every time I try to get close to someone, I get hurt in the end.

You wouldn't like me very much if you knew what I was really like.

In this way, the therapist's process comment has brought out important new concerns that can now be addressed in therapy. The therapeutic relationship and the content to be explored have deepened. *This process of helping clients to focus inward and to explore their resistance to looking within will provide some of the most important material to be addressed in therapy.* These important new concerns are not likely to be brought up for treatment unless the therapist uncovers them by using process comments and focusing clients inward. Thus, one of the most important reasons for adopting an internal focus is to evoke the central conflicts that are operating in the client's life and to do so in a way that makes them accessible for treatment.

Placing the Locus of Change with Clients

The first component of the internal focus is to help clients look within and become more aware of their own reactions and responses. A second component is that the client must begin to act from within, by adopting an internal locus for change. In this section, we examine how clients can gain a greater sense of *effectance* in their lives by becoming active agents in their own change process (Bandura, 1982; White, 1959).

First, we will see how the therapist can use the therapeutic relationship to foster the client's own initiative and self-direction. Second, we will examine specific types of therapeutic interventions that encourage the client's participa-tion in, and control over, the change process. Throughout, the therapist's goal is to respond to the client's feelings, interests, and initiative and to thereby support the client's growing sense of competence.

Using the Therapeutic Relationship to Foster Clients' Initiative

In the course of treatment, a therapist has the opportunity to convey to clients the broader message that they are able to take charge and effectively manage their own lives. Effective therapists of every theoretical orientation nurture clients' own sense of personal dignity and self-efficacy. Beyond verbal encouragement of clients' initiative and autonomy, therapists must provide clients with the experience of *acting* more effectively during the therapy session. Once clients are experiencing greater control in their relationship with the therapist, it is often relatively easy to help clients generalize this greater sense of personal power to other relationships beyond the therapeutic setting (Bandura, 1977). Let's see how therapists can help clients assume an active, initiating role in therapy.

The first way to help clients feel responsible for and capable of change is to repeatedly encourage clients to create their own agenda in therapy and to talk about whatever they feel is most important or relevant for them right then. Once clients begin pursuing their own concerns in this way, the therapist can be an active participant who helps clients to explore their own needs, wishes, and concerns more fully, to understand their problems, and to generate potential solutions and behavioral alternatives.

Why is it essential to encourage and respond to the client's own initiative, even in time-limited and short-term modalities? In one way or another, most clients are unable to act on their own wishes, preferences, and interests. In past relationships, these clients have not had significant others respond affirmingly or acceptingly to their own genuine feelings—especially of sadness and anger—or support their own interests. As a result, these clients will feel enlivened and encouraged but sometimes anxious, too, if the therapist supports their feelings, perceptions, and self-direction. Regardless of treatment length or the therapist's theoretical orientation, therapy becomes a more intense and productive experience when the therapist can successfully engage clients in pursuing their own interests and exploring their own concerns.

When the therapist can help clients explore and understand the material they produce, rather than merely direct them to the therapist's own agenda, change will usually begin to occur. When clients provide the momentum and direction for therapy, they often experience a corrective relationship where they can have their own opinions, act more effectively, and be more in charge of their own lives. Once clients have begun to initiate, the therapist can actively participate by providing the types of responses that this particular client seems to utilize most productively. Commonly, this will include interventions from a wide variety of theoretical orientations—contributing information, explanations, or interpretations; constructing behavioral alternatives; suggesting possibilities about what significant others may be doing or intending; providing interpersonal feedback about how others may be perceiving or reacting to the client; and so forth. *Whatever interventions or techniques the therapist uses, however, they will be far more effective when given in response to the clients' own agenda, interests, or concerns, rather than the therapist's, employer's, or spouse's.* This process dimension is one of the most important characteristics of the therapeutic relationship.

Avoiding a hierarchical relationship   We have already noted that clients may resist the therapist's attempts to make them active participants in treatment. Why? Frequently, clients will be conflicted about expressing their own wishes, acting on their own initiative, and achieving successes or being able to enjoy them. Because so many clients have not been supported in their own individuation and feel guilty or anxious if they exhibit self-direction, competence, or success, they continually elicit advice, direction, and explanations from the therapist. Therapy will not be productive if this helper-helpee mode comes to characterize the therapeutic process. This mode shifts responsibility away from the client and onto the therapist and creates the hierarchical relationship discussed in Chapter 2. Clients will not be able to feel greater efficacy and adopt a stronger stance in their lives as long as this dependency is fostered—their *compliance* is being reenacted in the therapeutic relationship rather than challenged.

Too often, therapists unwittingly comply with the clients' subtle or overt request to tell them what to talk about in therapy and what to do in their lives. It is flattering to think that we know what is best and can tell others what to do; it appeals to the narcissism in every therapist. As we would expect on the basis of client response specificity, there certainly will be circumstances in which strong directives, advice, and firm limits are facilitative for a client. However, where such initiatives characterize the ongoing process of therapy, clients will remain dependent on helpers to manage their lives. Because this problematic process occurs subtly (but frequently), we need to examine further this dependency-fostering approach to therapy.

Supporting clients' own autonomy and initiative   Effective therapy of any modality or treatment length should foster and enhance the client's self-efficacy. Therapists cannot just talk with clients about choice, responsibility, personal power, or autonomy, however; they must create a relationship where clients behave this way. To do this, the therapist must first give clients permission to follow their own interests and actively encourage them to introduce the material that is most relevant to them. As detailed in Chapter 2, the therapist must then be able to identify the subjective meaning and feeling in that material and must encourage clients to expand and elaborate this further. In response to clients' increasingly specific exploration, the therapist actively helps clients clarify and understand their experience; together, they can begin to generate more effective ways to respond. When this mutual interchange continues throughout several sessions, most clients become deeply committed to treatment and find that meaningful change has already begun to occur. *Facilitating this growth process may be the primary challenge and satisfaction of being a therapist.* Let's see how these ideas might sound or play out with a client:

Client:

I need more direction from you. What do you think is going on here?

Therapist:

I have a couple of ideas to share with you, but I think it would work best if I heard yours first.

Client:

If I knew what to do, I'd just do it, and wouldn't ask you or be coming here to see you!

Therapist:

All right, let's swap ideas. You tell me what you think is going on for you; and then I'll tell you what I see occurring; and let's see what we can put together.

Client:

Like I just said, I don't know—I don't really have any ideas.

Therapist:

Let's wait for just a minute and see if anything comes to you. If not, I'll be happy to go first and start us off.

Client:

Well, maybe I'm afraid of being left or alone or something like that.

Therapist:

That's very interesting, and fits with what I'm thinking about. Tell me more about what it means to be "left" or "alone"; it sounds important.

Client:

I think I've always been worried about that.

Therapist:

Well, it makes me think that maybe your parents were physically present for you, but that they cut off from you emotionally whenever you disappointed them. I wonder if that created the feeling of being all alone, even though others were present—which would have been very confusing to a child.

Client:

Yeah, I think that's right, and the worst thing about it was that it seemed like it was all my fault. They were going away or withdrawing because *I* let them down. What do you think?

Therapist:

I think you have very good ideas, but something often seems to hold you back from expressing them in the strong way you have just been doing with me. I have noticed that happening in here with me sometimes, and I hear it in your relationships with others as well. I wonder if there is some connection between holding yourself back in this way and being afraid of being left. What do you think?

This type of interaction, in which the therapist and client build on what the other has just produced, is a collaborative *independence-fostering* approach to psychotherapy. Clients have the experience of sharing responsibility for the course of treatment, as the therapist actively encourages them to define, address, and resolve their own concerns and conflicts. As therapists support clients in achieving this more actively involved stance, however, they are simultaneously contributing their own information and guidance in a way that creates a working partnership.

A common misconception is that longer-term, dynamic, or relationship-based therapies are dependency-fostering, whereas short-term, problem-solving, or strictly behavioral approaches are not. Actually, whether the therapy fosters dependence or independence is determined by the therapeutic process and not by the length of treatment or the theoretical orientation of the clinician. In short- or longer-term therapy, the client's dependency is inappropriately fostered when the therapist repeatedly directs the course of therapy, gives advice, and prescribes solutions for the client. Let's examine an effective middle ground between the nonproductive polarities of directive and nondirective control.

Shared control in the therapist-client relationship   As we have begun to see, therapy will be most successful when the therapist and client share control over the direction of therapy. It is overly controlling for the therapist to play the predominant role in structuring and directing the course of treatment, and it is ineffective to nondirectively abandon clients to their own confusion and frustration. In a more productive relationship, the therapist will encourage and enable the client to take the lead but will contribute information, clarifications, and alternatives to the material that the client has produced. It will be therapeutic for many clients to experience a relationship in which control is shared by both participants, rather than a relationship in which control is held by one party or the two parties compete for control. This therapeutic collaboration is the key to empowerment and provides the client with the support, freedom, and permission to act in new and stronger ways that expand old templates. For example, in Chapter 2, we stressed that in the beginning of treatment, the therapist should encourage the client to take the initiative and bring up whatever he or she would like to discuss.

Therapist:
I would like to begin each session by having you bring up what you want to talk about. I think it's important that we work on what is most important to you.

Some therapists will be frustrated by this approach because they feel an urgent need to direct clients toward solutions and action. However, a directive stance, although it will be helpful to certain clients at certain times, imposes several important limitations. First, as we have seen, a directive approach prevents the therapist and client from uncovering conflicts that were not evident in the client's presenting problem. Second, the goal of therapy is not merely to fix the client's presenting problem, but to teach clients that they possess the personal resources to resolve problems and effectively negotiate their own adult lives. Working collaboratively, the client gains an increasing sense of self-efficacy, which directive approaches cannot provide.

Responding to these arguments, the action-oriented therapist might ask, "What if you follow the client's lead and it takes you nowhere?" Certainly, a nondirective approach requires an unrealistic amount of time and patience from the therapist and often results in a disorganized and unfocused therapy. More-over, a purely nondirective approach is likely to reenact problematic relational patterns for clients who grew up in a permissive home, just as a purely directive approach may reenact such patterns for clients who grew up in an authoritarian home. As we have seen, however, there is an effective middle ground of shared therapist-client control, although it has not been delineated well in the counseling literature.

Therapists need treatment plans and intervention strategies with short-term, intermediate, and long-term goals, but these must develop out of collaborative interaction with the client, who will then feel ownership of the treatment process. In order to implement this interpersonal process, *the therapist must be able to tolerate ambiguity* and to refrain from subtly leading or controlling the interaction. Instead, the therapist must try to create opportunities for clients to voice their own concerns and act on their own initiative. The therapist must then be able to decenter, enter into the client's subjective worldview, and grasp the central meaning or feeling that the emerging material holds for the client.

This approach requires an active but noncontrolling therapist who is neither directive nor nondirective. In this model, therapists respond actively by communicating their understanding and affirming the client's experience; joining the client in expanding and exploring further what the client has produced; and providing additional structure and focus for treatment by highlighting the repetitive relational patterns, core conflicted affects, and pathogenic beliefs that recur in the client's narratives. Furthermore, the therapist is using process comments to address treatment impediments—such as resistance and an external focus—and to prevent relational reenactments by describing or talking about what is transpiring between them in their current interaction.

As we will see in Chapter 6, an authoritative parenting style that encompasses both parental warmth and firm limits reflects a more effective middle ground between the better known and more widely adopted authoritarian and permissive parenting styles. In parallel, therapists will find that shared control offers most clients a more productive alternative than either directive or nondirective approaches. Let's examine some therapeutic interventions that illustrate this middle ground of shared control.

Therapeutic Interventions That Place Clients at the Fulcrum of Change

As we have seen, the therapist's goal in the interpersonal process approach is to encourage the client's lead while still participating actively in shaping the course of therapy. The critical technique is for the therapist to intervene without taking the impetus away from the client. The following examples illustrate effective and ineffective ways to do this.

Ineffective interventions   Suppose that the client is filling the therapy session with seemingly irrelevant storytelling. The therapist cannot find a common theme to any of the client's stories or understand the emotional connection and personal meaning that the narratives hold for the client. It seems as if nothing significant is occurring. At that point, it is easy for the therapist to stop the stories and direct the client toward a specific topic that the therapist thinks would be more fruitful. Although this may be effective at times, therapists will succeed far more often if they revitalize the therapeutic interaction without shifting the impetus away from the client and onto the therapist. In the following dialogue, the onus for therapy comes to rest with the therapist, and the client loses an internal focus for change.

Therapist:
> I don't think this is taking us anywhere.

Client:
> Yeah, I'm not sure where I'm going with this either. What would you like me to talk about?

Therapist:
> You've had trouble asserting yourself in the past, and I think we need to look at that problem. Last week, you said you wanted to ask your boss for three weeks of vacation instead of two. How are you going to handle that confrontation?

Client:
> I'm not sure. What do you think I should say?

Therapist:
> To begin with, you need to arrange a face-to-face meeting with him. It is important that only the two of you are present, so that you can have his

full attention and there is less threat for either of you to lose face. Then use the "I" statements we have practiced to directly state what you want.
Client:
Sounds good, but what would you say to him first?

This exchange will not be productive, even though the therapist has moved the client to a more salient topic and provided useful information about effective confrontations. The fulcrum of therapeutic movement has tipped from the client to the therapist, and a hierarchical teacher-student process has been established. Most clients will not be able to incorporate and act on the therapist's useful information until they pick up the momentum and begin to actively participate again. Furthermore, because the client remains in a passive, following role as the therapist continues to explain and inform, the client does not gain the increasing sense of mastery that comes from participating in successful therapeutic movement. A more productive intervention might be to suggest or inquire about another topic that is more pertinent, and to wait until the client becomes actively involved again before offering additional information or further shaping the direction of treatment. If the client does not become actively engaged again, the therapist may simply observe this with a process comment.

Effective interventions   Let's look at how the therapist can reorient the client toward more productive material in a way that keeps the momentum for therapy with the client. Process comments that make the current interaction between therapist and client a topic for discussion are often effective in these circumstances.

Therapist:
Are you using our time the way you want to right now? I don't have the feeling that what you're talking about is really very important to you.
Client:
Yeah, I'm not sure where I'm going with this either. What would you like me to talk about?
Therapist:
I think that we should try to identify what would be most important to you, and talk about that. What would that be right now?
Client:
I'm not really sure.
Therapist:
Let's just sit quietly for a moment, then, and see what comes to you.

Beginning therapists often feel uncomfortable with pauses and silences and may find themselves filling or guiding them. If therapists can refrain from doing so, however, these are opportune moments for clients to establish their own agenda and/or introduce important and unexpected new material. In addition, this type of process comment invites the client to approach more substantial material but without taking the impetus away from the client. The therapist has directly intervened by sharing her observation and asking about the client's perceptions of their current interaction, but has still left the client an active participant in revitalizing the discussion. Routinely, clients produce far more significant material when they are invited to lead in this way rather than to follow.

Revealing new conflicts through effective confrontations   New client conflicts can be identified for treatment when the therapist intervenes without taking the locus for change away from the client. Imagine the following situation: A 19-year-old client in a college counseling center had been telling his therapist about his attempts in early adolescence to observe his mother undressing. The client had been detailing his voyeuristic efforts at great lengths, but the therapist did not feel that this was genuinely of much concern for the client. Unless the therapist could find the relevant meaning for the client, he wanted to move on to other, more salient issues in the client's current life. In the following dialogue, the therapist uses a process comment to refocus the client in a way that keeps the locus for change with the client and opens up an important new conflict area for therapy.

Therapist:
> Does this feel like an important topic that you want to discuss with me? As I listen to you, I don't get the feeling that you are really very interested in what you are telling me. Is that the case, or am I not understanding the meaning this holds for you?

Client:
> I thought therapists were interested in this oedipal stuff. I figured you would want to hear about it.

Therapist:
> I'm struck by the fact that you are telling me what you think I want to hear, rather than working on what is most important to you. Maybe that's something we could look at in therapy. I wonder if you find yourself doing this in other relationships as well—trying to sense other people's needs or unspoken demands at the expense of expressing your own interests or concerns.

Once again, the therapist has used a process comment to revitalize therapy without taking the impetus away from the client. In this vignette, the therapist has also used their current interaction to uncover a potential new problem area for the client. Because this new issue arose from their joint interaction and not from the therapist's own theoretical agenda or personal directive, most clients will be interested in exploring it. Thus, the therapist has effectively focused the client internally, identified a new conflict area to be explored, and left the client an active participant in directing the course of therapy.

As the client begins to work with the new problem areas that have developed out of the therapeutic relationship, important inroads will be made on the client's presenting problems as well. In most cases, significant progress cannot be made on the presenting problem until movement is made on other related conflicts that are not evident in the initial sessions. Clients often have been unable to resolve their presenting problems because they have not addressed other related issues. As we have seen, adopting an internal focus will identify salient conflicts that clients did not realize were relevant to their presenting problem. We will return to this issue of expanding and reframing clients' conceptualizations of their problems in Chapter 7.

Enlisting Clients in Resolving Their Own Conflicts

In this section, we bring the internal focus one step further into the therapeutic relationship. Once clients have begun to look within and share their inner world, the therapist can further clients' sense of efficacy by engaging them in understanding and resolving their own conflicts. A common misconception about psychotherapy is that the therapist is completely responsible for figuring out what is wrong and what clients should do. Many beginning therapists have unrealistic expectations that they must be very wise and possess insightful solutions to clients' problems; they then suffer under these performance demands. Here again, this misconception places the impetus for change in the therapist's lap and gives clients the role of passive recipients, waiting to be enlightened. A more effective approach is to extend the collaborative alliance into the problem-solving phase of therapy as well.

Therapists will be more effective if they elicit clients' active involvement in understanding and resolving their own conflicts instead of merely setting the stage for the therapist to announce interpretations, give advice and explanations, and provide behavioral alternatives. Clients' participation in a collaborative alliance frees the therapist to respond more fully to what clients produce and gives clients the opportunity to become more capable of managing their own lives. Rather than trying only to provide answers, therapists can develop the skills required to help clients explore their own problems and exercise their own strengths and personal resources. Real change has occurred when psychotherapy has not only resolved clients' presenting problems but has fostered their sense of competence as well.

To illustrate this third component of helping clients adopt an internal focus for change, let's look at two different ways of responding to a client's dream. Although the interpretation of the dream remains the same, the therapeutic approaches differ and enact very different interpersonal processes with the client.

Recapitulating Clients' Conflicts

Anna, a 22-year-old client, lives at home with her embittered and chronically embattled parents. For several years, Anna has been struggling with the developmental transition of leaving her family of origin. She cannot establish an independent adult life. Anna has few friendships, dates little, and has no serious educational or career involvements. Anna's mother and alcoholic father fight constantly, and Anna feels it is her responsibility to stay home and help her mother cope with her problematic marriage. Anna entered therapy complaining of depression.

After several sessions, Anna recounted to her therapist a dream that was of great importance to her. As the dream began, Anna was riding a beautiful horse across an open savannah. She felt as one with this graceful and powerful animal as they glided effortlessly across the broad grasslands. They sped toward a distant mountain, past sunlit rivers, birds in flight, and herds of grazing elk. Anna felt strong and free as she urged the tireless animal onward.

The distant mountains held the promise of new life amid green meadows and tall trees. Anna felt their promise quicken inside her as she urged the horse onward. But as the mountains drew near, Anna and the horse began to slow. The horse's legs became her own and grew heavier with each step. Anna desperately tried to will them on, but their footing became unsure and they began to stumble. At that moment, she was surrounded on all sides by menacing riders, ready to overtake and capture her. As she awoke, Anna choked back a scream.

"What do you think my dream means?" Anna asked.

Her bright, concerned young therapist offered a lengthy and insightful explanation. The interpretation focused on Anna's guilt over leaving home. The therapist suggested that, if Anna went on with her life and pursued her own interests, she would feel powerful and alive—just as she had in the dream. But before Anna could reach her goal and experience the satisfactions of having her own independent life, she would have to free herself from the binding ties of responsibility that she felt for her mother and her parents' marriage. Her loyalty to her mother threatened to entrap her in guilt and prevent her from having her own adult life.

Anna: *(enthusiastically)*

Yes, I do feel as if I am bad and going to get in trouble whenever I leave my mother. Is that why I dreamed that?

Therapist:

We've been talking about whether you are going to move into an apartment next month. I think the dream reflects your conflict over taking this big step on your own.

Anna:

I want to move out, but I can't leave my mother with my father. She says she will divorce him if I move out, and it'll be my fault. What should I do?

Therapist:

I can't make that decision for you. You have to be responsible for your own decisions.

Anna:

But I don't know what's best to do, and you always do. You're so much smarter than me. I thought about that dream a lot, and I didn't know that's what it meant.

Throughout the rest of the session, Anna continued to plead for advice and expressed her discouragement about being able to resolve her own problems. The therapist kept refusing to tell her what to do and gave a lengthy explanation about autonomy, independence, and the need for Anna to find her own solutions. At the end of the hour, Anna felt agitated and depressed, and the therapist was still explaining the need for Anna to make her own decisions.

Although the therapist was astute in linking Anna's dream to the current manifestation of her conflict over leaving home, this was an unproductive session. How did their therapeutic process go awry?

In this session, Anna metaphorically reexperienced the same conflict with her therapist that she is struggling with at home. In her family, Anna was trained to be dependent and to believe that she didn't have the wherewithal to succeed on her own. Further, she was made to feel guilty whenever she did act independently or on her own behalf. In the first 15 minutes of their session, the therapist acted as the knowing parent who gave all the necessary answers to the needing child. This interaction *behaviorally* communicated the message that Anna will be dependent on the therapist's superior understanding. On another level, however, this message was contradicted by the therapist's *verbal* message about independence. This mixed message about independence from the well-intentioned therapist immobilized the client. The therapist's direct interpretation could have worked well for many clients, but (recalling the concept of client response specificity) such an approach was problematic for Anna.

Anna reacted so intensely because she needed permission from the therapist to become more independent yet simultaneously expected that the therapist needed her to remain dependent—just as her mother did. When the relational pattern of having to remain dependent on the authority was reenacted with her therapist, Anna's fear that she was incapable of making her own decisions and having her own life was confirmed. Anna remained depressed until the therapist successfully reestablished a collaborative alliance. With the help of his supervisor, the therapist did this in the next session, by soliciting Anna's ideas about a problem they were discussing. The therapist expressed his genuine pleasure in watching her be so insightful and, following this, shared some related ideas of his own. This time, Anna readily picked up the ideas and used them to further her own thinking.

Providing a Corrective Emotional Experience

We now examine a different type of response to Anna's question, "What do you think my dream means?" Responses that engage clients in exploring dreams with the therapist will usually be more effective than any explanations provided by the therapist, no matter how accurate. Therapists can choose to respond in many different ways and can engage clients in exploring their own dreams and other material as follows:

> It sounds like a very important dream to me, too. Let's work on it together. Where should we begin?
>
> What was the primary feeling you were left with from the dream? Can you connect that feeling to anything going on in your life right now?
>
> What was the most important image in the dream for you? What does that image suggest to you?
>
> Let's exchange ideas. I'll give you a possibility I'm thinking about, and you share one with me. It'll be interesting to see what we can come up with together. Who should go first?

Each of these responses encourages the client to participate actively with the therapist. Ultimately, the therapist may want to give the same interpretation of the dream. Even though the therapist provides an explanation, however, *the process will be entirely different and far more enlivening for clients if it evolves out of their joint effort and is integrated into their continuing collaboration.* This gives clients a relationship in which they are not one down or dependent on the therapist but are encouraged to exercise their own abilities and resources. When this occurs, clients' motivation to work in therapy increases and their sense of competence is fostered. Facilitating such independence provides a corrective emotional experience for clients like Anna, who get the permission they need to grow out of their dependence and act capably in their relationship with the therapist—something they may not have experienced in the past.

Tracking Clients' Anxiety

Tracking clients' anxiety is a Sullivanian concept that will help therapists determine when and where to focus clients inward. Although the subjective experience of anxiety is always uncomfortable and may be intensely painful, anxiety is therapists' ally. It serves as a signal or signpost to help therapists identify clients' central conflicts. Therapists do not fully understand clients' problems—or what needs to occur in therapy to resolve them—until they understand what makes clients anxious. Why do certain situations or interactions make a particular client anxious, and how has the client learned to cope with recurrent anxieties? The answers to these questions assist greatly in the conceptualization of the client's dynamics and the formulation of effective treatment plans. As we will see, focusing clients inward on their anxieties will lead to the heart of clients' concerns.

Therapists can use a four-step sequence to track clients' anxiety and better use the guidelines for conceptualizing clients provided in Appendix B. In this sequence, the therapist (1) identifies manifest and covert signs of client anxiety, (2) approaches directly all signs of client anxiety, (3) observes the issues under discussion and the interpersonal process between the therapist and the client that precipitate the client's anxiety, and (4) focuses the client inward to explore the source and meaning of the client's anxiety.

Identifying Signs of Clients' Anxiety

Clients will become anxious in the session when the issues they are discussing touch on their central conflicts and activate their relational templates, pathogenic beliefs, or core conflicted affects. Clients develop certain interpersonal coping strategies that they employ over and over again, in different situations, to ward off the anxiety evoked by unmet developmental needs and conflicted relational patterns. If therapists can identify the interpersonal scenarios, subjective needs, or situational demands that make the client anxious, the therapist is better prepared to recognize the unifying themes or repetitive patterns that have triggered the client's anxiety. Thus, throughout therapy, therapists are continually asking themselves: "What makes this client anxious? When does the client feel threatened or insecure? Where does safety lie, and where does danger lie?"

In order to conceptualize the central conflicts or identify the repetitive patterns that are troubling clients, the therapist must actively attend to the anxiety (or anxiety equivalents) that they manifest. For example, the therapist must be alert for signs indicating that something clients are experiencing is making them anxious. There are a thousand different ways for clients to express their anxiety—nervous laughter, nail biting, hand gesticulation, agitated movement, stuttering, hair pulling, speech blockage, and so on. These anxiety equivalents will be expressed in endlessly varied ways across clients, and therapists will have to learn how each particular client tends to express anxiety. (This will be especially important when working with clients from cultures in which certain behaviors— such as gaze aversion—may be normative rather than a signal of anxiety.) When they become anxious, clients utilize the same interpersonal coping strategies in a repetitive or characteristic way. These strategies may include pleasing others, becoming critical or controlling, seeking help or acting confused, withdrawing from others, and so forth. We will examine these interpersonal coping strategies closely in subsequent chapters. For present purposes, it is enough to note that the first step in tracking the clients' anxiety is to observe and note when the clients become anxious.

Approaching Clients' Anxiety Directly

In Chapter 3, we suggested that therapists approach and draw out the feelings and concerns that emerge for clients as they enter treatment. In a similarly forthright manner, therapists are encouraged to approach clients' anxiety about an issue being discussed or about what is going on with the therapist at that moment. Therapists can focus clients on their experience in an open-ended way (for example, by asking, "What are you feeling right now?") or by labeling the anxiety more directly (for example, by saying, "Something seems to be making you uncomfortable right now. Any idea about what that may be?").

By focusing clients inward on their anxiety *as they are experiencing it,* and inviting them to explore it further, the therapist is leading clients closer to the source of their problems. As a result, some clients may be ambivalent about the therapist's request to look within and explore their anxiety further. On the one hand, clients may welcome the opportunity to share their anxiety with the therapist—and the confusion and pain that often follows close behind. On the other hand, clients may also recognize that approaching their anxiety will bring them closer to pathogenic beliefs about themselves and unwanted expectations of others that evoke acutely painful feelings. Thus, by approaching clients' anxiety, therapists may reassure them at some times or intensify their anxiety and arouse their resistance at others.

As we discussed in Chapter 3, the therapist must respond to the client's resistance whenever it occurs, by exploring why it is threatening or does not feel safe for the client to approach this anxiety-arousing topic. This approach will be far more effective than ignoring the client's reluctance or trying merely to push through it by persuading the client to talk about the conflict. Before the therapist and client can address the conflict itself, they must explore why it is threatening for the client to share it with the therapist. (Perhaps the client fears, for example, that the therapist will not like or respect the client anymore.) As we have already noted, the therapist must honor the client's resistance. The therapist does this by (1) trying to understand the original, aversive experiences that led the client to behave in this particular way; (2) helping the client appreciate how this "resistant" or "defensive" response was once a necessary and adaptive coping strategy; and (3) providing the client with a different and more satisfying response than the client has come to expect, as in the following example.

Therapist:

> No, I don't find you needy and demanding for asking me if I would call you before you make this big presentation on Tuesday. It would be easy for me to take a few minutes to do that, and I'm honored that you are willing to risk sharing your need with me.

Observing What Precipitates Clients' Anxiety

As we have seen, the therapist first observes when the client is anxious and then helps the client focus inward more directly on the experience. At the same time, the therapist has a third task: to try and recognize (1) what issue was just being discussed or (2) what type of interaction between the therapist and the client may have *precipitated* the client's anxiety. If the therapist can identify what it was that made the client anxious, the therapist will be better able to identify the pathogenic belief, maladaptive relational pattern, or core conflicted affect that may have generated the anxiety. Therapists must consider this precipitant carefully in order to formulate effective case conceptualizations and treatment plans.

The client will be able to discuss many different issues comfortably with the therapist. When the client becomes anxious, however, the therapist must try to identify the issue or theme that has just precipitated the anxiety. What was the client just talking about? Was it sex, death, intimacy, money, vulnerability, success, divorce, inadequacy? The answer will highlight the issues that are linked to the client's central conflicts and need to be explored further. As we know, therapists can help the client explore this more effectively if they have already formulated working hypotheses about the possible issues that may be generating the anxiety. Keeping process notes, as suggested in Appendix A, will help therapists formulate their working hypotheses more effectively.

More importantly, the client's anxiety is often a signal that the interpersonal process between the therapist and client is currently reenacting a developmental conflict or repetitive relational pattern for the client. For example, the therapist might observe client anxiety in the following circumstances:

1. As the therapist was expressing confusion about what was going on in the session at that point, the client became anxious, perhaps because her volatile, alcoholic father would start insulting and demeaning her whenever he felt uncertain or inadequate.

2. The client had just made an important insight and the therapist acknowledged his achievement. The client then became anxious, perhaps because he believed that he would always have to be so insightful and perform so well, as a parent had always expected of him.

3. The therapist had just been supportive, and perhaps the client felt anxious because strings had always been attached to what she had been given in the past.

Following her relational templates, the client may have become concerned that the therapist would make her pay for his support in some unwanted way, just as her caregivers always used to do.

The therapist should generate working hypotheses such as these about how the current therapist-client interaction may have aroused a developmental conflict or unwittingly reenacted a repetitive relational pattern for the client. These hypotheses can then be used to help clients explore the issues that evoked their anxiety and better understand the more primary concerns that generated it.

Focusing Clients Inward to Explore Their Anxiety

As we have seen, the therapist is alert for signs of client anxiety and approaches it whenever it occurs. Simultaneously, the therapist tries to identify what precipitated the client's anxiety—especially what has just transpired between them—and begins to generate working hypotheses about the relational patterns, pathogenic beliefs, or accompanying feelings that are the source of it. Finally, the therapist must focus the client inward, to explore more specifically what the threat or danger is for the client. If the therapist can help the client focus inward and express the thoughts, feelings, expectations, beliefs, or memories that are associated with the anxiety, the client's vague discomfort will usually become more concrete and specific. Consequently, the client's basic concern will be revealed more clearly, so that the therapist can provide a new and more satisfying response to it than the client has come to expect. The following dialogue illustrates this process.

Therapist:
> Right now, I'm wondering if you're afraid there's something I'll do, or something going on between us, that could be hurtful?

Client:
> You know, this doesn't sound very nice to say, but I don't think my mother really wants me to change very much. I don't think she's a mean person or anything, but I do think I'm getting better in therapy, and she's not completely happy about that.

Therapist:
> How so?

Client:
> Well, I haven't been depressed for a while now, and I've actually been feeling pretty good the last month or so. Maybe it's just coincidence, but it seems that, as I've gotten better, my mother has withdrawn and been harder to talk to. And I think this has gone on between us before.

Therapist:

Yes, I think you're right. You have been doing very good work in here and getting better, and I agree that may be kind of hard for your mother sometimes.

Client:

(*fidgets, begins picking at her nail, and starts talking about another topic*)

Therapist:

Did something just make you feel uncomfortable?

Client:

I don't know. How much time is left?

Therapist:

I'm wondering if something we're talking about, or something that might be going on between us, could be making you uncomfortable?

Client:

(*pause*) Maybe I'm afraid that you're going to go away or something?

Therapist:

That sure makes sense. We were just talking about how your mother seems to act hurt and withdraws when you are feeling stronger, and I just told you that I thought you were doing very good work in here.

Client:

Well, I guess so, but that doesn't really make me feel very good either, because you're not going to stay with me either if I get better.

Therapist:

What do you mean?

Client:

If I get better, then we stop. Isn't that the point of counseling anyway? Seems like I won't have my mother, or you, or anybody then.

Therapist:

I'm glad you can risk telling me this so clearly. I didn't understand at first, but you're right, the issue we are struggling with right now is just like what's happened before. To feel better and act stronger and be more independent meant that you have had to be alone. In the past, in order to be close to others, you have had to be sad and depressed and needy. I think you've been struggling with this dilemma all of your life and, right now, it feels like it has to be the same old story again for you and me.

Client:

Yeah, it kind of does.

Therapist:

But I wonder if we can do it differently for once. Can we work together and try to find a way to make it come out another way this time?

Client:

It's not working very well so far.

Therapist:
    Well, one difference I'm thinking about is that we're talking about it—and that hasn't happened before.

Client:
    What difference does that make?

Therapist:
    Maybe it's different because I can see what's happening for you and feel for you in this predicament. It's been very discouraging for you to be undermined in this way so many times.

Client:
    I suppose that's true, but these don't really seem like huge differences to me.

Therapist:
    OK. I see these differences as more significant than you do right now, but let's keep talking about this.

In this way, tracking the client's anxiety can help the therapist and the client focus inward and clarify the client's concerns and concerns. In particular, *it will often reveal how the client's problematic relational patterns have been activated or evoked within the therapeutic process.* By clarifying how aspects of the client's conflicts with others seem to be recurring between the therapist and client, the therapist can begin to *differentiate* the therapeutic relationship from the problematic patterns that have occurred in the past with others. We will return to other aspects of these basic orienting constructs in the chapters to follow.

Closing

The overarching theme in the interpersonal process approach is that clients will not be able to resolve their problems and change unless they enact a different and more satisfying solution to their conflict in their real-life relationship with the therapist. Helping clients to adopt an internal focus for change is one important way in which therapists provide clients with a different and more satisfying relationship than they have experienced in formative relationships and come to expect with others. By codetermining what topics and issues will be covered and by participating in finding solutions to their problems, clients gain an increasing sense of self-efficacy and uncover the interrelated constellation of feelings and beliefs that constitute their problems. Perhaps Bowlby (1988) has captured this interpersonal process most succinctly by saying that the basic therapeutic stance toward the client is not, "I know; I'll tell you" but, "You know; you tell me."

As clients begin to reflect on their own experience and become less preoccupied with others, not just their anxiety but their entire affective world opens up and becomes accessible in a way that was not possible before. This affective unfolding is a pivotal step in the process of change. However, the strong and sometimes painful feelings aroused by pursuing an internal focus can be threatening for clients and intimidating to therapists as well. The next chapter examines how therapists can help clients come to terms more successfully with the deep-seated emotions aroused by looking within.

Suggestions for Further Reading

1. Robert White introduces the concept of effectance motivation, an innate drive that leads individuals to actively explore and master their environment, in his classic paper, "Motivation Reconsidered: The Concept of Competence," *Psychology Review*, 66 (1959): 297–333.

2. Alan Wheelis discusses the internal process of change in his book *How People Change* (New York: Harper & Row, 1974). In particular, see the chapter "Freedom and Necessity," which uses the language of responsibility to clarify how clients must focus on themselves rather than others in order to change. Beyond the issue of responsibility, beginning therapists can learn much from other existential psychotherapists; see, for example, V. Frankl's *Man's Search for Meaning* (New York: Pocket Books, 1963); H. Kaiser's *Effective Psychotherapy* (New York: Free Press, 1965); and I. Yalom's *Existential Psychotherapy* (New York: Basic Books, 1981).

3. In a tour de force presentation, Albert Bandura argues that perceived self-efficacy is the underlying mechanism of change across all treatment approaches; see his article "Self Efficacy: Toward a Unifying Theory of Behavioral Change," *Psychological Review*, 84, no. 2 (1977): 191–215. He emphasizes that little enduring and generalized change results from verbal suggestion or persuasion. Instead, clients change when they have authentic experiences of enactive mastery—that is, when they have the experience of performing adaptive new coping responses with the therapist.

4. Therapists working with clients from cultures other than their own must be especially aware of normative verbal and nonverbal behavior. For a discussion of how to assess anxiety in Asian clients by focusing on nonverbal behavior, see S. Tseuneyoshi's case study "Rape Trauma Syndrome" in *Child and Adolescent Therapy: A Multicultural-Relational Approach*, edited by F. McClure and E. Teyber (Fort Worth: Harcourt Brace, 1996). Additionally, readers may wish to examine Chapters 1 and 2 of *Counseling the Culturally Different* (2nd ed.) by D. Sue and S. Sue (New York: Wiley, 1990), which address language differences among cultural groups.

CHAPTER FIVE

Responding to Conflicted Emotions

Conceptual Overview

Conflicted feelings lie at the heart of enduring and pervasive problems. Change in therapy is a process of affective relearning, and this change occurs when the therapist responds effectively to clients' conflicted emotions. When the therapist focuses clients on themselves, the troubling feelings that are central to their problems will emerge. When they look within, clients feel more intensely, and express more directly, the conflicted emotions that accompany their problems.

This affective unfolding is a pivotal point in therapy because it reveals the emotional basis of clients' problems. On the one hand, clients will welcome the promise of having therapists make contact with their problems on a deeper, more personal level. On the other hand, clients also want to avoid feelings that are too shameful, painful, or intense or that just seem hopelessly unresolvable. Thus, the therapist's response to clients' emerging feelings will significantly influence the eventual outcome of therapy.

Clients will not be able to change unless the therapist provides a more satisfying response to their conflicted emotions than they have received from others in the past. Unless the therapist is able to help bring out and then contain clients' emerging affect, the therapeutic relationship will lose its vitality, and therapy will be reduced to an intellectual pursuit. Thus, the purpose of this chapter is to help therapists respond to the full range and intensity of the feelings aroused when clients begin to look within. We will see that the therapist's ability to help clients change is largely determined by how the therapist responds to their emotions.

Chapter Organization

This chapter is divided into five sections. The first section provides basic intervention guidelines for responding to clients' affect. Therapists need to approach clients' feelings and help expand the affective component of their problems. The second section continues this theme of drawing out the clients' painful or conflicted affect and focuses, in particular, on working with the predominant or characterological affect that clients tend to reexperience in many different situations.

In the third section, therapists will be introduced to the interrelated sequence of emotions that comprise clients' conflicts. For many clients, three related feelings recur in an ordered sequence; therapists must identify and respond to each feeling in the sequence. Too often, clients cannot resolve conflicted emotions because therapists do not address each of the distinct feelings that make up this affective constellation. Once therapists are successful in evoking the client's conflicted emotions, they must have guidelines about how to respond to them. Thus, in the fourth section, we explore how the therapist can provide an interpersonal holding environment that helps contain conflicted emotions and thereby enables clients to integrate and resolve them.

In the fifth section, we examine personal characteristics that influence therapists' ability to respond to clients' emotions. This is one of the most important issues in clinical training, because therapists' own personalities, needs, and current life stressors can keep them from responding to all of the emotions that clients present. All therapists need to seek consultation at times, in order to manage their own emotional reactions to the material that clients present. Otherwise, they will be less capable of responding to the full range and intensity of clients' emotions.

As a final note of encouragement, readers should be aware that the material presented in this chapter is more complex, personally challenging, and harder for beginning therapists to employ than the material presented in other chapters. The issues discussed here often evoke significant personal concerns for therapists-in-training, and you should not expect to easily apply these principles with clients. Understanding these processes is far different from being able to utilize them effectively in ambiguous and emotion-laden therapeutic situations. Be patient: In a year or two many of these responses will be your own.

Responding to Clients' Conflicted Emotions

Many of the problems that clients present can be resolved by simply obtaining new information, trying out new coping strategies or behavioral alternatives, challenging faulty beliefs, or recasting problems in a new framework. These brief interventions are sufficient to help with many of the problems that clients present, and they constitute an important aspect of change for most problems. For the more enduring and pervasive problems that clients present, however, the conflicted emotions that accompany problems must be addressed as well.

At times, clients may be reluctant to work with the conflicted emotions that emerge in treatment. On the one hand, clients hope that the therapist will be able to help with their emotional reactions; on the other, they may want to avoid the painful feelings that they have not been able to master or resolve alone. There are many different reasons why clients may be reluctant to address these feelings with the therapist. They may be unwilling to reexperience the same searing pain when they have scant hope of change; to suffer shame or embarrassment by revealing feelings of inadequacy or need; to risk unwanted but expected criticism or rejection from the therapist for having feelings that seem unacceptable; or to struggle with their own fear of losing control. As a result of these and other concerns, clients sometimes resist the therapist's attempts to touch the emotional core of their experience, although at the same time they often long for this contact.

Once the therapist recognizes the client's relational templates and understands how significant others have characteristically responded to the client's pain, need, or vulnerability in the past, the fear or shame that motivates his or her resistance will make sense. As outlined in Appendix B, therapists must generate working hypotheses and try to conceptualize the clients' potential resistance to conflicted feelings and must attempt to work directly with those feelings rather than trying to bypass or avoid them. If not, therapy will be reduced to an intellectual exercise and will not effect enduring change that can be generalized beyond the therapeutic setting. The therapist must not comply with the client's resistance to approaching feelings but must keep in mind the client's simultaneous wish to be understood, cared about, and addressed in this deeply personal way.

Although conflicted emotions are central to most clients' problems, the therapist's affirming response to these emotions also provides the avenue to resolution and change. The corrective emotional experience that the therapist provides to clients—the experience of sharing their most profound feelings with a caring other who remains attuned, connected, and validating—loosens the hold of old templates, beliefs, and expectations, and empowers clients to make the lasting behavioral, cognitive, and affective changes that they desire. At this point, clients are significantly more receptive to interventions from all theoretical orientations.

In this chapter, we examine intervention guidelines to help therapists draw out or approach their clients' conflicted emotions and thus set in motion this process of change.

Approaching Clients' Affect

In their moment-to-moment interaction with clients, therapists must repeatedly choose how to respond to what the client has just said or done. For example, the therapist might respond by seeking more information, clarifying what the client has just said, relating this material to other issues the client has discussed in the past, or relating it to what is occurring between therapist and client at that instant. As a general guideline, however, *the most productive response is one that directly addresses the feeling that the client is currently experiencing.* The therapist's first priority is to acknowledge and approach the affective component of the client's response.

Imagine that the therapist and client have just sat down to begin their first session:

Therapist:

Tell me, Mike, what is the difficulty that brings you to therapy?

Client:

I'm having a lot of problems with my 15-year-old son. We disagree about everything and can't seem to talk to each other anymore. He doesn't do what I ask him to, and I don't like his values. I guess I'm pretty angry at him. His mother and I are divorced, and I'm thinking it may be time for him to go live with her. Do you think it's OK for a teenaged boy to live with his mother?

Therapist:

I don't think either of us understands what's going on between you two well enough to decide that yet. You said you were pretty angry at him. Tell me more about your anger.

The client has presented many different issues that the therapist could have chosen to pursue. Following the general guideline, however, the therapist approached the primary affect that the client presented. The therapist could have responded differently and inquired further about the issues that the father and son disagreed about, worked on values clarification with the father, obtained more background information about the father-son relationship or the divorce, provided research findings about the effects of mother-custody versus father-custody on boys, and so on. Although these and many other responses will often work well, *responding to the client's feelings will usually produce the most information and intensify the therapist-client exchange.*

In this dialogue, the therapist responded immediately when the client spoke about his anger. Therapists also need to respond to the covert or unverbalized feelings that clients experience. For instance, the therapist in our example might have chosen to respond instead to the more covert affect by acknowledging, "It sounds like you're feeling really discouraged about being a dad right now—maybe so discouraged that you just feel like giving up sometimes."

Strong feelings are aroused when the client discusses an important but conflicted issue. Such nonverbal affective signs as tearing, grimacing, or blushing are signals to the therapist that the client is addressing a significant issue. Responding to these nonverbal cues, the therapist draws out the affect that this situation holds for the client, as in the following example. Imagine that the client has been discussing her marital problems:

Client:

> I don't know if I should stay married or not. I haven't been happy with him for a long time, but I can see how hard he's trying to make our relationship work. And our 4-year-old son would be devastated if we broke up. I don't know what's right to do, and it's so important for everyone that I make the right decision.

Therapist:

> As you speak about this your face tightens. What are you feeling right now?

When the therapist responds to the client's feelings in this open-ended way, it often serves to clarify the client's central concern. Clients may respond by expressing more specific concerns:

> I'm so sad about hurting the people I love.
> I'm afraid that everyone will think I'm selfish for leaving and they'll blame me for the divorce. They'll think I haven't been a good mother.
> I don't want to be alone. I don't want to be married to him anymore, but I'm afraid I can't leave and make it on my own.
> I'm furious that I'm the one who has to make this decision. I'm responsible for every decision we make, and I always pay for it. It's not fair.

Throughout every session, clients have emotional responses to the issues and concerns that matter most to them. Sometimes the client's affect will be presented forthrightly; at other times, it will be subtle and elusive. Cultural, class, and gender factors will further shape how each client displays affect. For example, some clients may overtly state their reactions, whereas those from other cultural or familial backgrounds may simply avert their eyes or become noticeably quieter when certain feelings are evoked. However, if the therapist can find sensitive but direct ways to approach the client's affect and draw it out more fully, as in the preceding examples, it will evoke further disclosure from the client and clarify the client's conflict or concerns. This, in turn, will provide a narrower treatment focus.

Expanding and Elaborating Clients' Affect

We must look further at how the therapist can respond to clients' feelings. At times, it can be helpful to anticipate or label a client's emerging feeling (for example, by saying, "You must have been furious" or "You're frightened"). As a general guideline, however, a more effective response is to give clients an open-ended invitation to explore the feeling further. For example, the therapist may encourage clients to explore or express what they feel is central in their affective experience. A simple response—"What are you feeling right now?" or "Tell me more about that feeling"—is remarkably effective in eliciting the client's affect. An open-ended invitation is more effective than trying to label the client's feeling in an either-or fashion—for example, by saying, "Were you feeling _____ or _____ in that situation?" Such a choice may restrict the client rather than invite a free range of responses. The client often experiences something that the therapist has not anticipated; moreover, the client may have several feelings about the situation. The open-ended query is more effective in drawing out exactly what the client is feeling.

This type of open-ended response is also more effective than asking clients why they are experiencing a particular feeling. As a rule of thumb, therapists are well advised to avoid *why* questions. Many clients don't really understand why they feel, think, or do something. Asking them why makes them feel inadequate or resentful. The therapist will usually elicit more information from clients by offering an open-ended invitation—"Help me understand your anger better" or "Tell me about your anger"—than by asking, "Why were you angry about that?" By inviting the client to express more fully whatever feeling he or she is having, therapists are giving the message that they are interested in the client's own subjective experience and they are comfortable sharing whatever the client is experiencing. *Most clients have not had permission to share their emotions so freely or honestly in the past,* which is why this is one of the most important responses therapists can give their clients. Once they discover that they can be authentic with the therapist and remain connected and cared about, clients become more willing and able to take the risks necessary for the lasting changes they desire.

Clarifying clients' subjective feelings   Giving clients an open-ended invitation to explore their feelings will also help them to clarify what each particular feeling means to them. Too often, therapists wrongly assume that a particular affective word, such as *angry* or *sad,* means the same thing to the client as it does to them. Therapists should not assume that they understand a particular affective word without clarifying what it means to this particular client. Many therapists are able to understand this principle when there is a cultural difference between therapist and client; however, therapists should be aware that *every* client has a unique subjective worldview. Thus, therapists must actively encourage clients to articulate the personal meaning that a feeling holds for them. *This is especially important if clients repeatedly use the same affective word to describe themselves.* Finally, having every client elaborate affect in this way will safeguard therapists against becoming overidentified with the client and seeing the client's experience as being the same as their own, as many beginning therapists are apt to do.

Because they are often uncomfortable with many of their emotional reactions, clients themselves may be unfamiliar with their feelings. As a result, clients need to clarify the broad, undifferentiated feeling states that they often experience. Therapists can help clients learn more about the emotional reactions that comprise their problems by asking questions such as the following:

Can you bring that feeling to life for me and help me understand what it is like for you when you are feeling that?

Do you have an image that captures that feeling, or is there a fantasy that goes along with it for you?

Is there a particular place in your body where you experience that feeling?

Is this a familiar or old feeling? When is the first time you can remember having it? Where were you? Who were you with? How did the other person respond to you?

How old do you feel when you experience that emotion? Can you attach an age to it, such as 7 years old or 13 years old?

Does this feeling have its own sound or movement? Can you make a sound that would let me hear what it's like or a gesture that would help me see it?

Some of these exploratory questions will work well with a particular client and not at all with another. Following client response specificity, therapists will need to find what works best for each client. In general, however, these responses are all ways of approaching clients' emotions and clarifying their subjective meaning. These types of responses also tell clients that the therapist is concerned about their feelings—who they are and what is most personal and important to them. Further, they behaviorally inform clients that the therapist is different from many other people they have known, is comfortable sharing whatever emotions they may be experiencing, and wants to know *their* perspective.

When the therapist works with clients to clarify their emotions, it creates an opportunity for the therapist to see, understand, and care about the clients' most intimate and personal experience. A core aspect of most clients' problems is that significant others have invalidated, ignored, or disdained certain primary feelings in a systematic or repetitive way. As old expectations of ridicule or invalidation are disconfirmed by the counselor's care and affirmation, the client has an important experience of change. Later in this chapter we will examine how important it is for clients to receive new and more affirming responses that reassure them about the therapist's ability to respond to the most important concerns in their lives.

Experiencing feelings versus talking about feelings   Little change will occur until clients are able to stop talking about their emotions in an intellectualized or objective manner and actually experience their conflicted emotions in the presence of the therapist. The exploratory responses that we have been discussing will intensify clients' immediate affective experience; they will help clients subjectively experience more fully the emotional content of what they are discussing. Therapists can also help clients experience their feelings by *mirroring* their emerging affect. For example, when feelings are on the verge of coming through, clients may blink back tears, open their eyes wide in terror, clench their jaw in anger, tighten their lips to keep them from trembling, slowly turn their head from side to side in protest, rock or hold themselves for comfort, drop their heads and cover their eyes in shame, or hold their breath in fear. The therapist can sensitively mirror—nonverbally reflect—the clients' posture and expression as a way to clarify, amplify, and evoke the full intensity of the clients' emotional reaction.

However, beginning therapists need to observe how a role model effectively provides such affective mirroring; otherwise, they may misperceive it as trite mimicry. In fact, this highly refined responding may be compared to the sensitively attuned mutual cueing between Bowlby's securely attached mother and infant (Field, 1996) or to Kohut's (1977) empathic attunement. Think of how a very young child actively seeks his or her mother's eyes; the mother's eyes open wider in the delight of mutual recognition; the child responds by gurgling; the mother coos back in the same warm tone; the baby smiles, and the smile makes the mother laugh. By mirroring the client's affective expression in this respectful and deeply empathic way, therapists can communicate nonverbally that they see, recognize, and are comfortably accepting of the client's fear or sadness. This attuned responsiveness may not be effective with certain rigid, distrustful, or shame-prone clients or with clients who are not yet ready to deal with this level of intensity. Conversely, it may be highly effective with others—especially those who have suffered trauma so profound that words simply fail to encompass the depth of their pain.

Therapists may also draw out the client's affect by addressing any *incongruence* that they perceive between the client's narrative and the accompanying affect. In telling a story that the therapist finds sad, poignant, or disturbing, the client may display no feeling at all or an incongruent affect, such as laughter. Therapists are subtly but powerfully pulled to match the client's inappropriate affect, and often feel they are violating unspoken social rules if they do not do so. However, clients are usually relieved when the therapist risks violating the social rules, makes a process comment, and acknowledges this discrepancy—and, in so doing, implicitly invites the clients' true feelings or congruent experience.

Therapist:
> What you are saying is very sad and upsetting, yet you are smiling as you tell this to me.

It is important to draw out clients' affect in these and other ways, because the greatest opportunity to help clients change occurs when they are experiencing the full emotional impact of their problems. If therapists are available and compassionate to clients while they are experiencing their conflicted emotions, clients are able to move through their conflicts. Clients cannot do this when they are alone with their feelings or when they are insulated from the full intensity of them, as the following case example illustrates.

Throughout the session, Jean had been telling her therapist about her hopelessness and despair. She described her life as a merry-go-round of endless ups and downs that always returned to the same hopeless conflicts. This past week she had dropped out of school, as she had done "so many times before," and had given in to pressure from her critical and undermining boyfriend to let him move back into her apartment again. Jean described her hopelessness about ever being able to do or have anything for herself. At times, she acknowledged, the only way out of her misery seemed to be suicide.

The therapist could feel Jean's despair and was deeply moved by it. The therapist tried to work with Jean's feelings by acknowledging her pain and validating her experience.

Therapist:

I can see your hopelessness right now, and how painful it is to feel so defeated. You haven't been able to follow through and do what you wanted to do for yourself again, and that's so discouraging. As you've done so many times before, you felt that you had to meet someone else's needs rather than say no and do what you wanted for yourself.

As in all of their previous sessions, however, Jean was not able to let the therapist be emotionally connected to her while she was experiencing the sadness. Because this critical connection to the therapist was missing, Jean still felt helplessly alone with her feelings and hopelessly alone in her life.

The setbacks of the previous week had intensified Jean's distress. The therapist recognized that this crisis provided an opportunity to make contact with Jean's feelings for the first time:

Therapist:

It just seems hopeless.

Jean:

(*long pause, nods*)

Therapist:

You don't want to try anymore. You don't want to hurt anymore. You just want it to stop hurting, and sometimes suicide seems like the only way out.

Jean:

(*nods*)

Therapist:

You are so sad. You have been sad inside all of your life.

Jean:

    (*nods, looks at therapist*)

Therapist:

    I can see how very painful this is for you right now.

Jean:

    (*looks away, puts hands over eyes*)

Therapist:

    I want to be with you in your sadness. I can't take your sadness away, but you don't have to be alone with it anymore. I want to be close to you while you are feeling this.

Jean:

    (*looks at therapist sorrowfully, and slowly begins to cry*)

Therapist:

    Yes, I see your sadness there and how much this has hurt you. I feel very close to you right now.

Jean:

    (*begins sobbing and reaches to the therapist*)

For the first time, Jean was able to let the therapist be available to her while she experienced her despondency. The therapist acknowledged the risk that Jean had taken, and they discussed how it felt for her to share her sadness directly with the therapist. Later in the session, Jean volunteered that she felt lighter, and her depression lifted for the next few days. We will return to Jean later in this chapter and see how this issue continued to unfold for her over the next few sessions.

Identifying and Punctuating the Predominant Affect

Clients often feel confused, trapped, or defeated by their emotions; they may also feel ashamed, because their emotions seem irrational or inappropriate to the current situation. Clients often think that their feelings occur for no reason; they accurately perceive that their feelings are stronger than, or different from, what the situation calls for. Clients' inability to understand their own emotional reactions is the source of much distress and self-doubt. Helping clients make sense of their emotional reactions is an important pathway to their greater self-efficacy. Therapists can do this by working with the client's *predominant affect*.

    Clients usually enter therapy in response to one of two situations. First, a current life crisis echoes or recaptures an emotional conflict that originally occurred years before; the current stressor taps into an *old wound*. Second, clients may enter therapy in response to feeling overwhelmed by *too many stressors* in a short period. These stressors have overwhelmed the clients' usual coping strategies and defense mechanisms. As we will see, clients' experience often revolves around a central affect, such as being shame-prone or guilt-prone, and therapists need to identify and punctuate this predominant feeling.

An Old Wound

Many clients enter therapy because of an old wound. For example, a client who has suffered a painful loss through divorce or death will find it very difficult to cope with the stress of that loss if it evokes unresolved painful feelings from other significant losses in the client's past. The current loss may be associated with a predominant feeling—perhaps emptiness, betrayal, separation anxiety, or intolerable sadness. That feeling becomes both understandable and far more manageable when it can be linked to similar feelings that have been aroused by other losses in the client's life. Thus, *the therapist needs to identify the central affect that the current crisis has aroused and to link it to the original wound.* Linking this primary affect to the way clients felt in the past enables them to make sense of their seemingly irrational feelings, to feel more accepting and compassionate toward themselves, and thereby to gain greater mastery of themselves. Although this is essential in all treatment modalities, it may be the most significant action that therapists can take in crisis intervention.

Multiple Stressors

Most individuals have adaptive coping mechanisms that allow them to manage a single stressful event, unless it taps into a preexisting vulnerability. If a second or third stressful event closely follows the first, however, it becomes much harder for individuals to cope. The client's usual coping strategies fail, and the cumulative stressors precipitate symptoms and maladaptive defensive maneuvers.

When subjected to multiple stressors, clients often describe themselves as "overwhelmed," "defeated," or even "broken." They tend to repeat a *compacted phrase* that encapsulates their emotional responses to the stressful events:

I want to go away and just be alone.
It's too much for me; I can't stand it.
I don't care anymore; there's no point in trying.

As we noted in Chapter 2, therapists must look for *recurrent affective themes* and try to unravel the emotional reactions that are encapsulated in these compacted sentences. Underlying each repetitive compacted sentence are one or two central feelings that capture the basic impact these stressors have had on the client. Generally, the various stressors are arrayed around one or two integrating feelings, like spokes around the hub of a wheel. For example, in response to losing too many people or things, the client may have an overriding feeling of being despondent, empty, and alone or of being powerless and hopeless, or the client may be afraid of any further losses, changes, or new commitments. When therapists are able to articulate or capture the underlying feeling that *links* the impact of several different crisis events, clients are deeply reassured. They feel seen, heard, and understood, feel more in control of themselves, and invest further in the therapeutic relationship.

If counselors are to succeed in short-term and time-limited modalities, they will need this therapeutic skill, in order to make strong empathic connections to their clients in the initial session and to establish a collaborative working alliance early in treatment.

A Characterological Affect

As we have seen, it is important for therapists to respond to the client's moment-to-moment affect. Therapists can also listen for a recurrent feeling that pervades the client's life—a *characterological affect*.

As therapy proceeds, the therapist can often identify a predominant feeling that captures the client's current conflict and also characterizes his or her life; that is, if the therapist is accurately empathic and sees things from the client's point of view, the therapist can often find a central feeling that reflects the fabric of the client's life. Clients often have one or two core conflicted feelings that are familiar and continuous throughout their lives. When the therapist can identify these central feelings that capture their characterological conflict, clients often respond emphatically:

> Yes, that's how it has always been for me.
> That's what it's like to be me.
> That's who I really am.

Clients often seem to experience these central feelings as their fate, because they have always been there and it seems like they always will be. One of the most significant interventions therapists can make is to accurately identify and reflect these generic feelings to clients.

For example, after listening to the client for several sessions the therapist may respond:

> It seems as if you've always felt burdened by all the demands you feel you have to meet.
> I'm getting a sense that you've always been afraid that people are going to find you out and point up your true inadequacies.
> It seems as if you've always felt resentful that, no matter how much you do, it's never enough.
> I'm getting a sense that you've always felt wary that others are trying to put you down or take advantage of you.

Therapists must be prepared to respond to all of the varying emotions that clients experience in the course of therapy. However, *therapists will be most effective when they can identify the repetitive feelings that have recurred throughout a client's life and that the client considers to be central to his or her sense of self.*

Clients' Affective Constellations

It is rewarding but sometimes challenging to work with clients' conflicted emotions. Some clients are adamant in wanting to avoid their emotions; others talk endlessly about the same feelings without any change resulting. When clients are not progressing in treatment, one reason may be that therapists have only responded to the client's single presenting affect.

If the therapist acknowledges the client's current affect and invites the client to explore it further, a sequence of interrelated feelings will often occur together as a predictable, patterned sequence. This constellation of emotional reactions is central to the client's conflict; each affect in the sequence must be addressed. Although different variations or sequences occur for each client, the therapist can often identify a triad of interrelated feelings that repeat throughout the client's daily life. Making a similar point, Lazarus (1973) speaks of tracking a client's firing order. The process of change involves mastering each of the feelings in this affective constellation. In this section, we examine two affective constellations that commonly occur. We will see that therapists help clients change by responding to the entire sequence of feelings, rather than just responding to the first feeling in the sequence—the client's primary presenting affect.

Anger-Sadness-Shame

The first affective constellation we will examine consists of anger, sadness, and shame. The predominant feeling state for some clients is anger. These clients often experience and readily express angry feelings, including irritation, impatience, criticism, and cynicism. However, the client's anger is usually *reactive*; it is a secondary feeling that occurs in response to an original experience of sadness, hurt, or pain.

The therapist must acknowledge the client's anger, locate its source and target, and help the client find an appropriate way to express it. If the therapist stops at this point, however, the client will remain obsessively stuck in an angry, blaming, or critical feeling state and no change will result. Change will not occur until the therapist helps the client express the primary emotional response of sadness or hurt that led to the reactive feeling of anger. This primary feeling is more threatening, unacceptable, or painful for the client than the anger that it prompts.

To reach the original affect, the therapist might wait for clients to spontaneously ventilate their anger and immediately afterward invite them to examine what they are feeling. Following the expression of the reactive feeling, there is often an open window to the primary feeling of hurt or sadness. For example, the therapist might say, "Now that you've expressed how angry you are, what are you feeling at this moment?" The original feeling of sadness, hurt, vulnerability, embarrassment, or helplessness will often surface. By acknowledging and affirming this feeling as well, the therapist can often dislodge the client from the familiar reactive feeling of anger.

If this approach does not work, the therapist can inquire directly about the original experience that led to the client's anger.

Therapist:

Something must have hurt you very much for you to remain so angry. Tell me how you felt when that happened.

If the client again begins to express anger, indignation, or a sense of injustice about what occurred, the therapist focuses the client back on the original experience:

Therapist:

Yes, I can see how unfair that was for you, but how did it feel to you when . . . ?

If, in response to this query, the client's original feeling of sadness or hurt is expressed, and the therapist responds, the client comes in contact with the internal aspect of the conflict. It is this internal aspect of the conflict that has kept the client from being able to change.

As we have seen, when this type of client experiences anger, sadness often follows. Although the anger is easily experienced, the client has learned that it is not safe or acceptable to experience and/or share the original sadness, hurt, or vulnerability. To avoid the original painful feeling, the client defensively returns to the reactive feeling of anger in an obsessive or repetitive manner; this pattern may cause others to regard the client as an angry person.

The therapist's task is to help such clients experience the sadness, hurt, or vulnerability underneath their repetitive anger. However, as soon as the original hurt is activated, the client will defend against this feeling and reflexively return to the safer expression of anger. Thus, the therapist must explore the client's resistance to the original painful affect. *Often, the reason why clients resist the original affect of pain or sadness is that it arouses a third, aversive feeling of anxiety, guilt, or—most frequently—shame.*

For example, if the therapist inquires about clients' resistance to the original hurt or pain, they will often say something like this:

If I let the pain be there, it's admitting that she really did hurt me.
If I acknowledge that it hurts, it means he won.

If I feel sad, it shows that I really am weak.

Thus, if the therapist draws out the original pain that underlies the anger, a third feeling of shame or humiliation will also be aroused. To defend against both the original sadness and the shame associated with being sad, the client will reflexively return to the stronger anger.

Although shame may be the most common reaction, other clients will experience anxiety or guilt in response to feeling the original hurt. For example, some clients will say that, if they let themselves feel sad, hurt, or vulnerable, then "no one will be there," "others will go away," or "I will be empty and alone." These clients feel painful separation anxieties upon experiencing their hurt or pain and avoid this anxiety by reflexively returning to their anger. Still other clients will say that, if they let their sadness be real, they would be admitting that they really do have a need, which would make them feel selfish. Thus, for these clients, guilt is the third element of their affective constellation.

This triad of feelings exists for many clients: Frequent anger defends against unexpressed sadness, which, in turn, is associated with shame, guilt, or anxiety. *Significant and enduring change results when clients can experience, express, and contain each feeling in the triad.* More specifically, clients resolve their internal conflicts when they can (1) let themselves experience the full intensity of each feeling; (2) express the feeling and share it with the therapist so that they are no longer alone with it or hiding it; and (3) contain, hold, or tolerate the difficult feeling rather than having to move away from it, as they have in the past.

When clients can integrate their conflicted emotions in this way, they have mastered or internally resolved their conflict. They no longer need the maladaptive response patterns or defenses that they have employed in the past. Clients are then prepared to adopt new, more adaptive responses that they were previously unable to incorporate. This is a pivotal experience in the process of change.

Before examining another common affective constellation, we must pause to highlight the special role of shame. Although much has been written about anxiety, depression, and guilt, shame has long been ignored. Correspondingly, clients could be anxious, guilty, or depressed, but not ashamed. Shame is no longer a taboo topic, however; indeed, it is now regarded by many theorists as the "master emotion" ("Shame," 1995). The cardinal role of shame in most symptoms and defenses has been illuminated by the groundbreaking work of Helen Block Lewis (1971) and others.

New realms of understanding will open up to counselors as they allow themselves to begin hearing the family of shame-based emotions that routinely pervade their clients' narratives—self-consciousness, embarrassment, shame, self-contempt and self-hatred, shame anxiety, shame-rage cycles, and so on. When counselors are able to break cultural taboos and begin approaching shame reactions directly but compassionately, profound client movement often follows. Because most counselors have been socialized to avoid these shame-based affects, however, they will need assistance from their supervisors in order to begin responding to them.

Sadness-Anger-Guilt

Whereas some clients lead with their anger and defend against their hurt, others characteristically lead with their sadness and avoid their anger. This type of client presents with an undifferentiated feeling state of sadness, helplessness, vulnerability, and depression. These clients do not experience or express anger; they avoid interpersonal conflict and tend to respond to others' needs at the expense of their own. To illustrate this affective constellation, we return to the case of Jean.

Earlier we examined a dialogue in which, for the first time, Jean was able to let someone respond to her pain and be emotionally connected to her while she was experiencing it. Subsequently, the therapist explored why it had always been so difficult for Jean to share her pain and vulnerability with others. Jean responded that her painful feelings, memories, and needs were all sickeningly more real if someone else saw them. Jean had experienced these sad, empty, and hopeless feelings all of her life, but they had always been ignored or dismissed in her family of origin. Jean had internalized her family's taboo against unhappy and angry feelings. As a result, she had denied to herself the reality of the emotional deprivation and personal disparagement that she had suffered throughout her childhood. As often happens, Jean went on in adulthood to reenact the affective themes and relational patterns of her childhood conflicts in the problematic love relationships that she kept reconstructing.

When Jean returned to therapy the following week, she had made a significant change. For several months, she had been involved with yet another critical but irresponsible man who took her for granted and took advantage of her responsiveness. That week, when the boyfriend borrowed her car and arrived late to pick her up from work, she expressed her anger for the first time. Jean also told him that she deserved to be treated with more respect and that she had the right to have her own needs considered. Jean had remained in this relationship because she was afraid to be alone. At that point, though, she decided that the relationship was not good for her and ended it.

Once the therapist was able to touch Jean's sadness and hurt, her anger was activated and subsequently expressed in an appropriate manner. Jean had remained locked in her sad and helpless victim role, in part, because the direct expression of anger and the assertive expression of her own needs were both unacceptable. As often happens, Jean began to claim her own personal power once her feelings were expressed, and she received validating responses to them. We must look more carefully, however, at the sequence of feelings that unfolded for Jean.

Although Jean began the next therapy session by sharing her good news about ending the problematic relationship, she had difficulty staying with this feeling of strength. In the course of the session, Jean gradually retreated from her indignation at how badly this man had treated her, began to feel concerned about how she may have hurt him, and wondered if she had been selfish to end the relationship because he needed her so much. However, the therapist's working hypotheses had suggested that Jean would feel guilty upon experiencing her anger, setting limits, or responding to her own needs. With this awareness, the therapist was able to recognize Jean's guilt, question its validity, and help her explore it.

Thus, we see that, like the anger-sadness-shame constellation, Jean's affective constellation included a third, aversive feeling—guilt. Jean defensively avoided her anger because it aroused guilt, which served to reflexively return her to her characteristic affect of sadness and helplessness. As the therapist helped Jean identify and consider all three of these interrelated feelings, she was able to relinquish her victim stance for increasing periods.

With the therapist providing a supportive holding environment, Jean could acknowledge the reality of the painful deprivation and disempowering invalidation that she had originally experienced in her family. As Jean stopped avoiding the reality of these painful memories, she stopped reenacting them in her current relationships; she no longer recreated frustrating relational patterns that evoked the same old disheartening emotions. Jean also began to accept her anger at the hurtful and unfair ways she had been treated—as a child and in her adult relationships—and she learned to assert her own limits and preferences in the presence of others' competing needs or expectations, without becoming immobilized by guilt. Coming to terms with each component of this affective constellation eventually left Jean feeling stronger and more in control of her own life than she had ever been. Once again, significant life changes followed from Jean's internal resolution of her conflicted emotions. It is by responding to each feeling in clients' affective constellations that therapists help clients resolve their conflicts.

How does the predominant characterological affect relate to the affective constellation? The predominant characterological affect will typically contain the leading edge of the affective constellation or triad. For example, a client who presents affectively a predominant sense of helplessness can easily access and express sadness, which is first in their triad. Often, the first emotion in the triad is the one that was allowed in the client's family of origin; all other affects became subsumed under, or covered over by, this emotion. The second and third affects in the triad were typically taboo—considered too threatening to the functioning of the child or family—and could therefore not be acknowledged or expressed. Over time, the child came to see these legitimate feelings as unacceptable, and the child's ability to respond authentically to his or her experiences was impaired.

By validating the legitimacy of these suppressed affects and sharing them with clients while remaining supportively connected and caring, the therapist gives clients the experience of expressing themselves authentically without catastrophe resulting. Once clients realize that these real but deeply painful feelings can be expressed and shared without the familiar but unwanted consequences, they feel able to reveal other aspects of their functioning. From this experience of being connected in an authentic way, clients become more responsive to a variety of interventions from different theoretical orientations and begin to risk making the changes they desire.

Holding Clients' Pain

Having drawn out the full range of the client's feelings, the therapist's next concern is how to respond most effectively. The therapist must provide a safe holding environment to contain the client's feelings. In essence, such emotional containment requires the therapist to maintain a steady presence in the face of the client's pain and distress. Let's look at what this means.

As we saw in Chapter 2, therapists must be able to empathize fully with the emotions that the client expresses and to articulate that empathic understanding to the client. Therapists must also be able to communicate their compassionate acceptance of the client's pain nonverbally, by their presence and emotional availability. In the midst of the client's intense emotions, however, the therapist also maintains an *observing self*; this self conveys to the client that these emotions will not overwhelm the therapist, that the intensity of the pain will decrease, and that the upsetting event is not catastrophic but can be endured.

In this section, we explore how the therapist can provide such a holding environment and give clients a new and more satisfying response to their feelings than they have received from others in the past. When an emotionally responsive and affirming therapist provides this corrective emotional experience and contains clients' core conflicted feelings, clients are empowered to begin acting in new and more adaptive ways.

Clients' Resistance of Feelings to Avoid Interpersonal Consequences

As we saw in Chapter 3, clients can be resistant to treatment; likewise, they also defend against painful feelings at times. As before, the therapist does not simply want to push through these defenses in order to reach clients' sadness or other significant feelings. That would be winning the battle but losing the war. Resistance and defense are less about intrapsychic issues than about expectations of receiving familiar but unwanted responses from others that evoke fear, shame, or guilt. Therapists create a more affirming interpersonal process, with greater safety for clients, if they enlist clients' collaboration in trying to understand the very good reasons why it has been threatening to experience or share certain feelings. For example, rather than pressing the client to disclose a difficult feeling, the therapist may use responses such as these to explore the relational threat:

> OK, let's not talk about how you feel about this. Let's talk about what will happen for you if we did.
>
> If you cried or let me see your vulnerability, what might go on between us? Is there something that has happened before with others that you don't want to happen again here?
>
> What is the danger for you if you let me see your sadness? Help me understand what could go wrong between us or how you have been hurt in the past.
>
> How did your parents respond to you when you were feeling _____? What did they say and do? Can you recall the look on their faces or what they were feeling toward you?

Thus, the therapist does not push for the content—the specific feeling—but works with the client to clarify the defense—the reality-based reasons for not wanting to experience or share the feeling. As in Chapter 3, the therapist is honoring the client's resistance by collaborating with the client in trying to understand why this defense was once a necessary and adaptive coping mechanism. *The therapist's basic orienting assumption is that the client's fear or defense makes sense in terms of what has transpired in past relationships, although it is no longer necessary or adaptive in many current relationships.* Thus, if the therapist and client can clarify how the client's defense was necessary and adaptive at another point in time, and the therapist can make it clear that he or she will not respond in the same way that others originally responded, it will soon be safe enough for the client's conflicted feelings to emerge.

Let's look further at the interpersonal or relational aspect of clients' defenses. Though clients may tell therapists that they don't want to explore their underlying feelings because they are "too painful" or it "doesn't do any good anyway," this is not the whole story. Beginning therapists should always generate working hypotheses about the interpersonal consequences that clients may be expecting if they express certain feelings. Clients usually avoid conflicted feelings because of their relational templates and the corresponding expectations that certain unwanted interpersonal consequences will result if they share with the therapist feelings that have been unacceptable to significant others in the past. For example, on the basis of their experiences with others, clients commonly have expectations of being ridiculed; being invalidated and having their feelings discounted; being ignored and unheard; seeing that others are hurt or overwhelmed by their feelings; and so on. Such interpersonal threats are usually far more important than their own discomfort, and clients hold these expectations toward the therapist as well—even though the therapist has never responded in any of these ways! In sum, the discerning therapist can usually find very good reasons why clients learned that certain feelings were unsafe and can begin to differentiate the therapeutic relationship from other relationships that clients have experienced in the past.

One purpose of clients' defenses against their feelings is often to protect their parents. Many clients have learned to avoid or deny painful feelings in order to protect their caregivers from seeing the hurtful impact that they are having. This loyalty to the caregiver is carried out at the clients' own expense, however. As adults, these clients are still complying with delimiting familial rules and colluding with their caregivers in denying the impact of hurtful parental actions. By so doing, these clients are preserving insecure attachment ties at the cost of not acknowledging their own painful feelings. This protects the (internalized) parents from seeing what they are doing (or once did) that is hurting the child. If clients participate in this denial, avoid the internal conflict, and idealize the parent, homeostatic family rules are maintained and the clients remain emotionally connected to internalized parents who remain idealized. However, if painful feelings are permitted expression in therapy and the therapist affirms clients' experience without rejecting, stigmatizing, or demonizing the parent, chronic symptoms such as lifelong dysthymia often lift. In the process, though, familial rules are broken, and emotional ties to internalized parental figures are threatened. Consequently, separation anxieties and separation guilt are evoked.

To sum up, interpersonal themes motivate clients' defenses against their painful feelings. By clarifying that these unwanted interpersonal consequences have indeed occurred in past relationships but are not going to occur in the therapeutic relationship, the therapist provides the safety that clients need in order to experience, share, and ultimately resolve their conflicted emotions.

Providing a Holding Environment

As we have already seen, the best way for therapists to help clients resolve painful feelings is to provide a safe holding environment to contain the client's distress. When clients' core conflicted feelings are evoked, their relational templates or internal working models will be activated. In these affect-laden moments, clients *anticipate* that they are going to receive the same problematic response from the therapist that they have received in the past. It is often difficult for beginning therapists to grasp that clients fear this, even though the therapist has never responded to them in this problematic way. A corrective emotional experience occurs when the therapist responds in a new and safer way that resolves, rather than metaphorically reenacts, clients' original conflict. This is not always so easy to do, however. *Unless the therapist clarifies both what is occurring in the therapeutic relationship and what the therapist is thinking and feeling as clients experience these conflicted feelings, they routinely misperceive or misunderstand the therapist's benevolent response along old, problematic relational lines.* Without this clarification, for example, clients may believe that the therapist is privately contemptuous, judgmental, disappointed, frustrated, or burdened by their feelings, just as significant others were in the past. With these considerations in mind, let's examine further how therapists can provide a holding environment that contains the client.

While growing up, the client had certain developmental experiences that were painful, frustrating, confusing, frightening, disempowering, and so forth. Enduring conflicts arose, in part, because the adult caregivers were not able to provide a safe, affirming, and understanding context to help the child contain these emotional reactions. This containment occurs when the child is psychologically held in a close interpersonal envelope of empathic caring, and the child's concerns are heard, taken seriously, and understood by a caregiver who remains emotionally present and responsive. In other words, the child's distress registers with the caregiver and is taken seriously, without being diminished or exaggerated, and the child feels confident that the caregiver will remain available to help manage the problem, if only by expressing interest and concern. Before children can develop the capacity to manage feelings on their own, someone must first provide this holding context for them. Too often, however, this essential developmental experience is not provided. For example, suppose the child is sad or hurt. In some families, the child's sadness

- Arouses the parent's own sadness, and the parent withdraws, which leaves the child emotionally alone
- Makes the parent feel guilty, and the parent tries to cheer the child up and deny the child's feelings, which leaves the child confused and alienated from his or her her own authentic experience
- Makes the parent feel inadequate in the parenting role, and the parent responds derisively toward the child's sadness, which leaves the child ashamed of his or her vulnerability

In such scenarios, the child's sadness cannot be heard and supported; as a result, these feelings cannot run their natural course and come to their own resolution. Young children cannot contain a strong, painful feeling by themselves without support from an emotionally available caregiver. As a result, the child will have to devise ways to deny or avoid this painful feeling. Years later, when this child is an adult client and risks sharing the unacceptable or threatening feelings with the therapist, the therapist must provide the holding environment that was missed developmentally. We have already seen two examples of this in

Chapter 2: Marsha's second therapist heard and responded to her experience of emptiness, and the therapist of the depressed young mother affectionately articulated her mother's own unanswered cries. Both therapists provided a holding environment. Turning to specifics, we'll look first at what therapists should *not* do; then we'll examine further what therapists can do to help clients contain and resolve their conflicted feelings.

Too often, therapists avoid or move away from clients' painful emotions; in so doing, they reenact the clients' developmental conflict and leave clients unrelated or unconnected to others while they are experiencing their emotions. The most common reason why therapists do this is because *they assume too much responsibility both for causing the client's pain and for alleviating it.* Beginning therapists may think to themselves, "I got the client into this, so now it's up to me to get the client out of it." Both parts of this belief are false. By responding in the ways suggested in this book, therapists reveal feelings that are already present in the client; they do not cause the pain. Further, it is demeaning or patronizing for therapists to think they can fix clients, get them out of their feelings, or put them back together again. In actuality, the therapist never has the power to manipulate another's feelings in this way, and attempts to do so often recapitulate the parents' controlling or dependency-fostering stance with the client.

When therapists take on responsibility for the client's feelings, and place on themselves the expectation that they will fix or solve the problem, the usual outcome, understandably, is feelings of inadequacy. In addition, therapists may be threatened by their own conflicted feelings that have been evoked by the client. In such situations, therapists are likely to respond to the client's affect in ineffective ways:

- To interpret what the feelings mean and intellectually distance themselves
- To become directive and tell the client what to do
- To become anxious and change the topic
- To fall silent and emotionally withdraw
- To self-disclose or move into their own feelings
- To reassure the client and explain that everything will be all right
- To diminish the client by trying to rescue him or her
- To become overidentified with the client and insist that the client prematurely make some decision or take some action to manage the feeling

When these ineffective responses occur, there is no holding environment. Therapists' avoidance, as reflected in these responses, often dynamically recapitulates the client's developmental conflict. For example, by trying to talk her out of her feelings, Marsha's first therapist reenacted her experiences in her family. Marsha's first therapist went even further to avoid Marsha's feelings by, in effect, sending her away: When Marsha got near her sad, empty feelings, the therapist suggested that she join a group and talked to her about seeing a psychiatrist for antidepressant medications. Although referrals such as these are certainly appropriate and necessary at times, in this case they were used to distance the therapist from the client's feelings. This is, sadly, a common occurrence.

If these distancing and controlling responses are ineffective, what should the therapist do instead? The therapist needs to welcome the client's painful feelings and approach them directly. For example, the following would be a more effective response to Marsha's strong feelings.

Therapist:

I'm glad you're taking the risk of telling me about these sad and empty feelings. Tell me more and help me understand them better.

Therapists also want to find genuine ways to express their understanding, acceptance, and concern. Therapists can do this in words, of course, but it is equally important to communicate nonverbally in tone of voice, facial expression, and posture. In particular, therapists must communicate emotional connectedness to clients; therapists must be present with clients while they are experiencing their feelings. There are many ways to do this, and much of this communication is nonverbal or involves only a few words. We'll now consider some of the elements found in effective responses that create a supportive, holding environment. Of course, having assimilated these guidelines, therapists will want to tailor their responses to fit each particular client, as well as to be congruent with their own personal styles.

Initially, the therapist's primary goal is to stay emotionally present and connected to clients while they experience painful feelings. As discussed in Chapter 2, the therapist may also want to articulate or reflect what clients are feeling and to affirm the reality of their experience. This can be done simply by saying, "You're very sad right now." With some clients, it may be helpful to further validate their experience by saying, "Of course you are feeling sad right now. It hurt you very much when he did that." Therapists also need to find a way to express their own sincere compassion for clients and connectedness to them.

Therapist:

I can see how much this has hurt you and how sad you are feeling right now. I feel very close to you, and honored that you choose to share such a special part of you with me.

In addition, therapists must demonstrate behaviorally that they can tolerate the full intensity and range of clients' feelings. In other words, they must communicate by their manner—and overtly in words with some clients—that they are in no way hurt, burdened, or undone by clients' painful feelings.

Therapist:

I'm privileged that you choose to share this with me. I'm with you right now, and we're going to stay together in this until we have worked our way through it.

As we will see, if they are to undergo a new or corrective experience, clients must see that the therapist is not overwhelmed or threatened by their feelings, is not judgmental or rejecting of them, does not need to move away from them in any way, and is still fully committed to the relationship. Beginning therapists often fail to appreciate how clients' old templates or internal working models are activated in these moments and take precedence over the current reality.

As the client gathers his or her equilibrium from a strong emotion, the therapist may also want to describe succinctly what caregivers and others have done in the past when the client felt this way. The therapist can then clarify how their current relationship is different from the client's relationships with others in the past and emphasize that, unlike others, the therapist is comfortable with, or accepting of, the aspect of the client that has been revealed. It is important to debrief clients by asking them how it has been to share these feelings or personal disclosures with the therapist. Therapists often miss this important opportunity; genuinely touched by the client, they fail to appreciate that it is at these intense, emotion-laden moments that the client is most likely to misperceive the therapist's caring response in terms of earlier relational templates. To correct these expectable misunderstandings or transferential distortions, it is necessary to ask clients what they think the therapist was thinking, feeling, or going through while they were feeling so sad or hurt. Long-standing transference distortions can be identified and resolved at these moments. For example, the therapist might clarify the client's distortion by means of a self-involving comment:

Therapist:
No, I wasn't thinking that you "looked silly" and "sounded like a baby" at all. I was touched by your sadness. I think your feelings make sense and fit exactly with what happened to you.

Finally, when traumatic or abusive memories are being relived, it may be useful to share the comforting reality that the traumatizing event is over and the client has survived it and to emphasize that the client and therapist are just living through the emotional reactions to what has already happened and not the crisis itself.

When offered sincerely, simple human responses of validation and care mean much to clients and help restore their dignity and efficacy. Unfortunately, beginning therapists often believe that they have to do much more than this to respond adequately to the client's pain. These concerns may reflect the therapist's own unrealistic expectations that they have to respond in some precisely correct way to the client's pain. In reality, the client does not need the therapist to provide eloquent, crafted responses in a calmly self-assured manner. Such perfectionism is unnecessary and may simply immobilize the therapist. As Winnicott (1965) says, the child needs only "good enough mothering." Therapists provide a satisfactory holding environment when they find their own genuine ways to enact these guidelines. Clients will be thankful for sincere efforts and find them comforting, even if they are expressed in a fumbling or halting way.

Change from the Inside Out

Clients improve markedly when the therapist provides a holding environment that allows them to contain feelings that have been too painful to integrate or to come to terms with. How does this come about? A corrective emotional experience occurs when the therapist can be supportive and stay emotionally connected to clients' pain without denying, minimizing, intellectualizing, or otherwise moving away from it.

Many clients hold a primary irrational belief about their most central, conflicted feelings. In its most primitive expression, the fear about their feelings often is either "I will be destroyed" or "You will be destroyed." Some clients learned in their families of origin that their painful feelings are "too much" or "overwhelming" or that, if they express the feelings, they "won't be able to stop crying," "will go crazy," or "won't be able to breathe." In contrast, some clients believe that the other—the therapist or internalized parent—will be "over-whelmed," "hurt," "burdened," or "impaled" or will "go away" because of the intensity or unacceptability of their feelings. Other conflicted clients are double-binded and struggle with both pathogenic beliefs at the same time.

Therapists disconfirm these "grim, unconscious pathogenic beliefs" (Weiss & Sampson, 1986) and provide a corrective emotional experience when they can (1) compassionately accept clients' painful feeilngs; (2) stay emotionally connected to them; and (3) not be injured, burdened, or overwhelmed by them. If the therapist does this, some clients will be profoundly relieved to find that neither they nor others need be hurt by their feelings. How does this translate into real change? Clients' experience that their most painful or threatening emotions can be compassionately accepted—rather than producing the unwanted consequences they have come to expect—brings a more authentic emotional connection to themselves and to the therapist. It also frees them to take the steps necessary for enduring changes in their coping styles, behavior, and expectations of themselves and others. The clients now have greater freedom to live authentically, to identify their own needs and more actively choose how they want to shape their lives.

Moreover, in a supportive context, clients' feelings can run their course and come to a natural close, often in only a matter of minutes. Clients then recover by themselves and readily reestablish their own internal equilibrium. Through this corrective emotional experience with a caring and supportive therapist, clients disconfirm pathogenic beliefs about themselves and learn that their feelings—and they themselves—are not dangerous, disgusting, weak, or bad. As they have the experience that they can remain in a caring relationship, their feelings, although still a part of them, no longer compel them to be isolated or ashamed. As a result, clients no longer need to maintain long-held defenses, symptoms, or elaborate interpersonal coping strategies—the feelings that had to be sequestered away can now be experienced, understood, and integrated. When the therapist can hold clients' pain and allow them to tolerate experiencing and sharing it, their developmental conflict is resolved, and they will become more able to contain their feelings on their own. *This is change from the inside out.* This interpersonal process often propels a far-reaching trajectory of enduring behavioral change, and long-standing symptoms commonly abate at this point. Ironically, when clients risk exposing their pain, vulnerability, or shame, as long as the therapist responds with kindness and understanding, they feel more powerful. One of the most satisfying aspects of being a therapist is to facilitate clients' self-efficacy in this way. To illustrate the principles discussed in this section, let's look at a brief case study.

Sherry, a first-year practicum student, was struggling to assimilate all of the new ways of thinking about therapy that she had been learning. It was an exciting but difficult year; her entire worldview was shifting. Part of the difficulty was that her practicum instructor and her individual supervisor were suggesting very different ways to respond to clients. Having grown up with combative, divorced parents, she had conflicts about trying to please both sides, which were being aroused in this situation—making the already complex reality of clinical training even more difficult to manage. At times, Sherry was immobilized as she tried to sift through these competing ideas, find what made sense to her, and come up with something that she thought would be useful to say to her clients. Two-thirds of the way through her first practicum, she felt more self-conscious and inhibited about responding to people than she had been when she started the program!

Despite her insecurity, Sherry was learning a lot and had been responding quite effectively to her client, Maria. Near the end of one of their weekly sessions, Maria began to disclose a very painful memory. With tears in her eyes, Maria recounted the sickening experience of a date rape 15 years earlier, when she first went to college. Listening to Maria's painful story, Sherry felt sad and angry. However, as she began to think of all the things she should and should not say, and what her supervisor would want her to say, she felt frozen and could not find the words to say anything at all. Although it was less than a minute, the silence grew painfully awkward as Maria finished speaking and waited for Sherry to respond. Unable to speak, Sherry just sat there. Shaken by Maria's painful story and thinking of all the different ways she "should" respond, she couldn't find the words to say anything at all.

After an uncomfortable pause, Maria tried to take care of Sherry and started talking again, but this did not last long. After a few minutes, Maria chided, "Don't you understand! He hurt me a lot when he did that!" Sherry had made a mistake, but she was able to recover by validating Maria's anger and affirming her experience.

Sherry:

Yes. He did hurt you very much. It was frightening and painful for you. I'm very sorry that happened. And I understand why you are angry at me for not responding to you. I was touched by what you said, but I just couldn't find the right words for a minute.

Maria:

You couldn't?

Sherry:

No, I wanted to, because I felt for you. But I just couldn't find the right words for a minute, and the longer I searched for them, the worse it got. I'm really sorry this happened. I feel embarrassed about it, and I know it must have been hurtful to you.

Maria:

It's OK. I don't know what to say either sometimes. I don't know why, but this makes me think of my daughter when she was little. If I would take the time to stop and pick her up and really listen to her when she needed me, we would get through her upset pretty quick. My husband called it "collecting her." But if I was in a hurry and tried to ignore her or speed her up, she would escalate, and it would go on for a longer time.

Sherry:

You're a very good mom. I think you're remembering that right now because that's what you needed from me—to be attended to, cared about, collected. You needed me to give you what you gave your daughter.

Maria:

Yeah, I think that's just exactly what I needed. (*crying*) You know, I never dreamed of telling my mother what happened to me—you know, about the date rape thing. She never seemed to listen to me when . . .

Sherry recovered from her "mistake" and began providing the holding environ- ment that the client herself had articulated so well with her own young daughter. As in this example, validation of the client's dissatisfaction with the therapist often leads the client to clarify related conflicts or feelings in other relationships, to achieve meaningful insights that do lead to behavior change, or to bring up important new material for discussion. Moreover, by responding effectively to Maria's painful feelings at the second attempt, Sherry gave Maria a corrective emotional experience that also helped her to progress in treatment.

To sum up, when the pain associated with the client's central conflicts emerges, therapists have the best opportunity to help the client resolve problems and change. The most important corrective emotional experiences usually occur around the therapist's new, more satisfying response to the client's emotions.

However, *the client's maladaptive relational patterns are also most likely to be reenacted with the therapist in these intense, affect-laden moments.* Why? Clients' vulnerability activates their relational templates or internal working models, and, consequently, they expect the same hurtful response from the therapist that they have received in the past and readily distort or misperceive the therapist's benevolent response to fit the old expectations. To compound this, the therapist's own personal issues or countertransference reactions are likely to be aroused in these intimate moments. Thus, in the next section, we address countertransference issues and turn to factors in the therapist's own life that make it difficult to respond to clients' emotions.

Personal Factors That Prevent Therapists from Responding to Clients' Emotions

Just as it is easier for clients to externalize and focus on others rather than themselves, it is also easier for therapists to focus on the client than it is to look within at their own issues and dynamics. However, if therapists are not willing to work with their own emotional reactions to the feelings that clients present, they will tend to avoid clients' feelings or to engage them only on an intellectual level. The purpose of this section is to help therapists anticipate and manage their own reactions to the evocative material that clients present.

No enduring change occurs unless therapists express respect and compassion for the emotional conflicts with which clients are struggling. In practice, however, it is often difficult for therapists to respond to all of their clients' feelings, because every therapist brings his or her own personal conflicts, developmental history, and current life stressors to the consulting room. Let's examine how these factors can keep the therapist from responding effectively to clients' emotions.

Therapists' Need to Be Liked

Most students who are drawn to clinical practice are kind people; they are genuinely concerned about others and ready to give of themselves. Nurturing individuals by nature, they can readily empathize with, and respond to, others' hurt and pain. At the same time, many therapists also have strong needs to be liked. As a result, many therapists have difficulty providing a neutral, nonreactive response to the angry, controlling, competitive, critical, distrustful, demanding, or otherwise negative feelings and relational patterns that some clients will begin to construct or reestablish with the therapist (Strupp & Hadley, 1979). Other therapists have strong needs to nurture others who are dependent on them. These therapists enter clinical practice, in part, in order to fulfill these needs, and they may have difficulty supporting the client's healthy individuation at times.

Many counselors were caretakers in their families of origin. They are drawn to clinical practice, in part, by the unrecognized need to rescue their caregiver from alcoholism, an unhappy marriage, chronic depression, and so on. This prevalent countertransference issue makes it harder for therapists to address clients' painful emotions. Those emotions evoke for therapists the familiar predicament of needing desperately to fix their caregiver and, because this is an impossible task, threaten them with the prospect of once again feeling inadequate, a failure, or a disappointment to their all-important other.

All therapists need to assess their own motivations for becoming clinicians and to examine what needs of their own are being fulfilled by doing therapy. *The issue is not whether therapists have their own needs and countertransference propensities, but how they will deal with them.* All therapists have certain countertransference reactions, and self-examination is one way to manage them effectively.

Beginning therapists are taken aback when their well-intended attempts to approach the client's affect meet with anger, criticism, or disdain at times. Too often, beginning therapists absorb the client's rebuff, comply with the defensive side of the client's feelings, and stop approaching the client's conflicted emotions. This response is most likely from therapists whose needs to nurture or to be liked have been frustrated. To be effective, however, therapists must continue to work with the client's negative feelings that have engendered distance or discouraged others from responding in the past. They must find other ways to approach the client's conflicted feelings and change the maladaptive relational patterns that result.

To do this, therapists can adopt the strategies for working with the client's resistance that were presented in Chapter 3. Therapists should not push clients to express a feeling they are reluctant to share, but explore what the threat or danger may be for them if they do so.

What does it say or mean about you if you feel _____?
What could happen between us or what could go wrong for you if you let yourself feel _____?
How have people responded to you in the past if you felt _____?

Therapists must forgo winning the client's approval; they should not try to be a "nice" person who does not approach difficult feelings or awkward interpersonal situations. On the other hand, therapists must not demand that clients enter into, or stay with, a certain affect; instead, they allow clients to enter and leave painful emotions as they wish. Letting clients have it their own way avoids the issues of coercion and compliance that some clients had in their families of origin and that some people of color have experienced in the dominant culture. Therapists provide an effective middle ground when they address the client's resistance by honoring it and providing a safe place for it to be explored when the client is ready to do so. Feeling safe in this new and different relationship, clients usually are willing to experience feelings that needed to be held away in the past.

To sum up, therapists are most likely to have difficulty working with the client's angry, critical, and controlling feelings toward them. Therapists must learn to address these feelings, however, because therapy is likely to falter whenever they cannot respond to a particular feeling that the client is struggling with.

Therapists' Misperceptions of Their Responsibility

Too often, therapists believe that, if they respond to a client's feelings, they are responsible for causing the feeling or for taking the client's pain away. Assuming responsibility for the client's feelings is inappropriate and disempowers the client. Let's look at three aspects of this problem.

First, some therapists avoid their clients' feelings because they do not want to hurt them or make them feel bad. These therapists mistakenly believe that responding to a painful feeling in the client is the same as causing that feeling to exist. Therapists must keep in mind that they did not cause the client to have this feeling; it was present before the client met the therapist. The therapist has simply invited the client to express the feeling, acknowledged it, or responded to it. The misconception that the therapist has caused the client's painful feeling leads to feelings of guilt. Guilt over making the client hurt will immobilize therapists and prevent them from responding effectively. Beginning therapists would do well to realize that they are not so powerful as to cause these long-held beliefs and painful feelings that the client is presenting.

A second problem occurs when therapists confuse their legitimate responsibility to respond to the client's feelings with an inappropriate belief that they are responsible for *causing* the client's original feeling. If the therapist inappropriately accepts responsibility for causing the client's feeling, it follows that the therapist must also be afraid of the client's angry retaliation for having to reexperience the painful affect. Clients will hurt when painful feelings are expressed, and reactive anger or shame may be aroused. Thus, both guilt and fear of the client's retaliation may keep the therapist from approaching the client's conflicted emotions. Moreover, if responding to the client's pain does happen to evoke the client's anger or rebuke, the therapist will be unable to tolerate this reaction and work with it neutrally. Instead, the therapist may become either defensively angry or punitive toward the client or, most commonly, will overreact to the criticism and withdraw. Thus, therapists must learn to respond to clients' original sadness or reactive anger without accepting responsibility for causing these feelings.

A third problem arises when therapists labor under the impossible burden of having to "fix" the hurt they have "caused" the client to feel. When therapists take responsibility for causing the client's feeling, they typically assume responsibility also for making the client feel better. This is an impossible burden; therapists can never take away clients' hurt or fix their pain. Therapists can share clients' experience, however, and this sharing provides understanding and validation that are empowering.

Therapists who assume responsibility for clients' feelings are apt to avoid those feelings—to tell clients what to do, to reassure them, and to explain and interpret their emotions before they can experience them. In most cases, these approaches ultimately prove ineffective.

If therapists should not shoulder this inappropriate responsibility, what should they do for their clients? The therapist's appropriate response to the client's feelings is threefold. First, the therapist must help clients identify, experience, and express more fully whatever feelings they are having. Second, the therapist must be empathic or emotionally available and responsive to clients, so that clients can share their feelings with a concerned other rather than having to experience them alone, as they often have in the past. Third, the therapist needs to confirm the validity of clients' responses, by helping clients understand their feelings and make sense of why they are experiencing this particular feeling at a particular moment. If the therapist accepts this challenge but does not assume responsibility for causing or changing clients' feelings, the therapist will be freed up to respond effectively to whatever feelings clients present.

Even therapists who are well aware of the pitfalls of assuming responsibility for clients' feelings are likely to struggle with this issue when the client's affect becomes very intense. For example, a novice therapist was treating a young boy who was in the midst of a potentially life-threatening battle with leukemia. The therapist would sit with the boy in the hospital and, among other topics, sometimes talk with him about the possibility that he could die: "What would be the scariest thing about dying? What are you most afraid of?" The therapist approached the child's feelings in this direct way, and the boy responded, "I don't want to die, I don't want to die! I'm afraid to die, I want to live!" Feeling helpless and seeing the fear in this young child's eyes, the therapist questioned what she was doing and felt like a mean or sadistic person to be making this child feel such anguish.

But the child went on to say, "I don't want to be alone, I'll be all alone if I die!" Inviting the child's feeling revealed his concern that he would be all alone. Hearing the boy's separation anxiety and fear of abandonment, the therapist understood his primary concern and was able to talk with him in more helpful ways. More important, the therapist was able to help the child's mother understand his central fear; his mother was able to reassure him that, if he were to die, her love for him would last forever. By talking with the mother and son together about his fear of being left alone, the therapist allowed them to become much closer during his illness. This closer contact and deeper understanding between them measurably alleviated the child's distress, in general, and his intense separation anxieties and fears about dying, in particular.

When working with such profound emotions, therapists almost inevitably feel to some extent that they are responsible for causing the client's pain. As a result, they tend to avoid these intense feelings. To keep this from happening, therapists must consult with a supervisor or colleague to help them manage their own exaggerated feelings of responsibility. Otherwise, these countertransference reactions will prevent therapists from responding effectively and giving clients the holding environment they need. In the example of the boy with leukemia, the therapist could not have stayed with the intensity of the boy's fear or the mother's potential loss without regularly scheduled consultations with a supervisor to help her manage her own emotional reactions.

Family Rules

Therapists may not be able to respond to certain feelings that the client presents because of *family rules* about emotional expression that they learned in their family of origin. Every family has unspoken rules that govern emotional expression: A rule-bound homeostatic system prescribes which feelings can be expressed and which are unacceptable. This system also prescribes when, how, and to whom certain feelings can be expressed. The permitted intensity or degree of emotional expression is also controlled by homeostatic family mechanisms. For example, if too much conflict, independence, or closeness is expressed in the family, a corrective homeostatic mechanism automatically brings the level of emotional expression back within acceptable limits. The following example illustrates how homeostatic mechanisms operate to keep family emotional expression within prescribed limits.

Suppose that in the Smith family a moderate degree of conflict and anger can exist between the older sister and the younger brother. A lesser degree of frustration and irritation can be expressed between the father and both children. However, anger or even overt disagreement is rarely expressed between any family member and the mother and never expressed between the father and the mother. What is the homeostatic mechanism that serves to keep this system operating within these boundaries? Whenever a child begins to express angry feelings toward the mother or to have direct conflict with her, the mother stops the child's emotional expression by looking sad and hurt. If the child's anger continues, the father intervenes and says, "Don't talk to your mother that way," or else the mother employs the father as a regulatory influence by saying, "I'm going to tell your father about this when he comes home."

In contrast, if anger, conflict, or differences of opinion begin to escalate between the mother and father, the children serve as homeostatic regulators and respond in predictable, patterned ways to terminate the parental conflict. For example, the oldest child, 8-year-old Mary, serves as a go-between in the parental relationship. She tries to make peace between her parents by carrying messages back and forth between them. This mechanism usually succeeds in terminating parental conflict, except in periods of exceptional family stress. If the parental conflict continues, the younger son may then provide a diversion to the parental conflict by acting out in some predictable way—by breaking something or having an accident. When the son makes himself a problem that mother and father must jointly address, their rule-breaking level of conflict is terminated, and the family's emotional equilibrium returns to acceptable levels.

This example can be linked to our case study of Jean. Imagine that we have just turned the clock forward 20 years. The oldest child in the Smith family, Mary, who used to serve as the mediator to assuage parental conflict, is now a graduate student in clinical training. Mary is an intern at a county mental health agency and has been seeing Jean in therapy for the past four months. The family rules that Mary has learned in her family of origin will influence her ability to respond to Jean's emotions—just as the family rules that all therapists have learned in their families will influence how they respond to their clients.

Earlier in the chapter, we observed the critical incident in which Jean was able to let Mary respond to her pain and share her sadness for the first time. Things had gotten somewhat better for Jean since then, but her problems continued. Jean was able to end her problematic relationship with her boyfriend but she soon became depressed again and fell back into a victim role. Jean was still unable to say no to the demands of others or to follow through and act on her own wishes and interests. Frustrated by her continued depression, Jean angrily criticized Mary for the first time: "I keep coming here every week, but I'm not getting better. You make me have these awful feelings and it just doesn't help. I wonder if you know how bad you're making me feel sometimes!"

Mary was taken aback by Jean's angry outburst, even though it was short-lived and Jean quickly became apologetic. Mary was especially surprised because she thought they had been working together more effectively in recent weeks. Throughout the rest of the session, Mary backed away from Jean's anger by offering reassurances. She tried to convince Jean that she really did care for her, and that Jean would get better if she would continue to come to therapy and work with the difficult feelings they had been sharing. Mary felt terrible that Jean had been critical and anxiously sought help from her supervisor as soon as the session was over.

Mary's supervisor was responsive to her distress over Jean's anger and was able to help Mary understand what went on for her during the session. In reviewing a videotape recording of the session, the supervisor helped Mary recognize how she had tried to assuage Jean's anger with reassurances, rather than allowing Jean to feel angry with her or disappointed about her lack of progress. The supervisor explored with Mary why she had responded in that way.

With her supervisor's help, Mary realized that she had been unable to tolerate Jean's anger. She had indeed tried to assuage it, just as she had always done with her parents. Mary's role in her family of origin was to take care of the conflict, rather than to allow others to express their dissatisfaction and take their concerns seriously. In addition to clarifying Mary's old response pattern, the supervisor also helped Mary to generate working hypotheses about what this critical response may have meant for Jean. The supervisor noted that Jean expressed her anger after sharing her hurt, and hypothesized that the anger may be a part of her affective constellation that needed to be addressed. In other words, Jean needed the experience of being able to express her anger at someone who would take her concerns seriously and would not punish or abandon her. In part, Jean may have been testing to see whether she could be angry or critical of Mary and still remain in the relationship with her—which Jean had not been able to do with her primary caregivers.

Without this helpful consultation, Mary and Jean may have become stuck in their relationship, and therapy might not have progressed. Jean could not move forward and make progress with her relational conflicts as long as Mary could not respond to the anger component of Jean's affective constellation. In their next session, however, Mary was able to recover and invited Jean to discuss her frustration with treatment and anger toward her in more depth. Although Mary still felt an anxious compulsion to reassure Jean and move on, Mary was able to contain her own discomfort enough to be able to discuss Jean's concerns, better appreciate what was not working for Jean in their relationship, and suggest some accommodations to make things work better for her.

Clients will not be able to resolve long-standing conflicts unless the therapist can respond to each component of their conflicted affective constellation. Therapy often stops progressing at the point at which the therapist cannot respond to a particular feeling that the client is experiencing. With this in mind, *all therapists should examine the rules that governed how emotions were dealt with in their own families, and reflect on how they are bringing their own familial roles and rules for emotional expression to their clinical work.* For many therapists, their own therapy will be the best way to change ineffective family rules that still govern how they act with their clients.

Situational Conflicts in Therapists' Own Lives

Situational stress in therapists' own lives may also prevent them from responding effectively to clients' emotions. For example, when personal problems in therapists' own lives are similar to those their clients are experiencing (perhaps marital conflict, parenting problems, or conflicts over caring for aging parents), therapists may have difficulty approaching their clients' affect or, more likely, therapists will not be able to allow clients to experience fully the feelings they are struggling with.

Therapy is a very human interchange. Although the therapist's personal availability is the greatest asset in helping clients change, it will also be an obstacle at times. When the therapist fails to accurately identify the feelings that the client is presenting, avoids rather than approaches the client's feelings, or becomes personally invested in changing how the client responds to or feels about an issue, the therapist's own countertransference issues are operating.

Countertransference reactions will occur at times for every therapist. The issue is how they are dealt with. When countertransference reactions occur, therapists must consult with a colleague or supervisor to better understand and manage their own reactions. When therapists find that they are repeatedly having difficulty with the same type of affect, or when supervision does not free them up to respond to clients' feelings, therapists should seek therapy for themselves. Because change is predicated on the relationship that the therapist creates with clients, therapists must make an ongoing, lifelong commitment to working with their own dynamics. Without this willingness to work on their own personal reactions to clients, the help therapists can provide will be limited. Furthermore, therapists who are unwilling to work on persistent countertransference reactions are those most likely to have a negative therapeutic impact on their clients. It is a privilege to be able to help people change, and therapists honor that privilege by acknowledging their own limitations and personal involvement in the therapeutic process.

Closing

As we saw in Chapter 4, clients' conflicted emotions will not emerge in therapy unless the therapist focuses clients inward. An externalizing focus will lead therapists into a problem-solving and advice-giving mode that precludes clients' affective exploration. As clients talk about the problem out there with others, the therapist's complementary role is to give advice, reassure, or disclose what the therapist has done to solve similar problems. Although these responses can be helpful at times, limited change will occur for most clients if they characterize the ongoing course of therapy.

But if the therapist does not tell clients what to do, what does the therapist really have to offer? The most important way to facilitate change is to help clients master their conflicted emotions by reexperiencing and integrating feelings that have been too painful, shameful, or unacceptable to tolerate in the past. Clients have experienced tragedies in their lives that will be entwined with the symptoms and problems they present. The therapist helps clients resolve their presenting problems by sharing the full intensity of whatever feelings they are experiencing. The most important aspect of therapy is to respond to clients in these deeply personal ways and thus to provide a new, corrective experience as old expectations and relational patterns are disconfirmed. Yet therapists must also be aware that their biggest obstacle to providing clients with a relationship that can produce change will be their own discomfort with certain client feelings. Thus, therapists must make a lifelong commitment to working with their own personal issues and examining how their own personhood influences their clinical practice.

In the next three chapters, we will see how the therapist can conceptualize clients' dynamics on the basis of their conflicted emotions and the problematic relational scenarios they begin to reconstruct with the therapist. Then, in the final two chapters, we will show how this conceptual formulation can be applied to treatment plans and can help clients resolve their conflicts and change.

Suggestions for Further Reading

1. For information on basic communication skills to help therapists approach or inquire further about clients' feelings, see Chapters 5 and 6 of G. Egan's *The Skilled Helper* (Pacific Grove, CA: Brooks/Cole, 1994). These two chapters also help counselors to listen without judging and to become more aware of their own cultural biases or filters in listening and responding to clients.

2. One way for therapists to increase their effectiveness with clients is to explore their own motivations for becoming therapists and to examine how their own dynamics are expressed in their clinical work. It is just as important to focus on the personhood of the therapist as it is to focus on client dynamics and intervention strategies. Two sources that can help therapists explore these issues are: Chapter 21 of J. Bugental's *The Search for Authenticity* (New York: Holt, Rinehart & Winston, 1965) and A. Guggenbuhl-Craig's *Power in the Helping Professions* (New York: Springer, 1971). See especially the concept of the wounded healer.

3. At times, techniques can be useful to elicit and clarify the client's affect. A book that describes Gestalt techniques for intensifying client affect is W. Passons' *Gestalt Approaches in Counseling* (New York: Holt, Rinehart & Winston, 1975).

PART THREE

Conceptualizing Client Dynamics

CHAPTER SIX
Familial and Developmental Factors

CHAPTER SEVEN
Inflexible Interpersonal Coping Strategies

CHAPTER EIGHT
Current Interpersonal Factors

Familial and Developmental Factors

Conceptual Overview

The previous chapters have focused on the interaction that occurs between the therapist and the client. In the next three chapters, we step back from this intensive tracking of the therapeutic process and consider how therapists conceptualize their clients' dynamics. In order to be effective, therapists must be able to formulate a clear conceptualization of the client's personality and problems, and this case conceptualization must lead to specific treatment plans and intervention strategies. Because novice clinicians have had little training or experience, however, it is difficult for them to conceptualize their clients' dynamics or apply this conceptualization to treatment. Although no single book, supervisor, or theory is sufficient to teach therapists how to recognize and respond to the relational patterns that organize their clients' personality and problems, helpful guidelines are available.

How can therapists understand their clients' problems and use this conceptualization to guide their treatment plans? One way to understand how significant and enduring problems develop is to examine what occurs for offspring in families that function well. Once the basic familial processes that produce healthy offspring are identified, it is easier to understand how normal development can go awry and lead to the enduring personality conflicts that clients present. Thus, this chapter helps therapists recognize the familial and developmental antecedents of many clients' problems. Family systems concepts that clarify the genesis of many characterological problems will be introduced in order to aid therapists in conceptualizing their clients' dynamics. The discussion of developmental issues in this chapter will be integrated with an interpersonal model for further conceptualizing clients' functioning in the next chapter.

Chapter Organization

In this chapter, we examine three fundamental dimensions of family life. In the first part, we discuss the structure of family relations and the nature of the parental coalition. In the second part, we examine the separateness-relatedness dialectic that was briefly introduced in Chapter 1. In the third part, we explore child-rearing practices. Problems in each of these areas lead to developmental conflicts that are central to many of the symptoms and problems that clients present. By understanding how these familial factors shape personality and development, therapists will be able to respond more effectively to their clients. Therapists will be even more effective when they can broaden this perspective and take into account the cultural context in which family relations and child-rearing practices are embedded. Cultural values and beliefs often circumscribe acceptable behavior and exert considerable influence on family functioning.

Structural Family Relations

Structural family relations comprise the relatively enduring patterns of alliances, coalitions, loyalties, and alignments that exist in the family. These relationships are the basis for family organization and define how the family operates as a social system (Minuchin, 1974). Such family patterns, in addition, are often restricted or bounded by cultural values and beliefs. Thus, structural family relations and the culture within which they are delineated shape family communication patterns and the roles that family members adopt. Among the different structural family relationships that exist, the parental coalition is the pivotal axis of family life and shapes much of how the family functions (Teyber, 1981). In particular, the nature of the parental coalition will be very important in determining how well children adjust.

The Parental Coalition

In healthy families, the marital relationship is the primary two-person relationship in the family. Each spouse has a primary loyalty commitment to the other, and the marital coalition cannot be divided by grandparents, children, friends, employers, or others, although the relative influence of extended family members may be more pronounced in some cultures (for example, in Native American cultures). Parents will still have differences and conflicts between themselves, and they will be committed to others as well, but the marital relationship is a stable alliance; its cohesiveness is a fundamental feature of healthy families. In contrast, two-parent families in which the primary emotional bond or involvement is not between mother and father often produce offspring with enduring personality conflicts. In these problematic families, the primary alliance is between a grandparent and a parent, between a parent and a child, or within some other dyad. Although it is important to recognize that cultures vary in how strongly they emphasize child care responsibilities and family loyalties with extended family members, family system theorists emphasize that the parental coalition is the primary emotional axis in healthy families.

How does a primary parental coalition develop? Both parents must have been able to individuate within their families of origin and to achieve a sense of their own personal identity, based on the ability to experience and act on their own authentic feelings, interests, and values in early adulthood. In this developmental transition of forming their marital coalition, they must also be able to establish a new balance of family loyalties. This complex process involves maintaining ties and responsibilities toward the previous generation, while shifting their loyalties to a primary commitment to a peer relationship. Thus, as young adults, these offspring have been able to transfer their primary loyalty commitment from their parents to a spouse. This process, accompanied by increased maturity, self-esteem, and sense of competence, permits these young adults to define and elaborate their own sets of values and beliefs. This does not mean losing their sense of connectedness to their own parents but signals increased flexibility in their relational self and involves transferring their primary relational commitment to a peer relationship characterized by mutual empathy and empowerment.

If young adult offspring have not been able to shift their primary loyalty commitment from their family of origin to their spouse, a primary marital coalition will not develop. The primary loyalty of one or both spouses will remain with their parent. When this occurs, the primary dyadic relationship in the family is a *cross-generational alliance* between a parent and the adult offspring, rather than a primary marital coalition between the husband and the wife. With the birth of children, this cross-generational alliance (for example, maternal grandmother-mother) usually repeats in the next generation and results in a primary parent-child coalition as well (for example, mother-child). Before going on to see why these cross-generational alliances are so problematic for offspring, we must better understand the important developmental processes that are being enacted in forming a primary parental coalition. To do this, let's examine the wedding ceremony as a cultural rite of passage.

The wedding ceremony is a familiar ritual; its central purpose is to publicly mark a transition in loyalty bonds from the parents and family of origin to the new spouse. How does this occur? In the standard Christian ceremony, once family, friends, and clergy are all assembled, the father walks the bride down the center aisle to the front of the church. In front of all, the father symbolically gives the bride away by placing her hand in the hand of the waiting groom. The father then leaves the couple by stepping back from the front of the church and sitting down beside his own wife. The bride and groom, who have now been separated from the parental generation, turn away and step forward to be married. In this ceremony, the bride and groom are shifting their primary loyalty commitments and publicly defining themselves to family and friends as an enduring marital couple.

This shifting of primary loyalties from the previous generation to the new spouse is a major developmental task. According to Erikson (1968), this transition is a normative developmental crisis that arouses difficult feelings of loss and requires a complex redefinition of individual identity, as well as familial roles, membership, and boundaries—even in healthy, well-functioning families. This developmental challenge becomes far more conflict-laden and leads to symptom formation when the young adult offspring are trying to emancipate from families of origin that have not had a primary marital coalition. The question of who comes first, parent or spouse, may be painfully enacted in conflicts that surround planning for the wedding day. Offspring embroiled in cross-generational alliances often find that their wedding day, and the preparations for it, are a wrenching battle between conducting the ceremony as the couple being married would wish and doing it the parents' way. Usually, the underlying conflict is over the shift of loyalties from the previous generation to the spouse, but it is played out in arguments over who will be invited to the wedding, where it will be held, how it will be conducted, who is financially responsible, and so on. In contrast, young adults who enjoy their wedding day have been allowed to switch their primary loyalty from parents to each other. *These couples are not being pulled apart by competing loyalty ties, with the associated feelings of guilt.*

How does this model of a primary parental coalition apply to single-parent families? Are they inherently problematic because there is not a primary marital coalition? Absolutely not. However, this structural model can still help us understand why some single-parent families function effectively, whereas others have problems.

In single-parent families that function well, as in nuclear families that function well, there are clearly defined *intergenerational boundaries* that separate adult business from child business. By necessity, children in single-parent families usually need to take on more household duties, and there is often more emotional sharing between parent and child than in two-parent families. Even so, in healthy single-parent families, roles and responsibilities are clearly differentiated between adults and children. This means that the single parent still performs executive ego functions for the family, such as providing a well-organized household with predictable daily routines for children, making decisions and plans for the family, and setting limits and enforcing rules. Further, although parents certainly have legitimate needs for companionship and support, these are met primarily in same-generational peer relationships rather than through the children.

In contrast, single-parent families will have more problems when adult and child roles are not clearly distinguished, when intergenerational boundaries are blurred, and when too many adult needs are met through the children. With this general background in mind, we now examine how the parental coalition influences child development and why primary marital coalitions are more effective than cross-generational alliances.

How the Parental Coalition Influences Child Adjustment

Why is the nature of the parental coalition so crucial for family functioning and child adjustment? To answer this question, we look at four common problems of adjustment among children from families in which the primary coalition is not between mother and father.

1. *Problems with limits and authority.* When mother and father have a stable alliance, children cannot come between their parents and play one parent against the other. For example, if a primary coalition exists between the father and an 8-year-old son, the mother will have trouble setting and enforcing limits with her son. If the mother tries to put her son to bed at 9 P.M. or to have him pick up his room, he can often defy these attempts at discipline by appealing to his father. The son learns that he does not have to do what his mother says because his father is likely to take his side. When children can play one parent against the other in this way, they learn that rules do not apply to them. Having learned that they do not have to conform to adult rules in the home, they will often disobey teachers and try to manipulate other authority figures in their lives. These problems with limits and authority will carry over into adulthood and are especially likely to lead to problems in the work sphere, where adults must be able to conform to rules and cooperate with others. A tragic illustration of this principle is found in Woodward's (1984) biography of the late comedian John Belushi.

When Belushi was an early adolescent, his authoritarian father would work long hours in the restaurant business. Belushi was supposed to do chores for his mother at home. By making up humorous but hostile imitations of his father and performing these routines for his mother, he could get her to laugh at his father and manipulate her out of enforcing household duties. With his mother and siblings in stitches, Belushi would escape his chores and head out the door to meet with his friends. Among other problematic messages here, Belushi learned that, if he could successfully play one parent against the other, limits and rules did not apply to him. Despite his talents, his life was a study in compulsively pushing limits and being out of control with food, money, sex, and especially drugs. After years of chronic drug use, he died from an overdose of cocaine and heroin.

Therapists will not see large numbers of clients with this background unless they work in alcohol and substance abuse programs. Clients who learn that they can manipulate controls by dividing the marital coalition are more likely to develop impulsive acting-out symptoms. When therapists do see them in treatment, these clients will push limits, avoid responsibility, and be demanding of the therapist. For example, they will often arrive late for appointments; expect to be able to stay beyond the scheduled time; be delinquent in, or try to avoid, payment of fees; or be insistent on calling the therapist at home. These challenging clients receive a corrective emotional experience when therapists can tolerate the clients' disapproval, set firm limits, and not be manipulated out of their usual treatment parameters. At the same time, however, therapists must do this in a nonpunitive way that also communicates their continuing personal commitment to the client and compassion for the very real fear and loneliness that has been engendered by these developmental experiences. When therapists are able to do this, they will help clients address more directly the fears and needs that underlie this behavior and help them find more healthy and effective ways to get those needs met.

2. *Exaggerated self-importance.* Children in families with a primary marital coalition are more likely to gain a realistic sense of their own power and control than are children raised in cross-generational alliances. When children form the primary emotional bond with a parent, they become too important to the parent and exert too much influence over the parent's well-being. As a result, these children gain an exaggerated sense of their own importance and a grandiose sense of their own ability to influence others. Children cannot gain a realistic sense of their own limits and capabilities when they are encouraged in the illusion that they can prop up a parent's sagging self-esteem, maintain their parent's emotional equilibrium, or make important decisions for the parent.

It is highly reinforcing for children to feel so powerful vis-à-vis their parent, and they will be extremely reluctant to give up their special role. These clients will presume this special status with the therapist and are often effective in reestablishing this relational configuration. In particular, such clients may offer the therapist a subtle but all-too-easily accepted invitation to establish a mutual admiration society. These sophisticated clients are often adept at flattering the therapist and establishing an elite sense of mutual, shared superiority; in this cozy relationship, the client is very responsive to nuances in the therapist's mood, is skilled at making the therapist feel special, and, in turn, expects to be treated as special by the therapist. Once this unspoken deal is struck, however, the therapist feels constrained because confronting or disagreeing with the client or focusing the client inward on internal conflicts seems like a betrayal of the special relationship. Such a reenactment, of course, will keep the therapist and client from contacting the client's real problems and relational conflicts.

In parallel with their grandiosity, such clients also have a strong sense of inadequacy. This inadequacy is a pervasive source of anxiety that arises because, as children, they were never capable of meeting their parents' needs. These clients often suffer from extreme performance anxieties and feel deeply inadequate to meet the exaggerated demands they now place on themselves. As before, the therapist's task is to remain compassionate toward both sides of their dilemma. On the one hand, these clients struggle with profound feelings of inadequacy and experience intense anxiety over all of the things they must control. On the other, if they relinquish the demands they are struggling to meet, the clients surrender their family role and identity as special; in so doing, they dissolve the illusion of safety or secure attachment that this interpersonal strategy originally provided. Further, to become healthier and relinquish control also threatens to undo or destroy their (internalized) parent, who has convincingly communicated for decades that she or he cannot survive without them. Thus, getting better in treatment arouses two intense conflicts: guilt over leaving their parents to manage their own lives; and anxiety over disrupting the illusion of a secure attachment in their relationships.

The therapist must not overreact to such clients' compliments, criticisms, or distress. The therapist needs to be attuned and responsive in order to be appropriately affected by clients, but (recalling client response specificity) these clients will get worse if the therapist's own personal equilibrium can be too readily influenced. In the past, these clients were able to exert too much control over their parents' well-being, and the therapist does not want to reenact this relational pattern in therapy. In other words, these clients feel safer and are deeply reassured when the therapist is responsive to them but does so without becoming overreactive—even though they may try to elicit overreactions from the therapist.

3. *Emancipation conflicts.* Many children caught in cross-generational alliances have problems with emancipation. If the mother and father have a primary marital coalition, children are free to grow up. As successful young adults, they can leave home while still preserving close ties and involvement; they do not have to emotionally disengage or physically break off contact with the family. This is possible because the parents do not need the child to remain dependent on them in order to fulfill their own lives. In contrast, if the primary coalition is between parent and child, these offspring may feel *separation guilt* over leaving their parent alone or forsaking their parent to the unfulfilling relationship with the other spouse. As a result, they are especially likely to feel dissatisfied with their achievements, to be unable to establish or sustain fulfilling relationships with others, or to be chronically depressed. Guilt over emancipation frequently underlies academic failure in college, as well as many other symptoms and problems that students present in college counseling centers (Teyber, 1983). Thus, cross-generational alliances impose binding loyalty ties that make young adults feel guilty about leaving home, successfully pursuing their own career interests, and establishing satisfying love relationships in adulthood. Although emancipation issues are expressed differently in varying cultures and will reflect a different balance of

separateness and relatedness, symptomatic guilt and depression often ensue when offspring are not permitted culturally sanctioned avenues of individuation and coupling.

Many clients, like Anna in Chapter 4, struggle with binding separation guilt and survivor guilt. These clients may feel guilty about being happy, succeeding in life, or even getting better in therapy. It is important with all clients, but with these in particular, that *therapists enjoy clients' happiness and express pleasure in their success.* Some trainees are unsure of how to respond when clients begin the session by saying, "I feel really good today; I don't have any problems to talk about right now." When that happens, clients with separation guilt are apt to assume, on the basis of their old relational templates, that their happiness, success, or independence has hurt or wounded the therapist. The therapist *disconfirms* such clients' separation guilt and faulty relational expectations—and expands the healthier ways in which they can be connected to others—by enthusiastically responding, "That's great! What's the best thing going on for you today?" Without such affirmation and encouragement of the successful aspects of clients' lives, therapy will bog down and become stuck as these clients repetitively share deprivation, emptiness, sadness, fear, or pain without any progress or resolution.

As we will explore in Chapter 10, conflicts over independence and success will be activated for many clients around the time of terminating therapy. In particular, clients who struggle with separation guilt will feel especially bad about no longer needing the therapist and heading off successfully on their own. Therapists must find multiple ways to give these clients permission not to need them, to leave when they are ready to go, and to enjoy their own successful lives. Guilt over growing up and becoming stronger, succeeding in work and love, or perhaps even surpassing the parent or therapist with more happiness, success, or a better marriage is a common issue, although few clients will be able to recognize it as such on their own. Fortunately, it is relatively easy to help many clients dispel these binding guilt-related beliefs by communicating in words and behavior that the therapist enjoys their competence or success, takes pleasure in their stronger functioning, and is in no way hurt or threatened by their bolder stance.

Finally, cultural factors play an especially important role in these issues of individuation and family loyalty. Family connectedness or close family ties may be of special concern to some Asian, Latino, or other clients from strong relational cultures. Entering the subjective worldview of these clients, therapists must learn to appreciate the complex and subtle balance between being loyal to family and culture and possessing their own authentic self. In order to become familiar with the culturally sanctioned avenues for individuation that exist, therapists will find it helpful to consult with others who are knowledgeable about the client's cultural context. With such consultation, therapists can help clients find ways to individuate that are acceptable within their own culture. Although the balance of separateness and relatedness will be different in each familial and cultural context, counselors who are open to these differences can help clients find culturally acceptable outlets for expressing their own individuality and forming love commitments.

4. *Parentification of children.* When the primary coalition is between a parent and a child rather than between the mother and father, the result is frequently the *parentification* of children. A role reversal occurs: Rather than the parent responding to the child's needs, the child takes on the role of meeting the parent's emotional needs. When certain of the parent's emotional needs are not met by his or her spouse, the parent inappropriately turns to one or more of the children to meet his or her own adult needs for affection and intimacy, for approval and reassurance, or for stability, direction, and control. Therapists should be alerted to the possibility of such problems when they hear adults describe their children as their "best friend," "lifeline," or "confidant" (Teyber, 1992).

It is problematic for children when they become responsible for, or take care of, the emotional needs of their parent. The problem with this role reversal is that the natural flow of nurturing from parent to child is reversed. Children must give to their parents rather than receive, and the children's age-appropriate dependency needs go unmet. These parentified children grow up to feel overly responsible for others, afraid of depending on others, and guilty about having their own needs met. As adults, parentified offspring often describe themselves as feeling empty or "having a hole inside" as a result of having given rather than received throughout their childhoods. Parentified children initially enjoy having such a special and powerful role vis-à-vis their parent. As adults, however, they ultimately come to resent having been used by their parent and having been deprived of their own childhoods.

Parentification is common in the background of clients and therapists alike. Perhaps as many as one-half of all clients seeking outpatient psychotherapy have been parentified to some extent. Because their basic relational orientation is to take care of others, parentified children select careers as nurses, teachers, and therapists. Although kind and capable, they often are guilty about saying no, setting limits, and meeting their own needs. Because they do not draw boundaries well, they tend to become overidentified with others' problems and are prone to experience burnout. Since they grew up having to take care of their parent, it is now threatening to relinquish this control in their current lives. For example, it will be hard for them to let others share in meeting demands, although they may also resent having to do everything themselves. These control issues may also be evidenced in symptoms such as airplane phobias, for example, where these individuals must temporarily relinquish control to the pilot. These clients also have problems in close personal relationships, because it is too threatening to relinquish the control necessary to be intimate with someone.

Note, however, that in some families—for economic or other reasons—one child may temporarily assume a parental role with younger siblings. This is not the same as parentification, because it is always clear that the parent is in charge when he or she is home and that this child's temporary role is not to care for the parent's emotional needs but to assist in the functioning of the home in the parent's absence.

Clients who have been parentified will be highly attuned and responsive to the therapist. It can feel great to work with these clients in the short run, because they astutely discern what they need to be or do so that the therapist will feel competent or secure. If the therapist does not collude in reenacting this relational pattern, however, clients can begin to explore the consequences of having missed being allowed to be a child and have a childhood because they had to switch roles and take care of the parent. It will be relieving yet anxiety-arousing for these clients when the therapist begins this exploration by making process comments.

Therapist:

> Here again, it seems as if you are taking care of me. I really do appreciate your concern for me, but I also see how this puts you back in your old predicament. In fact, it leaves me wondering what happens to your needs—if there is *any* place where you can ever be responded to.

To illustrate this important concept, let's consider a case study. Carol was a highly regarded psychiatric nurse. An utterly dependable and take-charge person, she could seemingly handle every situation that arose. In a hospital emergency setting where she had to deal with very disturbed patients in crisis, her rapid and accurate assessments, good judgment, and compassion for highly disorganized patients had earned her the respect of the entire staff. Although considered a superstar at work, Carol sought therapy for her recurrent depression and loneliness.

Carol was a bright and engaging client who kept trying to get the therapist to lead, talk about herself, and become involved in interesting discussions about abstract clinical issues. The therapist was usually effective in resisting Carol's strong pulls to externalize in these ways, however, and repeatedly focused Carol inward (by saying, for example, "What were you feeling when he . . . ?" "What were you afraid was going to happen if you . . . ?"). Carol expressed clearly that it was new for her to attend to herself and her inner experience in this way, saying that it made her feel uncomfortable. The therapist was genuinely sympathetic to her concern and did not press for this internalizing focus when Carol did not want it, but instead asked Carol on occasion how it was for her to look within at her own experience. It soon became clear to both of them that guilt over being "selfish" and shame over being "the center of attention" were evoked by the internal focus. Thanks to the therapist's accepting and affirming responses to these reactions, however, both soon agreed that important new material was emerging.

Only seven weeks into treatment, Carol disclosed something that she had never told anyone: Her stepfather had sexually assaulted her in her early adolescence. Carol successfully fought him off, although she was scratched, had her blouse torn open, and suffered a bloody nose when she literally pushed him off her. Carol recounted her ordeal in detail but without emotion, and the therapist responded effectively in caring and validating ways. The therapist, herself a mother of two daughters, soon asked what was for her the burning question.

Therapist:
> You have kept this awful secret for almost 20 years. I'm glad that you can share it with me now, but I'm sorry that you had to be alone with this much pain for so long. What kept you from telling your mother?

Carol:
> I didn't want to put more on her. She couldn't have done much anyway, and I didn't want her to worry.

Carol's poignant response illustrates the plight of the seriously parentified child. As in most aspects of her relationship with her mother, Carol's own profound needs for protection and comfort were set aside in order to meet her parent's need.

To sum up, there are many ways in which children are better adjusted in families with a primary marital coalition than in families with cross-generational alliances. Therapists will see the consequences of these structural family relationships operating with most of their clients. As we will see next, the nature of the parental coalition also influences two other basic dimensions of family life: the separateness-relatedness dialectic and child-rearing practices.

The Separateness-Relatedness Dialectic

The second basic dimension of family life that informs therapists about the nature and source of client conflicts is the *separateness-relatedness dialectic*. At times, the separateness side of this dialectic is misunderstood. In this context, autonomy, separateness, or individuation does not imply cut-off, selfish, or uninvolved individuals who are insensitive to their relational contexts or responsibilities to others. Rather, autonomy connotes being connected to our genuine self—our inner voice and our own feelings, needs, and wants. It is the ability to make choices based on awareness of our own experience within our own relational contexts. Thus, the autonomy side of this dialectic connotes *connection to our authentic self and the ability to respond purposively and choose, instead of merely complying with what others want us to do*. With this clarification in mind, let's see how this family systems term refers, in part, to the family's ability to respond to the child's need for both closeness and autonomy.

Recall the dialectic of separateness-relatedness that was introduced in Chapter 1. As children develop, each family faces the task of having to provide both intimacy and individuation—to be close and separate at the same time. Infants and very young children must have basic attachment needs met by responsive caregivers. Both of these basic, developmental behavior systems—of attachment, on the one hand, and of exploration, curiosity, and mastery on the other—are present at every stage of the lifespan. Although there will be varying expressions and timetables within different cultural contexts, strivings for authentic self-definition and expression figure predominately during adolescence for securely attached offspring. Whereas parental caregivers must respond to young children's basic attachment needs, they must exercise very different parental skills with adolescents; they must be able to encourage and facilitate the adolescents' emancipation. Parents who have a healthy sense of self are able to be sensitive and flexible in their relatedness; they give their children permission to grow and develop meaningful relationships outside the nuclear family, which is particularly important during adolescence. Thus, healthy families are characterized by *experiential ranging* between these two poles of intimacy and individuation (Farley, 1979). Effective caregivers have the breadth and flexibility to range back and forth along this separateness-relatedness continuum and meet children's changing needs both for closeness and for autonomy.

In contrast, less effective families do not have the interpersonal or emotional range to function well at both ends of the continuum. These less effective parents cannot meet the changing needs that children present at different developmental stages. Enmeshed families cannot support older children's developmental growth toward individuation, whereas disengaged families cannot provide a secure base for young children and meet their needs for closeness.

This conceptualization provides a useful model for therapists, who similarly need a broad interpersonal range to respond flexibly to clients' needs. This interpersonal range is a critical component of client response specificity; depending on their developmental history, clients will need different degrees of relatedness-separateness. Further, these needs will change during the course of treatment, as clients develop an increased sense of competence and mastery. With this in mind, let's see how the separateness-relatedness continuum relates to the parental coalition. The following example shows how families without a primary marital coalition are more restricted in their experiential range.

Following Solomon (1973), suppose that a young married couple has not made much progress with Erikson's identity and intimacy tasks in late adolescence and early adulthood. As a result, they have not been able to psychologically separate from their families of origin sufficiently to shift their primary loyalty ties from parents and establish a successful marital coalition with each other. Such newlyweds are likely to have a high degree of conflict that they are both unable to resolve. In many cases, the couple will attempt to deal with their relationship problems by having children. As Bowen (1966) illuminates, these parents often experience the children as extensions of themselves. Without being aware of this identification, they tend to project their own unresolved developmental and marital conflicts onto the children—in an attempt to distance and control those conflicts. Such parents select a certain child who matches their own templates— perhaps an oldest or youngest child, a boy or a girl—to carry or express certain unmet needs, unacceptable feelings, or unwanted aspects of themselves. Through this "family projection process," described by Murray Bowen, Ivan Boszormenyi-Nagy, and other multigenerational family therapists, children are scripted into roles, and delimiting family rules and myths are established to help parents defend against personal conflicts that have been activated by marriage and child-rearing.

To illustrate, if the mother's need for closeness or intimacy cannot be met by the father, she may turn to the child to have her emotional needs met. This solution will work well in the short run, but it establishes the basis for long-term family problems. In many cases, the mother could not emancipate psychologically from her own parents, and she is made anxious and/or guilty by individuation. As long as the child remains dependent on her, her need for closeness is met, and her anxiety or guilt over individuation is managed. On the other hand, the father in this prototypic scenario is uncomfortable with closeness; it makes him anxious. The primary mother-child coalition allows him to gain greater distance from the anxiety-arousing demands of closeness from both the spouse and the child. In this way, the child's role is to hold the marriage together and, at the same time, to ensure distance between the couple.

This situation eventually poses a dilemma for the child's continuing development, however. The child's innate push for growth will destabilize the marital relationship, and this will arouse each parent's own internal conflicts. Thus, these parents need to prevent the child's growth, independence, or competence in order to maintain the marital status quo. At the same time, the parents will express anger and frustration to the child about the child's refusal to grow up and act responsibly. This family's limited experiential range is most likely to come to a head during adolescence, when offspring face the developmental hurdle of leaving the family of origin. The mixed messages that the child has received about growing up and remaining dependent will often erupt in psychological symptoms and acting-out behavior at this time. Furthermore, these underlying conflicts have been internalized by the child and often continue to be expressed in symptoms and problems throughout adulthood.

The separateness-relatedness dialectic is a useful orienting construct to help therapists conceptualize basic familial and developmental processes. It also helps therapists intervene more effectively with the process dimension. We will return to the separateness-relatedness dialectic later and see how therapists can use it to better understand the interpersonal process that therapists and clients are enacting.

Child-Rearing Practices
Three Styles of Parenting

Child-rearing practices comprise the third dimension of family life that helps therapists understand how personality conflicts develop. Although further subgroups have been identified, most families employ one of three parenting styles: *authoritarian, permissive,* or *authoritative.* Our examination of these three styles of childrearing will show how each is related to the problems that clients present in therapy. Much of this discussion draws on research by Baumrind (1967, 1971, 1983) on the effects of child-rearing practices.

Authoritarian parents are firm disciplinarians. Children are given unambiguous prescriptions for acceptable and unacceptable behavior. Parental rules are clearly explained, and the consequences for violating them are consistently enforced. Authoritarian parents also hold high expectations for their children to behave in a responsible and mature manner. Children are expected to perform up to their abilities and to be competent and contributing family members.

Authoritarian parents do not give children reasons or explanations for the rules they set, however. They do not discuss compromises or alternatives with children. Children of authoritarian parents cannot ask why a rule is set; they must simply obey. Furthermore, although authoritarian parents discipline their children and expect much of them, they do not provide their children with much warmth or affection.

In contrast, permissive parents are lax disciplinarians. Children often do not know what behavior is expected of them, and they do not know what consequences will occur if they violate parental norms. Most importantly, permissive parents do not consistently enforce the rules that they do set. As a result, children of permissive parents learn that they do not have to obey because parents will not follow through and enforce the rules they set anyway.

Thus, authoritarian parents exert a high degree of control over their children. However, these limits and rules are not coupled with warmth and affection, with parental explanations for why the rules have been set, or with opportunities for the children to discuss compromises and alternatives with their parents. In contrast, permissive parents may be more expressive, loving, and communicative with their children, but their children are not disciplined, nor are they expected to behave in a mature, responsible manner. Before looking at the problems that result from these two ineffective child-rearing styles, let us contrast them with the more effective, but less well-known, authoritative style.

The authoritative child-rearing style produces the most healthy, well-adjusted children. Authoritative parents possess a wide range of parenting skills and are able to combine firm discipline with nurturant child care. Although authoritative parents set and enforce firm limits, they also communicate with their children. Authoritative parents give reasons and explanations for the rules they set and they will negotiate compromises and discuss alternatives with their children. While authoritative parents expect mature, responsible behavior from their children, they also provide a great deal of physical affection and spoken approval. In short, authoritative parents combine the most effective features of the other two parenting styles. They are nurturing parents who communicate with their children, but they are also firm disciplinarians who place high demands for maturity on children. Research has found that authoritative parents produce children who are independent, self-controlled, successful with peers, and generally have a positive, happy mood.

We have seen that the authoritarian parent sets limits but does so harshly, whereas the permissive parent cannot take charge and place appropriate controls on children. Most parents are unable to both set rules and do so in a caring way; they believe that they must be either strict or loving. In contrast, authoritative parents are more effective because they have a wider experiential range and can be both firm and loving. Despite the best of intentions, however, balancing these two domains is hard for most parents. Authoritative parenting requires that parents possess the diverse personality strengths to be nurturing and emotionally available to children, yet still able to set clear rules and to tolerate children's disapproval when they firmly enforce these rules. Authoritative parents also make an effort to talk with children and explain what the rules are and why certain behavior is encouraged or discouraged, to entertain alternatives and compromises, and to follow through and enforce the rules that have been set. Further, authoritative parents must also possess the ability to be highly discriminating, in order to accurately assess the upper level of their children's abilities. Because children are growing rapidly and constantly changing, it is difficult for parents to accurately discern what demands for mature and responsible behavior will challenge children yet allow them to succeed. Thus, it is understandable—but unfortunate—that only a minority of parents provide such effective authoritative parenting.

As trainees listen to their clients' narratives and learn more about their developmental histories, they will often find that their clients have been reared by authoritarian parents. Offspring from permissive parents have substantial problems but, as we will see, they are less likely to seek help. What about children of authoritative parents? These healthier offspring will not usually be seen in longer-term treatment for enduring personality conflicts, although they may seek help in crisis situations (such as a child's major illness) or when negotiating developmental hurdles; for example, they may seek premarital counseling. In the following section, we examine more fully how child-rearing practices are reflected in the conflicts that clients present in therapy.

Consequences of Child-Rearing Practices

In socialization, young children gradually give up their own immediate demands for gratification in order to maintain parental approval. It is the threat of parental punishment and parental disapproval that leads young children to inhibit their own impulses and conform to parental standards. If all goes well, children's conflict between the desire to gratify their own impulses and the fear of parental punishment will be successfully resolved. *A successful resolution is one in which children learn to obey, but without sacrificing their own initiative and positive self-regard.* Healthy children can become self-controlled and self-reliant without inhibiting their own initiative or losing their sense of being prized by their parents (Wenar, 1994).

Children of authoritarian parents, however, must give up too much of their initiative and positive self-regard to maintain parental approval because authoritarian parents provide too little nurturance and affection. These children are obedient and achieving, but they are also anxious and insecure; they comply with their parents out of fear.

By the time the child is of school age, this conflict with the authoritarian parent has been *internalized:* What was originally an interpersonal conflict becomes an internal or intrapsychic conflict. These well-behaved, insecure children become harsh, critical, and demanding toward themselves. Many of these offspring will seek therapy as adults; they present with symptoms involving guilt, depression, unassertiveness, anxiety, and low self-esteem.

Offspring of permissive parents will also have adjustment problems, but their symptoms are different. These offspring are more likely to develop acting-out or externalizing problems, and they are less likely to seek therapy as adults. Children of permissive parents learn that they can avoid the consequences of their own behavior by manipulating others. Rules and limits do not apply to them, and wishes can be gratified without delay. Research has found these children to be dependent, immature, demanding, and unhappy. They have little self-control, low frustration tolerances, and do poorly with their peers. As adults, these offspring tend to be impulsive and to avoid taking responsibility for their own behavior. As a result, they are more likely to develop acting-out disorders such as alcohol and substance abuse. They also tend to be more self-centered and demanding in their interpersonal relations and less capable of making commitments and following through on obligations.

Authoritarian Parenting, Love Withdrawal, and Insecure Attachment

Authoritarian parents typically employ discipline techniques based on love withdrawal. Instead of communicating that they disapprove of the child's behavior, parents respond with such anger or rejection that they communicate essential disapproval of the child's basic self or personhood. This communication is often nonverbal; it involves inflection, gesture, and facial expression as much as it involves words.

To punish or shape the child's behavior, or to communicate their anger—often accompanied by contempt or disgust—these parents withdraw their warmth, care, and emotional connection to the child. They may say, for example, "Get out of here! I don't even want to have to look at you." Many moderately authoritarian parents may respond in these ways occasionally, when they are tired or upset, and may later try to clarify that it is only the child's behavior that they disapprove of, not the child's personhood. For highly authoritarian parents, in contrast, withdrawal of approval, warmth, and emotional contact is a *frequent* technique, and the child's emotional ties to the parents are profoundly disrupted on a regular basis. The child understandably develops symptoms and defenses to cope with the intense separation anxieties and the resulting shame-based sense of self that these painful, repetitive disruptions engender. For these children, *the attachment affect of shame is evoked because their legitimate attachment needs for the caregiver—about which they feel intensely exposed and vulnerable—are rejected.*

A continuum of authoritarian responses Authoritarian child-rearing and love withdrawal techniques occur on a continuum of severity. In families that are only somewhat authoritarian, disruption of ties from love withdrawal may not be severe, and may occur only when parents are stressed, tired, or frustrated. If love withdrawal does not occur routinely, if other opportunities for emotional connection between parent and child are available, and if the break in relatedness is not too severe, ties can soon be restored. The anxiety may still be intense but usually, through compliance and assuming responsibility both for causing the disruption and for restoring it, the child can reestablish contact. As therapists increasingly develop familiarity with these attachment configurations, they will come to recognize the characterological interpersonal strategies their clients have learned to employ in their efforts to restore ties or relatedness to others, especially the pervasiveness of *compliance* in the very diverse symptoms and problems such clients present with.

In families that are more severely authoritarian, however, the disruption will occur more often and more severely. As we will explore later, children in these families will be exposed regularly to experiences of interpersonal loss, emotional isolation, helplessness, and blame in daily interactions around discipline and control. As we move further along this authoritarian continuum, verbal abuse, ridicule, and debasement of the child will occur and may erupt into physical abuse. In this situation, the child experiences his or her parent's anger, contempt, and emotional withdrawal as assaults on his or her emotional needs and sense of self. It leaves the child feeling deeply vulnerable—psychologically alone, emotionally unworthy, and bad. Concluding that they are to blame for their parents' anger, such children feel ashamed of themselves. This sense of responsibility and intense shame are inevitable if children are to experience some sense of control and order in their lives. The alternative would be to view the parents as malevolent and punitive, which would leave the child feeling even more powerless and ineffective— living in a malevolent world inhabited by malevolent people. Thus, when the child experiences this severely authoritarian style of parenting, it is the genesis of an essentially shame-based sense of self.

Moving even further out on the continuum of authoritarian responses, some parents are so completely removed psychologically that they become dissociated. For a few moments, usually while in the midst of their own shame-rage cycle, these parents lose all vestiges of emotional connectedness to their own experience. More importantly, they lose any awareness of their child's experience and all feeling for the frightening and humiliating impact they are having on their child at that moment. It is terrifying for children when angry, threatening parents become dissociated in this way; they may describe their parents in these moments as being "possessed," "not there," or "somebody else." Most clinical trainees will not be familiar with these dissociative ego states and will be understandably upset by the pain and fear they engender. However, by exploring the basic dynamics of these dysfunctional families, therapist's can learn important principles for working with clients from less disturbed backgrounds.

Parents may abuse or mistreat children for many different reasons. Some abusive parents were indulged when they were children and experienced no consistent limits or consequences for their behavior. Others mistreat their own children in the same ways they were once mistreated. *This is especially likely if the relationship was highly ambivalent and the parent who hurt them was also caring or warm at times* (Rocklin & Levitt, 1987). Abuse of this sort tends to occur when parents are in dissociated ego states, which often have been facilitated by alcohol or other substance abuse. In those states, parents reenact with their children abusive events from their own childhoods. In many cases, these parents can remember and talk about the abusive events that happened to them, but the fear and shame that accompanied their experience is not available to them; they relay the story without affect. By reenacting the abuse, rather than reexperiencing it, they defend against the painful feelings evoked by their own mistreatment. Typically, one of their own children comes to represent themselves or the unwanted or disowned aspects of themselves (the emotional needs, vulnerability, or anger that originally evoked the abuse); which child is chosen will depend on age, gender, birth order, temperament, or other characteristics.

This disturbing reenactment has received different labels within different theoretical systems: projective identification (object relations theory); identification with the aggressor (psychoanalytic theory); the family projection process (family systems theory); turning passive into active (control/mastery theory).

However labeled, it is a primitive attempt to manage trauma and gain control of feelings that once were intolerable. This intergenerational transmission of pathology stops if the abusive parent can not only remember the events, but also tolerate experiencing the fear and shame they originally felt and dissociated from in their own past. This psychological task is not easy to accomplish.

Let's review these complex dynamics again. Object relations theorists and family systems theorists both tell us that the more conflicted, developmentally arrested, or unintegrated we are, the more likely we are to instill unacceptable aspects of ourselves in others—especially in our children. Evoking or instilling our own conflicts in others is a way of seeking an external solution to an internal problem (Chapter 4). Thus, in this type of abuse, the abusive parents' reenactment with their own child is a primitive psychological defense—an attempt to gain some control over the painful feelings associated with their own past abuse. By externally reenacting some version of the trauma over and over again with their children, abusive parents do not have to remember or experience internally their own painful feelings. That is, by evoking the same terrified or otherwise unacceptable feelings in their child that they were once made to feel—by turning passive into active—abusive parents do not have to experience the frightening trauma and shame-laden defeat of abuse as their own; their fear and shame is expressed vicariously through the child. For children in extremely authoritarian families, such abuse is a constant threat that organizes their daily experience and, ultimately, their psychological adaption to life.

Effects of severe love withdrawal Severe love withdrawal as a means of discipline is primarily associated with authoritarian parenting, but it occurs in other families as well. For example, variations of this dynamic can be seen in parents who act like martyrs. These parents communicate their hurt and disappointment nonverbally, by turning away and withdrawing emotionally or by means of a sigh or long-suffering look. Other parents also demand perfection from the child, exert excessive control over the child in order to obtain it, and withdraw their love when the child does not fulfill their perfectionistic demands. In these moments of parental love withdrawal, however they occur, the child in effect loses the parent; the child's emotional connectedness to the parent is temporarily broken, and the child's ties are situationally disrupted. This engenders separation anxieties and attachment shame until the child can comply and restore the tie. Of course, all children's ties to their parents will be threatened at times. *Significant problems occur when disruption of such ties is so intense and frequent as to characterize the relationship, and when these interactions are disavowed by the parent rather than acknowledged or validated.*

A constellation of significant emotional reactions occurs when authoritarian parents angrily cut off their emotional connection to young children. Even though the caregiver is physically present, the child is psychologically alone and suffers painful separation anxieties. This withdrawal of warmth, approval, and relatedness may also give rise to helplessness and hopelessness and, consequently, to chronic dysthymia and a shame-based sense of self as ineffectual. Thus, the child is made to feel bad and alone and desperately wants to restore the relationship and renew emotional ties, but is ultimately helpless to do so until the parent reengages. Moreover, the child's inefficacy is further exacerbated by being made to feel responsible for and deserving of the parent's anger, rejection, and withdrawal. In this immobilizing double-bind, the child believes that his or her behavior (being bad or imperfect, crying, feeling angry, disagreeing, needing help, and so forth) has caused the parent to go away, which is the response the child fears most because it threatens already insecure ties.

The child is angry at being abandoned and wishes to protest, of course, but this would only elicit further threats from the authoritarian parent. Through power assertion, the authoritarian parent does not allow the child to disagree, let alone find appropriate means of expressing anger. (The child may be told, for example, "I'm your father. You are never angry at me. Do you understand that? Look at me and say, 'Yes, sir.'") Thus, the child cannot protest interpersonally, or even experience anger internally, because such reactions will further threaten already tenuous ties to the parent. As a result, the child often turns the anger inward; it is expressed through self-deprecation, dysthymia, and having a shame-prone self. *This tendency toward self-blame is exacerbated as the child comes to identify with the parent and adopts the same critical or contemptuous attitude toward himself or herself that the parent originally communicated.* Just as this type of parent loses touch with the child's feelings or experience in these angry moments, the child in turn loses the clarity and authenticity of his or her own internal experience; such children lose touch with important aspects of themselves. Repression, denial, and splitting defenses result when these processes have been intense and pervasive. Poignantly, when they enter treatment as adults, such clients often describe their parents in idealized, conflict-free terms.

Low self-esteem, internalized anger, inefficacy, and loss of the caregiver's love constitute a prescription for depression. They also engender anxiety symptoms and the control conflicts found in eating disorders. In addition, these offspring develop identity conflicts because their feelings have been so pervasively invalidated. Because highly authoritarian parents are so rigidly demanding of conformity and obedience, children soon lose touch with their own internal experience. They may not know what they like and dislike; they may even be confused about what does or does not feel good to them. Their own experience can be so completely overridden at such a young age that they have no basis for later developing their own belief systems, differentiating their own values, or formulating occupational interests in late adolescence and young adulthood.

Therapists often find themselves working with the types of emotional conflicts and family dynamics just described. With these clients, the therapist's initial intervention strategy is to validate or affirm their subjective experience; to encourage their initiative—for example, by following their lead in therapy whenever possible; and to provide a treatment focus by helping them recognize, articulate, and for the first time act on their own feelings, interests, and preferences whenever possible. To better understand these dynamics, let's look further at the developmental experiences of children in highly authoritarian families when emotional ties are disrupted by severe love withdrawal.

The need for constancy of parental relatedness   There is already too little affection or nurturance in authoritarian families, and these thin threads of connection are repeatedly disrupted when angry, demanding, or rejecting parents emotionally disconnect from children. When young children are unable to secure and dependably maintain parental affection, they will not grow up to feel loveworthy. These children have missed the essential developmental experience of *constancy* in their attachment bonds. This developmental deficit has two aspects.

First, these children missed the experience of someone actively reaching out and choosing them. They do not feel loved, cherished, or even wanted in most cases. If the parent's love and caring is dependent on the child's efforts, and the child feels responsible for eliciting the caretaker's love, the child will not develop a secure attachment. In authoritarian families, children must find ways to cope with the intense anxiety generated by their insecure ties. To establish more secure bonds, they attempt to control or manipulate parental feeling for them; they learn to employ interpersonal strategies aimed at earning a parental response—for example, by being compliant and pleasing, by compulsively striving for achievement, or by perfectionistically trying to be good. As we will examine in Chapter 7, this attempt to win approval and love often becomes a characterological coping style that is integral to clients' symptoms and problems. Because of their childhood experiences, these clients often believe they have to elicit, or be responsible for, the therapist's interest in them.

Second, whereas secure children do not feel they have the power to destroy the parent-child tie, insecurely attached children are often induced to believe that their bad behavior—angry, demanding, or needy—disrupts the parent's loving commitment to them. Children in authoritarian families learn that their anger, their tears, or even their questions can provoke their parent to disrupt the parent-child tie with threats such as these:

> Wipe that angry look off your face. You will do what I say, and like it, or else!
> Stop those tears right now or I'll give you something to really cry about!
> Never ask me why. Just do what I say, when I say it.
> Now you've really done it. This time I've had it with you for good.

Even as adults, the sons and daughters of authoritarian parents may be subjected to the same threats: If they do not do what parents demand, they will (seemingly) destroy the relationship.

If you marry him, we won't come to the wedding or visit you anymore.
If you get a divorce, we are going to disown you and take you out of the will.
If you do that, no one in this family will ever speak to you again.

Authoritarian parents who use severe love withdrawal techniques to control their children are not just setting limits on unacceptable behavior; they are threatening to cut off fundamental relational ties. As attachment-seeking children try to cope with the intense anxiety this arouses, compliance becomes a generalized trait, pervasive personality constriction and inhibition occur, and obsessive/compulsive defenses often develop.

A diversity of attachment configurations   In their initial caseloads, most beginning therapists will work with some clients from authoritarian backgrounds. For clinicians who have had better developmental experiences, it may be hard to appreciate the emotional severity of highly authoritarian parenting. These therapists often wonder how the ostensibly normal and, in many other ways, decent parents of these clients can be so rejecting on occasion and can have caused such profound insecurities or even self-hatred in their clients. Other trainees may have difficulty with these client dynamics because they evoke the therapists' feelings about their own authoritarian backgrounds. To help therapists cope with this sometimes challenging material, let's look at a range of secure and insecure attachment configurations.

Molly experienced secure relational ties without fear of love withdrawal. In therapy, she recalled an incident when her mother was angry about something Molly had done when she was 7 years old. Her mother made very strong eye contact with her, reached out and touched her on the shoulder, and said in a calm but firm voice, "I love you, but I don't like it when you act this way and I don't wish to be around you." Molly felt confused, and protested, "But you *have* to want to be with me. You're my mom!" Her mother went on to explain that it was Molly's behavior that she did not like, and Molly recalled that she began to understand that concept.

Her mother was angry, and she got her point across that she did not like what Molly was doing, but Molly also felt secure in her mother's love. Looking back, Molly thought that her sense of secure attachment in this conflict was maintained primarily by the nonverbal messages that accompanied her mother's restrictions and explanations. Because of these secure relational ties, Molly was able to internalize stable, loving self-object relationships and to develop object constancy, as discussed in Chapter 1. As an adult, Molly is able to establish friendships in which she feels respected and affirmed and to have a marriage in which aspects of this same loving affect are present.

Often, children will be secure in their emotional ties with one parent but struggle with a lack of constancy with the other parent. For example, Ellen recalled that she always felt secure with her mother—even when her mother was mad at her. Her father was highly inconsistent, however; at times he was affectionate and available, but at other times he was overtly rejecting. Ellen recalled happy memories of her father patiently painting ladybugs on her roller skates, as well as painful memories of her father angrily yelling, "Get out of here. I can't stand having you around!"

Ellen's secure base with her mother allowed her to cope well with this intensely ambivalent relationship with her father. With the constancy of emotional contact provided by her mother, she could learn to anticipate her father's moods and stay away from him when necessary. Her feelings about herself were not based on the unstable fluctuations of her father's moods. As an adult, Ellen is especially perceptive and interpersonally aware; personality strengths often develop from such conflict-driven demands for coping or adaptation. Ellen's brother, however, was not so fortunate. Ellen's mother seemed to like girls better than boys, and her brother did not receive the secure emotional base with his mother that Ellen enjoyed. As a result, he was left to bob about unconnected at the mercy of his father's stormy emotional seas. Ellen describes her brother as always feeling very bad about himself and being depressed a lot. Now in his late 20s, he has been unable to make commitments to relationships or a career and cannot find a life for himself. Ellen says she worries about him a lot.

Let's look more closely now at insecure attachments. With certain clients, therapists may wish to ask directly about the constancy of parental affection:

> How did your parents respond to you when they were angry or disciplining you, and how did that make you feel about yourself?
>
> Tell me, as specifically as you can, what each parent said, did, and felt toward you when they were upset with you.
>
> When your parents were angry with you, were you more likely to feel that you had just done something wrong and disappointed them, or did you feel instead that you were unwanted, unloved, or bad in some basic and irreparable way?

In response to such queries, many clients will describe disciplinary techniques of love withdrawal that threatened their ties to their parents. Often, the only way for them to become emotionally reconnected was to accept their familial role of being bad, demanding, too sensitive, and so on. The following are some sample comments or actions directed at clients by their authoritarian and/or rejecting parents, along with the clients' statements about the impact of these parental messages. Although clients from authoritarian families can usually remember many hurtful comments and talk about the interactions of which they were part, the deep pain accompanying such exchanges is often unavailable to the clients.

Parental Comment or Action	Impact on Child
"Stop it, or I'll send you to live with your father!"	The client, whose father was an alcoholic, said this threat made her feel that her mother didn't really want her or care about her. As a child, the client was good all the time and stayed away from her mother to avoid this threat; because of her vigilence, it was not often voiced.
"How could you do this to me? Can't you think of anybody but yourself? What's wrong with you?"	The client said she always felt guilty and tried to earn love by figuring out what her parent wanted and trying to provide it.
"I've had it with you! Just you wait until your dad gets home!"	After this comment, the client's mother would withdraw and not speak to her for the rest of the day. The client said it made her feel very alone.
"Look what you did! Get away from me! I don't want anything to do with you!"	The client said he felt he was a terrible person.
"You did it wrong again; you always do it wrong. You're ruining my life!"	The client said she hated herself.
A father would wordlessly slap his son's face for "eating wrong."	The client said, "I don't remember having any feelings about it."

Highly authoritarian parents threaten their children, break emotional contact with them, and—in words and, especially, through tone—may communicate contempt or disgust for them. Children from such families typically struggle with anxiety and depression throughout their adult lives, and often report feeling guilt, loneliness, and self-hate when they enter treatment but without understanding why. *These clients have also developed elaborate interpersonal strategies to cope with this trauma, such as being good, taking care of the parent, or being quiet and "going away inside." As we will examine in the next chapter, the therapist needs to highlight these interpersonal coping strategies and focus on them with clients, or else they are likely to be reenacted in the therapeutic relationship.* These interpersonal defenses originally helped to protect the client from experiencing the painful feelings that resulted from repetitive hurtful interactions. If therapists acknowledge or inquire about these characterological coping strategies when they see them occurring, this accompanying pain will soon emerge. When therapists respond with affirmation and compassion, however, clients can tolerate reexperiencing these feelings, benefit greatly from the therapist's validating response, and begin integrating these sequestered feelings for the first time. Only then can these loyal clients stop protecting their parents at the expense of their own defenses and symptoms, stop denying their own feelings, and stop maintaining family myths of happiness and togetherness.

Clinical implications for working with disrupted ties Parenting is probably the most challenging task in life. The influential family therapist Salvadore Minuchin says forgivingly that parenting has always been more or less impossible. Almost all parents are trying hard to do the best they can for their children. Even many highly authoritarian parents, who indeed caused significant lifelong problems for their children, are not psychopathic, sadistic, or bad people in most cases. In parenting, as in other aspects of personality, people are uneven in their development. Most of these parents do other things well for their children, live by certain moral standards, believe they are doing what is right and best for their children much of the time, and usually are treating their children as well as or even better than they themselves were treated. Although children certainly are hurt by such child-rearing practices, almost universally they still love their parents and seek their approval. Therapists who fail to appreciate adult clients' willingness to give parents another chance or their lifelong efforts to improve or repair these all-important relationships will not do well. Therapists need to help clients assess whether caregivers have changed and to what extent they are capable of responding better now than before, and to help these current relationships between parents and adult offspring become as good as they can be. It is important that therapists allow and invite the client to express the range of their feelings toward their caregivers. Therapists who foreclose on this process by quickly echoing, and remaining stuck on, the negative aspects of the caregivers will rob their clients of the very important experience of integrating both the good and the bad parts of their caregivers. Over the long term, clients who cannot integrate the good and bad aspects of their parents will also have difficulty accepting and integrating the good and the bad parts of themselves (and of the therapist), which they need to do if they are to function in a more authentic way.

In sum, the therapist's role is not to foster splitting defenses by bashing parents who have been hurtful and making them bad, by encouraging clients to reject them or break off contact, or by replacing the parents with idealizations of the therapist. Nor is it to deny or in any way to minimize the real impact of hurtful interactions. Instead, the therapist helps clients to realistically come to terms with the good news and the bad news in their family of origin, to change their own responses to problematic others in current relationships, and to facilitate clients' current attempts to establish new relationships that are more affirming and responsive.

Therapists provide a corrective emotional experience when they remain emotionally available to their clients in a consistent manner; they provide clients with a continuously available point of contact. Such consistent emotional availability and responsiveness, week in and week out over the course of treatment, usually has more effect on client change than do more dramatic but isolated incidents of compelling insight, important self-disclosure, or other significant therapeutic interventions. As we will see, however, providing this consistent presence in the face of the client's maladaptive relational patterns is often difficult in practice.

The clients we have been discussing share the experience that caregivers withdrew from them in one way or another or were not emotionally available to them at important times of need. This developmental deficit—and the maladaptive relational patterns that result from it—can readily but subtly be recapitulated in the therapeutic relationship. For example, some therapists may have trouble being emotionally available to clients' pain and vulnerability; they may not be able to remain present with certain affects, such as the client's raw shame or intense sadness. The client's individuation, emancipation, or achievement may make other therapists uncomfortable because success or individuation was not supported in their own development or because the client's satisfaction evokes the lack of fulfillment or limits in their own work or marriage. Some therapists may also feel sad about losing the caretaking role they were scripted to assume in their family of origin. Others may compete with clients' success or even undermine it by highlighting the potential problems it might bring them, rather than taking pleasure in clients' strengths and celebrating their accomplishments. When such common countertransference reactions occur, clients are again left alone and unconnected in their experience. Clients' pathogenic beliefs that being sad, mad, or successful causes others to be hurt or angry or to abandon them emotionally are confirmed rather than resolved by this interpersonal process. When such reenactments occur without being addressed and rectified, they generate further insecurity for clients, as old templates or problematic relational expectations are reinforced.

In addition to these and other countertransference issues, factors in the client also make it difficult for therapists to remain consistently available. For example, many clients will report that they like and trust the therapist, but they may also believe that, if the therapist really knew them, he or she would not respect or care about them. These clients believe they have deceived the therapist or manipulated the therapist's positive feeling for them. This is why it is necessary for the therapist to help clients adopt an internal focus and to draw out the full range of clients' feelings toward, expectations of, and reactions to the therapist—including the client's perceptions of the therapist's reactions to the client.

Unfortunately, clients often hold the pathogenic belief that their despised, bad, weak, or otherwise unacceptable emotions constitute their real selves. The clients' conflicted emotions are resolved when the therapist sees the vulnerable, dependent, shameful, or other parts of themselves that the clients believe to be unacceptable and still remains affirming and emotionally connected to them. The clients' encounter with compassion and understanding, rather than the emotional withdrawal or rejection that they have learned to expect, is a powerful learning experience. Rogers (1951) emphasized that clients do not change until they accept themselves, and that they begin to accept themselves when they feel such acceptance from the therapist. Thus, by providing a safe holding environment in which clients can contain these core conflicted feelings, as they are experiencing them, the therapist allows clients to resolve those feelings. This pivotal experience not only permits the integration of those warded-off feelings, but also opens up a variety of therapeutic opportunities.

For example, in the next minute or two, clients often disclose important new material or make significant links to other related problems they have been working on. Clients may risk trying out previously threatening new behaviors with the therapist; they may be more assertive or risk bringing up problems between them that had been too threatening to broach. In the week that follows, clients may act in these stronger ways with significant others in their lives. Clients may also express more self-acceptance, resolve some ambivalence, or make an important life decision they had been unable to act on. Further, therapists can often successfully reframe problems or help clients see a situation in a different way—a new perspective that did not seem realistic or relevant previously. Because they feel safer now, clients may be able to hear the therapist challenge a pathogenic belief about themselves or identify a recurrent relational pattern. In short, clients often can make use of a wide variety of therapeutic and educational interventions that they could not utilize before.

Finally, some clients are adept at eliciting the same problematic responses from the therapist that they have received in the past. As we have already emphasized, a basic tenet of the interpersonal process approach is that most clients' conflicts will be temporarily reenacted with the therapist along the process dimension and need to be identified and reworked. It is easy to be empathic, warm, and genuine with clients who are usually cooperative, friendly, and respectful. However, other clients are more challenging because their interpersonal coping strategies—for example, being hostile, dominating, or controlling—alienate, intimidate, or frustrate the therapist. With these clients, it is particularly important for therapists to keep process notes and conceptualize the clients' core conflicted relational patterns— to hypothesize what the clients have tended to elicit from others in the past and what they are likely to evoke in the therapist now. When therapists prepare such conceptual formulations, they increase their chances of providing a corrective experience rather than responding automatically or reflexively to the client and merely reenacting the problematic scenario with which the client is familiar (see Appendixes A and B). The following case study illustrates how the therapist was able to resist reenacting the client's maladaptive relational scenario and instead provide a corrective relational experience, by remaining emotionally connected to him during a challenging disagreement.

John, a 13-year-old client, was being treated in a strict, physically punishing manner by his authoritarian stepfather. His mother passively accepted her husband's harsh corporal punishment and hostile derision of the boy; she often walked away and left the room when the stepfather was angrily deriding him. In treatment, John defied his female therapist, responded contemptuously toward her, and repeatedly tried to push her away. Near the end of an especially frustrating session, in which he had repeatedly pushed every limit, John was unwilling to speak. Trying to find some way to remain in contact with him, the therapist agreed that they would not have to talk together if he did not want to and that he could just throw the ball as he wished. Trying to join him in his chosen activity, she asked if she could silently play catch with him. He agreed but, of course, began throwing the ball too hard. This process continued throughout the session: John tested, and the therapist set limits by saying, "You have to throw the ball below my knees," while still working hard to try and find some way to connect with him.

Near the end of the hour, the therapist invoked their standing rule and asked him to help her pick up the room. John refused; the therapist insisted. Exasperated, she held him firmly by the arms, whereupon he recoiled. At this moment, the same conflicted emotions that John struggled with at home were evoked toward the therapist and were about to be reenacted in the session. The therapist had been sorely pushed by John for weeks; but even though she was frustrated with John, she was still trying to find another way to relate. She did not withdraw in resignation as his mother did, although she felt like it, and she did not physically dominate him, as his stepfather did. Instead, she met his eyes and slowly said, "You're a good boy. I still like you. I want us to work this out together." Even though she was upset, the therapist could still communicate that she felt for John and was remaining emotionally connected to him. This was a turning point in treatment: John was not able to reenact with the therapist the same painful scenario of domination by stepfather, withdrawal by mother, and angry but hopeless isolation for John that had been played out so many times at home. John grudgingly helped the therapist pick up the room; at their next session, however, John was more respectful of and communicative with the therapist.

In closing, clients will present in therapy with an extraordinary diversity of developmental experiences, family dynamics, and interpersonal coping strategies. Many clients' dynamics will be unrelated to the issues discussed here. However, authoritarian child-rearing practices, insecure attachments, and love withdrawal techniques will be among the most common experiences clients report in therapy.

Relating the Three Dimensions of Family Life

Let's now examine some relationships among the three basic dimensions of family life: the nature of the parental coalition, the separateness-relatedness dialectic, and child-rearing practices. Authoritative parents have the emotional flexibility and range of parenting skills necessary to meet children's needs for both closeness and individuation. The experiential range of authoritarian and permissive parents is more limited—specifically, to opposing halves of the separateness-relatedness continuum. Thus, the range of authoritarian parents does not include intimacy, nurturance, and mutuality. These parents tend not to satisfy young children's attachment needs and older children's age-appropriate dependency needs for warmth, attention, and understanding, although their emphasis on mature, responsible behavior often allows children to accomplish and achieve. In contrast, some permissive parents may meet children's emotional needs better, but they do not foster their children's individuation by expecting competent and independent behavior from them. Moreover, permissive parents are more likely to embroil children in cross-generational parent-child alliances, which further stifle their individuation. Of the three parenting styles, permissive parents are least likely to establish a primary marital coalition, most likely to embroil children in cross-generational alliances, and most likely to parentify children.

The family dynamics presented in this chapter offer a parallel to the relationship between the therapist and the client. Therapists, like authoritative parents, must develop their own capacity to respond at both ends of the separateness-relatedness continuum. On the one hand, effective therapists are those who have a broad experiential range that allows them to nurture and care for clients. On the other hand, therapists must also be able to set limits with clients, challenge them when necessary, and tolerate clients' anger and disapproval at times. For example, therapists need to be able to respond to the deprivation and emotional needs of clients who have been reared by highly authoritarian parents, but they must also be able to set clear rules and limits with the provocative, demanding, or testing behavior of clients who have been reared by permissive or inconsistent parents. Thus, the personal challenge for therapists is to examine their own experiential range, which has been shaped in their families of origin. Therapists must identify their own limitations and work to develop a broad interpersonal range so that they can respond to the diversity of problems that clients present. In order to achieve this, therapists must do their own family-of-origin work.

Such work is an integral part of training for those who intend to specialize in family therapy. In particular, Murray Bowen (1966) has written extensively about how clinical trainees can systematically study and sometimes change repetitive patterns of interaction in their own families of origin. By achieving greater understanding of their own family backgrounds and often greater empathy for all family members, therapists make an important contribution to their clinical skills. Any reader who would like to become a better therapist is advised to read about family-of-origin work (Boszormenyi-Nagy & Spark, 1973; Bowen, 1966) and prepare a family genogram—a three-generational map of family roles and structural family relationships (Hartman, 1978; Minuchin, 1974). By learning about their family of origin in this way, therapists learn about their own counter-transference propensities, their clients' dynamics, and the profound influence of familial experience on adult life.

Closing

In this chapter, we have examined three basic dimensions of family functioning: structural family relations and the parental coalition; the separateness-relatedness dialectic; and child-rearing practices. These three aspects of family life will help therapists understand the genesis of many clients' personality and emotional problems. For many readers, however, it is disruptive to study this material because it violates familial and cultural rules to question or examine the interaction and communication processes that transpired in their families of origin. Also, many trainees are parents or grandparents themselves. As they learn more about child development, family functioning, and parenting practices in particular, much guilt may be evoked as they recognize the limits of their own parenting abilities and the consequences for their children. Such readers need to relinquish unrealistic expectations of themselves, forgive themselves for not being a perfect parent, and pardon themselves for not knowing more about child-rearing than they did at the time.

Suggestions for Further Reading

1. Basic information on child development and child psychopathology will help therapists understand their adult clients' personalities and problems. One outstanding text is *Psychopathology from Infancy through Adolescence: A Developmental Approach* (3rd ed.) by C. Wenar (New York: Random House, 1994); see especially Chapters 4–8.

2. Structural family relations provide an illuminating road map for understanding family functioning. A seminal work in this area is Salvadore Minuchin's *Families and Family Therapy* (Cambridge, MA: Harvard University Press, 1974); see especially Chapters 3 and 5. As was mentioned in this chapter, Bob Woodward's *Wired* (New York: Simon & Schuster, 1984), a biography of the late comedian John Belushi, illustrates the effects of cross-generational alliances on child development; see Chapter 2.

3. *Parentification* is an important concept for therapists to understand because so many clients and therapists have been scripted into this familial role. Further information about this important concept can be found in Chapter 9 of E. Teyber's *Helping Children Cope with Divorce* (San Francisco: Jossey-Bass, 1992).

4. The family systems concepts of *separateness-relatedness dialectic* and *experiential range* are useful and readily applicable to individual psychotherapy. One informative article that can also lead to further reading in this area is J. Farley's "Family Separation-Individuation Tolerance: A Developmental Conceptualization of the Nuclear Family," *Journal of Marital and Family Therapy* (January 1979): 61–67.

5. *For Your Own Good*, Alice Miller's book on authoritarian child-rearing (New York: Farrar, Straus & Giroux, 1984), and especially *Becoming Attached*, Robert Karen's outstanding overview of attachment (New York: Warner Books, 1994), will help clinicians better understand many of their clients' formative developmental experiences. David Elkind's book *Ties That Stress: The New Family Imbalance* (Cambridge: Harvard University Press, 1995) describes the changing family system and parenting styles.

CHAPTER SEVEN

Inflexible Interpersonal Coping Strategies

Conceptual Overview

This chapter provides a model for helping therapists further conceptualize how client conflicts are expressed, maintained, and resolved. This model is based on the client's *generic conflict* and his or her *interpersonal strategy* for coping with this conflict. The client's generic conflict is a central conflict that pervades the client's life and links together the different problems that the client presents. The generic conflict has arisen out of the client's repetitive interactions with significant others, especially parents or primary caregivers, and usually represents a basic and pervasive unmet need, such as an insecure attachment. Clients' unsatisfying experiences with caregivers often give rise to pathogenic beliefs about themselves and others, and to faulty relational templates that shape their views of themselves and their expectations of others, and what will occur in relationships. To avoid or defend against the pain, frustration, or unmet needs that accompany these formative relational patterns with the caregiver, clients develop a repetitive or characterological interpersonal style. This coping strategy—such as always pleasing others or being perfect; repeatedly taking charge or being in control; routinely taking care of others and meeting their needs; consistently remaining invisible or avoiding all conflict— originally helped them to cope. However, it is no longer necessary or adaptive in most current relationships and actually causes many of the clients' presenting symptoms and relational problems. Therapists will therefore be far more effective when they can (1) identify the formative maladaptive patterns and resulting painful feelings that comprise a particular client's generic conflict; (2) understand the interpersonal coping strategy that the client has adopted to defend against this conflict; and (3) recognize how the generic conflict and interpersonal coping strategy are expressed in current problems and relationships, especially in the current interaction with the therapist. As beginning therapists formulate the client's generic conflict and interpersonal adaptation to it, this conceptualization will lend structure and organization to the disparate material that the client presents and will help to guide therapists' intervention and treatment strategy.

As we noted in Chapter 2, client conceptualization is an ongoing process that begins with the initial client contact. General hypotheses about client dynamics are formulated early in treatment, and these tentative working hypotheses are further refined or discarded as the therapist learns more about each particular client. It is not until the client's conflicted emotions emerge, however, that the central conflicts underlying the client's problems becomes clear. Having initially focused the client inward, the therapist can use the client's emotions that begin to emerge as guideposts to signal generic conflicts, which can now be addressed effectively.

Chapter Organization

In the first section of this chapter, we look at a model for conceptualizing the client's generic conflict—how it originally developed, the client's interpersonal adaptation for coping with it, and how it is being expressed in current symptoms and problems. This model for conceptualizing the client's conflicts includes five components:

1. The client's unmet developmental needs
2. The original environmental block that created the conflicts
3. The client's intrapsychic defenses against his or her own generic conflict
4. The client's interpersonal strategy to rise above or overcome the generic conflict
5. An interpersonal resolution of the generic conflict

Adapted from Horney's (1970) interpersonal theory, these five components provide a useful schema for conceptualizing client dynamics (see Figure 7.1).

This schema is an integrated system composed of numerous, complex psychological processes. After discussing each component of this model, we consider an extended case study in the second section and two case summaries in the third section, as illustrations of the model. After reading this chapter, therapists should turn to Appendix B for a structured outline that they can use to write case conceptualizations and to formulate treatment plans and intervention strategies.

A Conceptual Model

In this section, we examine an interpersonal model for further conceptualizing how client conflicts are expressed, maintained, and resolved. Our discussion does not emphasize how therapists can intervene with clients but instead gives therapists further guidelines for conceptualizing clients' conflicts and adaptation to them. Much of the discussion that follows is derived from the interpersonal theory of Karen Horney (1966, 1970). The five components of this model are

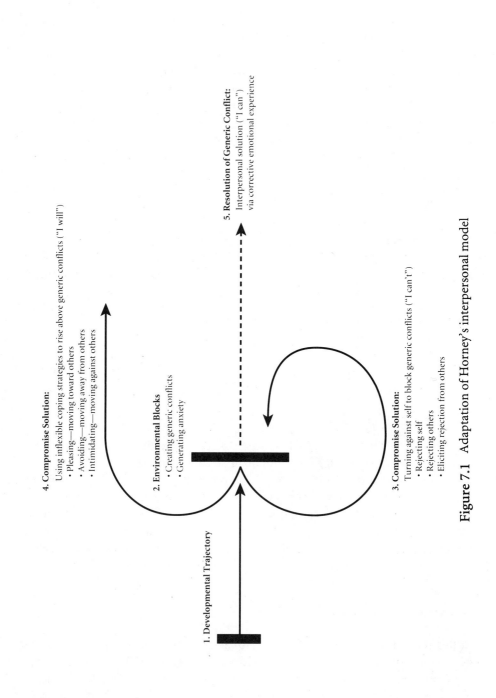

4. Compromise Solution:
Using inflexible coping strategies to rise above generic conflicts ("I will")
• Pleasing—moving toward others
• Avoiding—moving away from others
• Intimidating—moving against others

2. Environmental Blocks
• Creating generic conflicts
• Generating anxiety

5. Resolution of Generic Conflict:
Interpersonal solution ("I can")
via corrective emotional experience

3. Compromise Solution:
Turning against self to block generic conflicts ("I can't")
• Rejecting self
• Rejecting others
• Eliciting rejection from others

1. Developmental Trajectory

Figure 7.1 Adaptation of Horney's interpersonal model

schematized in Figure 7.1.* Let's begin by examining how enduring personality problems result when one or more of a child's basic developmental needs go unmet.

Clients' Developmental Needs

This conceptual model begins with children's basic needs for secure relational ties. Component 1 in Figure 7.1 refers in the broadest sense to young children's dependency needs on their parents and, in particular, to children's need for a secure attachment. If parenting figures are consistently responsive to the children's bids for attention, affection, and comfort, children will be able to freely experience and express this need. Of course, as children develop, they will have to learn to accept limitations on parents' ability to respond and to tolerate delays in parental response. If parents are punitive or unresponsive, however, anxiety will soon become associated with the children's needs. Children are then forced to fashion a solution; the *compromise solution* that they construct often alienates or removes them from their authentic needs and authentic response styles.

Consider children of preschool age whose need for warmth and affection goes unmet from authoritarian parents. When such children approach their caregiver for comfort or emotional refueling, their need is usually rebuffed in one way or another:

What do you want now? Why are you always needing something!
Big boys don't want to be held all the time. That's sissy stuff.
Leave me alone, honey.
Can't you see I'm busy.

If this type of interchange occurs repeatedly, children will soon learn to anticipate rejection and feel anxious whenever emotional needs for affection, attention, understanding, or reassurance are aroused. Children are biologically organized to continue experiencing attachment needs, but their direct expression is blocked by the parents' cold response (component 2 of Figure 7.1). The children are then compelled to find some way to cope with this painful set of circumstances.

Although clients will present with a wide range of problems that result from varied developmental backgrounds and environmental blocks, enduring psychological problems often begin when basic childhood needs for secure relational ties are not met. Thus, the source of most generic conflicts involves a failure to provide children with nurturance, clear communication, and consistent emotional access, on the one hand, and a predictable environment with consistent limits and discipline, on the other. Although the specific developmental needs

---

* The author wishes to thank Robert Hilton and Karen Humphrey for their help with this schematization.

and challenges, the types of blocks (for example, lack of parental response; responsiveness on the parent's timetable rather than when the child initiates contact and seeks comfort or stimulation; punitive or inconsistent response), and the specific coping strategies developed may vary, all clients tend to experience anxiety when their needs are blocked. As a result, children begin to construct a compromise solution designed to prevent continued anxiety engendered by these unmet needs. At the same time, however, children also find indirect ways to express and partially gratify this need. Components 3 and 4 of our model elaborate on how this compromise solution works to block or minimize the felt need that has now become anxiety-arousing and, simultaneously, to implement different inter- personal coping strategies in an attempt to rise above the need and indirectly gratify it. We now discuss the complex psychological maneuvers that individuals adopt to cope with these generic conflicts and examine how these relational templates and inflexible interpersonal coping styles are replayed in current relationships and reenacted in the therapy setting.

Clients' Compromise Solutions

In Chapter 2, we proposed three domains for conceptualizing clients' conflicts: maladaptive relational patterns, faulty beliefs about self and unrealistic expectations of others, and core conflicted affects. Although all three domains are inextricably intertwined, the environmental block in component 2 of Figure 7.1 principally evokes the core conflicted affect. As we will see, the faulty beliefs that arise from these generic conflicts and the interpersonal strategies used to cope with them are highlighted in components 3 and 4, respectively.

Clients employ two psychological mechanisms in order to defend against the anxiety associated with expressing, or even experiencing, their blocked needs. First, we will examine three ways in which clients defend against their generic conflicts by responding to themselves in the same way that others originally responded to them (component 3 of Figure 7.1). These can be thought of as *cognitive schemas* that clients adopt; that is, beliefs about themselves and others that, once adopted, regulate their behavior. Second, we will examine three ways in which clients adopt a general *interpersonal coping style* to try and rise above their unmet need (component 4 in Figure 7.1). In therapy, this style will often quickly become apparent in the way the client interacts with the therapist. Taken together, these two mechanisms of *blocking* and *rising above*, which represent attempts to minimize anxiety and provide a modicum of control, are clients' compromise solutions to their generic conflict. We will see how this complex defensive system works.

Blocking the unmet need   We have already described children who try desperately to get their parents to respond to their emotional needs, with the limited skills
and means available to them, but are unable to stop the parental rejection or to elicit
the nurturing and support they need. In order to gain some active mastery over their
helplessness and to ward off the anxiety aroused, such children begin to block
their own need in the same way that the environment originally blocked it. For
example, children may deny it as insignificant and regard themselves as unimportant or may reject it as too much and view themselves as shamefully demanding. Thus, such clients adopt the same hurtful responses toward themselves that
others originally expressed toward them; they behave toward themselves in ways
that make the original environmental blocks seem reasonable. In turning against
themselves, children are able to ward off the intense anxiety that would be
engendered if they felt that their true person was being seen and that their
developmental needs were reasonable and legitimate but still viewed as unacceptable to those they need and love most—their caregivers. As well as preserving ties,
this adaptation provides a sense of control and allows them to continue to view
the world and their parents as reasonable and just. This then becomes their
cognitive schema for viewing themselves and others; by acting accordingly, they
elicit responses consistent with the template. This adaptation has been referred
to as identification or modeling; some clinicians speak colloquially of "using the
devil's tools to fight the devil." Component 3 in Figure 7.1 reflects three ways in
which children begin to do to themselves what was originally done to them.

First, clients block their own need *intrapsychically* and respond to themselves
in the same hurtful, rejecting ways that others have responded to them. When
the children of nonresponsive parents feel a need for reassurance or affection,
they will block their need by feeling critical or contemptuous of themselves and
their own need. As adults, when their own unacceptable emotional needs are
aroused, they will often say to themselves the same critical things they heard from
others years ago—often adopting the exact words and tone—but without recognizing the source of these internalized templates or tapes. Furthermore, *the
critical, shaming, judgmental, or rejecting affect that clients feel toward themselves
as they replay these internalized tapes is the same affect that parenting figures
originally expressed toward them or their need years before.*

Second, clients block their anxiety-arousing need by reenacting in current
relationships the same conflict that was originally experienced in earlier, formative relationships. Thus, on an *interpersonal* level, clients will say and do to others
what was originally done to them. For example, the child of unresponsive
authoritarian parents may grow up to be a parent who disparages emotional
needs in one of his or her own children—most likely the child who most resembles
the parent in gender, birth order, temperamental characteristics, and so forth.

Especially in the arena of parenting, clients who have not integrated or come to terms with their own conflicted feelings are prone to reenact the same relational patterns and hurtful messages that they received in their own childhoods. As parents, many clients feel helplessly dismayed as they watch themselves respond to their children in the same hurtful ways that their parents responded to them. Despite sincere pledges that "I will never do to my children what my parents did to me," variations of the old themes occur with clocklike regularity. Here again, in the interpersonal sphere, adult offspring gain some control over their own anxiety-arousing feelings and defend against their own conflicts by reenacting what was originally done to them.

Third, clients block the experience and expression of their own conflicted need by *eliciting* the same unsatisfying response from others in current relationships that they received in their childhood. For example, unless the children with problematic developmental experiences have other corrective or more satisfying relationships to help them resolve these concerns, they are likely, as adults, to select a marital partner who cannot respond well to their needs for affection and intimacy. In the event that their spouse *is* capable of responding to this need, they may not be able to accept this emotional responsiveness. Why? The possibility of having a response to their old, unmet needs—even a well-intended, benevolent response—will often arouse too much shame or anxiety. The spouse's affection is likely to arouse the pain of the original deprivation and, following their relational templates, such clients will anticipate that their spouse will reject their needs as others have done in the past—even though the spouse has never responded in hurtful ways. Operating on schemas or internal working models that fit past relationships better than current ones, such clients may simultaneously elicit and reject nurturance from their spouse. This will certainly frustrate and confuse the spouse, especially if these mixed messages reenact or elicit the spouse's own problematic relational templates. In that case, the likely outcome is significant conflict and perhaps marital counseling or divorce.

To sum up, clients defend against their unmet needs by (1) intrapsychically responding to themselves in the same rejecting ways that parental figures did; (2) identifying with the rejecting parent, usually in an attempt to overcome their powerlessness, and then responding to others in the same hurtful way; and (3) choosing to become involved with others who provide the same hurtful response to them that they originally received as children. Although one of these three modes may be predominant for a particular client, most clients with long-standing developmental conflicts will employ all three mechanisms at different times in order to defend against their anxiety-arousing generic conflicts.

At the same time as clients are blocking their needs on each of these three levels, they are simultaneously trying to get them met in an indirect way.

Rising above the unmet need   As we have seen, when clients have had a primary need blocked throughout their development, they employ three strategies to defend against it. In all three maneuvers, clients do to themselves and others what caretakers originally did to them, by turning a passive experience that originally happened to them into an active experience over which they now have some control. Counselors need to focus on how clients take up the same hurtful affect that was originally expressed toward them in childhood and actively turn it against themselves to block their own unmet need. This turning against the self in component 3 of Figure 7.1 (often expressed as variations of "I can't") is only half of the conflict, however. Paradoxically, while defending against the anxiety of reexperiencing the unmet need, clients simultaneously try to rise above it and have it met indirectly. Component 4 of Figure 7.1 shows the other side of clients' compromise solution—clients' attempts to rise above the need and indirectly gratify it.

If they do not find reparative experiences with significant others, children's most important unmet developmental needs usually do not dissipate or go away on their own as the children grow up. These needs may be denied, warded off, sequestered away, or defended against, but they continue to seek expression. As a result, clients do not give up trying to have the needs met. At the same time, however, the unmet needs still remain too anxiety-arousing to be expressed directly. So clients try to rise above the unmet needs and indirectly fulfill them by adopting various interpersonal coping styles. Horney (1970), Beck and Freeman (1990), Strupp and Binder (1984), and others have utilized different theoretical frameworks and terms to describe similar interpersonal coping styles that clients repetitively employ. To illustrate these interpersonal coping strategies, we will focus on Horney's three lifestyle adaptations: moving toward, moving away, and moving against others.

Moving toward, moving away, and moving against   Clients adopt a fixed interpersonal style both to reduce the anxiety associated with an unmet need and to find some partial, indirect gratification of it. For example, in their families of origin some clients learned to cope with their unmet needs and problematic familial interactions by *moving toward* or pleasing people. These clients could earn some needed approval and diminish the threat of further rejection by complying with their parents and being unfailingly nice or good. Generalizing from their formative relationships with caregivers, they develop a primary mode of relating to others in which they submit, accommodate, or please others and never assert their own wishes. Thus, these clients have learned to defend against the anxiety engendered by their legitimate needs, to win some approval and to indirectly gratify these needs, by consistently moving toward others in a pleasing, servile way.

Other clients have learned that aggressiveness and resistance to parental wishes, if pursued long enough, will ward off the pain of not having their needs affirmed in a loving way. Among the responses they may have needed were to have the parent act more strongly, take charge of the family effectively, and provide the safety that would allow them to feel secure as children so they could come to view their needs and vulnerability as legitimate and reasonable. These individuals cope by *moving against* others and expansively seeking to be in control of themselves and others. They approach life and relationships with the orienting attitude that they must win what they deserve; they assert themselves, often aggressively, for this purpose. Thus, these clients adopt a repetitive interpersonal style of moving against others in order to protect themselves from experiencing the threatening or unwanted emotions evoked by their generic conflicts or developmental experiences.

In Horney's third interpersonal style, clients have learned that the best way to reduce their anxiety and defend against their unmet needs is by *moving away* from others, through physical avoidance, withdrawal, and self-sufficiency. Although clients may employ all three of these interpersonal coping styles at different times, one style usually becomes predominant or "characterological."

To review, we have seen that clients use three ways to cope with their nonsupportive, nonaffirming developmental experiences; these defenses are similar to the ways in which their original needs were blocked by parents or others in the environment (component 3 in Figure 7.1). We have also discussed the ways in which clients try to rise above their blocked needs and indirectly meet them (component 4 in Figure 7.1). These include the interpersonal styles of moving toward, against, or away from others that are employed pervasively or inflexibly in all types of situations. We are now ready to examine further how clients use their characterological interpersonal style to try to rise above their problematic developmental experiences. As we will see, however, even though clients derive significant secondary gains from their interpersonal coping style, it remains a defensive adaptation that avoids their real conflict and does not help resolve it.

Being special to rise above their nonaffirming developmental experiences
When the original conflicts were pervasive or severe, clients often adopt one of the three interpersonal styles—moving toward, moving against, or moving away—a pervasive or habitual way of relating to others, and this inflexible coping style is reflected in vocational and other life choices. Although there are many exceptions, individuals who move toward may find that careers in nursing or counseling are a good match for their caretaking skills and needs; individuals who move against may find that careers in law, medicine, or money management are congruent with their interpersonal style of taking charge and needing to control; and individuals who move away may feel comfortable as research scientists or artists or in a meditative lifestyle. Many clients will find some way to turn this defensive coping style into a virtue and use it to feel special and rise above their

need. Thus, clients do not experience their interpersonal strategy as a defensive coping system to avoid conflicts, but as a virtuous way of relating to others that may make them feel superior. However, this unrealistic sense of being special actually reflects the low self-esteem and shame that comes about because their developmental needs were unmet and blocked. These clients' only sense of worth or value comes from their ability to rise above their unmet needs by pleasing others, by achieving and succeeding, or by withdrawing and feeling cynically superior. Thus, clients convert their defensive interpersonal coping style into a virtue and invent an aggrandized false self. Although moving-toward clients are compliant and submissive, they do not experience themselves as servile. Instead, they often feel special or secretly better than others. They accomplish this by perceiving themselves as selflessly and endlessly loving, always sensitive to the needs of others, and committed to the ideals of peace and harmony. In this way, the defensive interpersonal style becomes a source of self-esteem—a virtuous quality that wins needed approval from others and wards off anxiety and shame.

Similarly, moving-against clients do not see themselves as angry, competitive, self-centered, or demanding, although they frequently behave in these ways. Instead, they see themselves as heroes or heroines—strong leaders entitled to direct others, who are seen as inferior and less capable. Finally, moving-away clients glorify their aloofness. They are often cynical and may feel better than the ordinary masses, whose involvement in everyday happenings is seen as mundane and small-minded.

Thus, to lessen their anxiety, clients develop an inflexible, pervasive interpersonal style that allows them to feel special; gain a sense of identity and self-esteem; and rise above their unmet needs, unacceptable feelings, and pathogenic beliefs about themselves. Their ability to turn a defensive coping style into a virtue or talent that testifies to their specialness is part of Horney's *neurotic pride system*. Clients' self-esteem is brittle and vulnerable, however. It is based on being special, controlling others' feelings and reactions, and rising above, rather than approaching and learning to value, understand, and integrate their authentic feelings.

We now have the two sides of clients' conflict. First, clients block their own needs and respond to themselves in the same critical or rejecting ways that adult caregivers originally responded to them. Clients come to internalize and feel toward themselves the same affect that the caregiver originally communicated toward them (for example, disappointment, contempt, resentment, or indifference). This affect will usually be evidenced as low self-esteem when clients enter therapy, and clients are often seeking treatment to be rid of this intrapunitive feeling. Usually, it is relatively easy for counselors to recognize how clients are turning against themselves and to respond in supportive and affirming ways.

Second, clients rise above their unmet needs and find ways to interpret their defensive strategy as evidence that they are special. This aspect of their compromise solution is usually more covert and is harder for therapists to address, primarily because the clients' neurotic pride, false self, or wounded narcissism is typically bound up with shame. Let's examine some other reasons why clients are reluctant to let go of their attempts to rise above their conflict.

Clients' interpersonal coping style has indeed granted some relief from the anxiety of unmet needs and unacceptable feelings by providing a sense of control over them. It has also allowed clients to earn some of the much needed affection, attention, or respect they want from others and, to some extent, from themselves. In this way, *clients' interpersonal coping style has become the primary source of their self-esteem and their principal avenue to achieving mastery and has provided orderliness to their lives. Further, clients have often developed genuinely impressive abilities and admirable competencies in the service of their coping strategies.* For example, many moving-toward clients have not only been able to make others like them, but have also become highly skilled at caring for others and responding with great sensitivity. Likewise, moving-against clients have often been able to achieve a great deal of success and power and may develop exceptional technical or other abilities along the way. Similarly, although moving-away clients have often been able to become safely aloof, they may also have developed a rich inner life or been able to develop creative artistic abilities.

Therapists need to help clients recognize the significant emotional price they pay for their coping style and to highlight as concretely as possible how it is creating stress and limiting their lives now. At the same time, however, therapists need to appreciate the genuine strengths that clients have developed through these coping strategies. We will see later how clients can retain the skills and strengths they have developed in these compensatory modes, but without utilizing them in the same narrow, delimiting, and overdetermined ways as in the past. Therapists again need to help clients *honor* their interpersonal coping styles and appreciate rather than pathologize them. Although they were absolutely necessary and represented the best possible adaption they could have made at that point in their development, they are no longer adaptive in most current situations. In this regard, clients will need to learn and appreciate that flexible interpersonal coping styles will serve them best in the present and in the future.

To sum up, by turning their defensive interpersonal adaptation into a sense of being special, clients often gain significant secondary benefits and genuine personality strengths. It's no wonder, then, that some clients may be reluctant to explore this adaptation with the therapist; it has become the basis of their identity, the primary source of their self-esteem, and the only way they have had of defending against their seemingly unresolvable conflicts. Once therapists understand and fully grasp all that this defense has provided for clients, and help clients to appreciate what it has accomplished, both therapists and clients will stop being judgmental or impatient with the clients' symptoms, will have genuine empathy for the clients' feelings of being threatened by change, and will be able to work together to develop more adaptive, flexible coping strategies. Understandably, clients will not want to risk giving up what they have without having something better to replace it. As we will see, however, clients never really succeed in rising above the original conflict through these compensatory lifestyle adaptations; by acknowledging the conflicts and addressing them more directly, they can generate a range of responses that will be more satisfying for the present and future.

Shoulds for the self, entitlement from others   Clients both gain and lose something by their compromise solutions. As a result of feeling special and virtuous, clients gain what Horney terms *false pride* and a sense of *entitlement* from others. For example, the self-effacing, moving-toward client expects unrealistic adulation and approval from others and often demands constant reassurances. The expansive, moving-against client claims leadership, respect, and allegiance from others. And, although it may look as if detached, moving-away clients have no claims, they demand that others never criticize them or make requests of them. These neurotic claims leave clients exposed to continuing frustration and conflict, however, as others frequently do not accept such unrealistic demands and unspoken expectations.

Accompanying clients' excessive demands on others is a harsh and uncompromising set of "shoulds" that they place on themselves; these recurrent cognitive messages reflect their impaired self-esteem and impoverished sense of self. For example, moving-toward clients suffer under the self-imposed demands that they should be the perfect lover, teacher, spouse, and so forth. They must always be calm, caring, sensitive, and responsive to the needs of others and never feel angry, be critical, or act selfishly. Moving-against clients demand of themselves that they should be able to quickly overcome all difficulties and obstacles. They must be able to control all of their feelings at all times and overcome bad moods simply by an act of will. Moving-away clients pay for their specialness by believing that they should be able to work tirelessly and always be productive. They demand that they should be able to endure anything without becoming ruffled or upset and that they should never need help or reassurance from anyone. As we will see, such individuals will not always be able to meet these demands. When the inevitable failures occur, they may suffer great shame and may undergo major life crises, such as anxiety attacks, suicidal feelings, and depressions; at such critical times, it is hoped that individuals will also seek help.

Thus the *tyranny of the shoulds* is an exacting price to pay for the client's sense of specialness. One of the most reassuring interventions therapists can offer is to communicate their compassionate understanding that this interpersonal coping strategy is core to clients' sense of self and the basis for whatever self-worth they may possess and that it is the organizing theme of their relationships and their life. Therapists also need to help clients appreciate that these interpersonal coping strategies were once necessary and adaptive, while clarifying the unnecessary price in symptoms and pain that the client is paying.

Resolving the Generic Conflict

Component 5 in Figure 7.1 illustrates how the client's generic conflict can be resolved. Ultimately, the client's compromise solution to the original conflict will never succeed. The client can block the anxiety-arousing need and try to rise above it in the ways described here. However, these compensatory maneuvers will not release the client from the conflict. Furthermore, not only will the lifestyle adaptation of moving toward, away, or against never allow the client to resolve the original conflict, this defensive adaptation will expose the client to continual frustration. Let's examine three reasons why the client's compromise solution will not succeed, and then explore what can be done to help the client successfully resolve the generic conflict.

First, the tyranny of the shoulds places unrealistic and unrelenting demands on the individual. These can be heard, for example, when clients enter therapy exclaiming, "I can't do all of this anymore! I can't be a perfect parent, a loving spouse, and a successful professional. I can't be everything to everybody anymore." Individuals' quality of life is diminished by these harsh and fatiguing expectations on self, even if they do not provoke a crisis that brings them to therapy.

Second, clients' sense of entitlement and being special will usually fail as well. Self-effacing clients will not be able to win the love or approval of everyone they need; expansive clients will not always succeed or be able to make others defer to their demands; and detached clients will encounter criticism from others or situational pressures that force them to compete. Inevitably, clients' attempts to become special and rise above their conflict will lead to repeated interpersonal conflicts, failure, and frustration. Clients usually enter therapy when situational life stressors, developmental transitions in adulthood, or aging have caused their interpersonal coping strategies to fail.

Therapists must grasp the profound significance of the pain that clients suffer when their attempts to be special and rise above their conflict fail; Kohut (1977) speaks of the *narcissistic wound* that is so painful to clients' core or most central sense of self. Even though the demands that clients place on themselves and others are unrealistic, it is often excruciating when their interpersonal adaptation fails. Tragically, some individuals will even attempt suicide in response to such a failure, and many others will contemplate it. Why are some clients driven to such an annihilating response? As we have seen, the interpersonal coping style that many clients have adopted is the basis of their identity and their primary source of self-worth, but it is a brittle substitute for the genuine self-esteem they lack. From these clients' point of view, they *are* their ability to please, to achieve, or to remain superior and aloof. Such clients overreact to seemingly insignificant events (for example, gaining 3 pounds over the holidays; losing their patience with a trying 2-year-old) because their entire sense of self is threatened when their interpersonal coping strategy fails—and, following their templates or internal working models, their shame-based sense of self is revealed by their imperfection or mistakes.

When such students get an A minus, rather than the A they demand of themselves, they may feel completely undone or defeated. Often, therapists respond initially to this disappointment by trying to provide a more realistic appraisal of the situation or by challenging the maladaptive overreaction to the grade. However, this well-intended, reality-based response is generally an empathic failure that misses the deeper meaning for the client. Such a reality check is certainly useful later, but a more productive response initially is a bid for an empathic connection: The therapist acknowledges that this grade seems to indicate once again that the client is not perfect and cannot always get it right—despite Herculean efforts. In other words, the therapist tries to communicate an understanding that this experience really is devastating for the client, and a sense of compassion toward the burden of perfectionism that the client has suffered under for so long. Empathic bids such as these demonstrate that the therapist grasps the core meaning of this prototypic experience, which occurs over and over again for the client, and truly understands the client's subjective worldview. Therapists will become important to their clients at this point and provide them with a sense of safety to explore further the significance of this adaptation. By understanding clients in a way they have never experienced before—through appreciating and articulating the full impact of this failure, defeat, or imperfection for clients, an impact that has always seemed so irrational to themselves and insignificant to others—therapists help clients free themselves from the tyranny of the shoulds and their sense of entitlement from others. In this way, they provide the holding environment that allows clients to enter deeply into their disappointment and shame, to grieve their losses, and to come to terms with or integrate the various parts (good, bad, lonely, angry, loving, and so on) of who they are.

The therapist must recognize that it can be shattering for some clients when their interpersonal coping style fails because it arouses the original helplessness, vulnerability, and shame that was evoked by their nonaffirming childhood experiences, which left their needs exposed but unmet. Eventually, the therapist must respond to the emotional deprivation beneath clients' compensatory attempt to obtain safety, security, and acceptance by moving toward, moving against, or moving away. As noted before, therapists can do this only after they have first responded to the anxiety, shame, or depression aroused by clients' pathogenic beliefs that—if they are no longer able to successfully please, achieve, or remain aloof—they are not loveworthy, do not matter, cannot protect themselves, are inadequate, and so on.

Third, the most important reason the client's compromise solution cannot succeed is that the client can never rise above the original conflict or unmet need. No matter how saintlike or loving, how successful or accomplished, or how self-contained and independent, the client is still defending against the original conflict, rather than resolving it. All these interpersonal maneuvers are attempts to *avoid* the problem by blocking it or to *manage* the problem by trying to rise above it rather than address it directly. This is clearly illustrated when the client's increasing success in life is not met by pleasure or satisfaction, but by increasing depression, cynicism, or guilt.

Let's review this complex sequence (Figure 7.1). When the child's basic developmental needs are blocked by the environment in one way or another, the individual will adopt a compromise solution that temporarily relieves the anxiety of having basic emotional needs unmet but does not resolve the underlying conflict. On the one hand, clients try to block their own needs, just as they were originally blocked by significant others. Component 3 in Figure 7.1 illustrates how clients turn against themselves, losing their self-efficacy and developing self-punitive feelings and a sense of helplessness and hopelessness ("I can't").

On the other hand, clients try to rise above the old unmet need by becoming special, turning their defensive lifestyle adaptation into a virtue, and obtaining some indirect gratification of the original need. Clients will be highly invested in trying to overcome and rise above their generic conflict; typically, to have to come to terms with it will feel like a shameful defeat. Thus, component 4 in Figure 7.1 indicates clients' grandiose sense of being special and defiant attempts to rise above the conflict. (They might say, for example, "I will never let them hurt me again.") This rigid, defensive stance reflects an act of will to overcome these basic developmental needs and rise above them, rather than trying to resolve them by addressing them directly.

In order to resolve these generic conflicts, clients must relinquish their compromise solutions. Attempts to block the original need and to rise above it are fragile solutions that cannot be sustained; failure experiences that precipitate crises and symptoms are inevitable. Thus, instead of avoiding the conflict in these two ways, clients will have to approach it, by allowing themselves to experience the original unmet need in a more supportive relationship that does not meet the original need, but does give clients the understanding and validation they need to tolerate, contain, and live through the feelings that were unacceptable or intolerable before. The interpersonal solution provided by the therapist's new and more effective response also allows clients to integrate other reactive feelings, helps to disconfirm pathogenic beliefs about themselves, and alters relational templates or faulty expectations about what may occur in relationships. Now, as adults, clients are able to understand, integrate, and resolve the original hurt and vulnerability by reexperiencing it with the therapist. In this way, what was once dreaded can now find compassion, validation, or understanding, and clients can integrate these feelings rather than continuing to try to defend against them. In other words, clients can now approach and resolve their conflicts. When clients allow the therapist to help contain the painful feelings, the old relational patterns are broken and new templates become possible. The therapist's behavioral disconfirmation of these faulty beliefs, evidenced by his or her compassion and understanding, also allows clients to be more self-accepting and forgiving. Thus, clients can now acknowledge and integrate the previously unacceptable parts of themselves, and can better acknowledge and accept both good and bad in significant others (component 5 in Figure 7.1). In other words, once the original hurt or need begins to emerge in therapy, the therapist's task is to provide a more satisfying response than clients have received in the past. The therapist does this, in part, by holding and being compassionate toward clients' pain, as discussed in Chapter 5. As we will see more fully later, this provides clients with a corrective emotional experience that can produce lasting and meaningful change.

If clients receive a more effective response from the therapist than they have received from others in the past, they will not need to block their need or attempt to rise above it anymore. The therapist's more accepting and affirming response to the old need and the conflicted feelings that accompany it will alter the old relational pattern and alleviate the anxiety that the original need aroused. As this occurs, clients will be able to begin responding in new and more adaptive ways that improve their relationships and expand the limited interpersonal style that they have adopted to cope with their conflict. In Chapter 9, we will examine more fully how the therapeutic relationship can be utilized to help clients resolve their generic conflict, and how this emotional relearning with the therapist can be generalized to other relationships in clients' lives. For now, an extended case study of a moving toward client, and two case summaries of the other interpersonal coping styles, will help therapists apply this conceptual model to therapeutic interventions.

Case Study of Peter: Moving toward Others
Developmental History and Precipitating Crisis

Peter was an insecurely attached child throughout his childhood and his authoritarian parents left his age-appropriate dependency needs for affection, warmth, and caring profoundly unmet. In addition, Peter's parents divorced when he was 6 years old, and his father did not take an active parenting role after the divorce. Peter's infrequent contacts with his father were superficial and inconsequential.

Peter's mother was overburdened by the demands of raising three children on her own, working full time, and trying to make some kind of personal life for herself. She was involved with several men in the years following the divorce, but she was never able to establish an enduring love relationship. Frustrated by the many demands and few pleasures in her life, she often made Peter the target of her resentment. She was irritable toward Peter, criticized him when things went wrong in her life, and felt resentful of his needs. Although she tried her best to be fair to the children and give them a good home, she could not be emotionally responsive and affirming toward Peter.

After his father left, Peter quickly learned that taking care of his mother was the best way to ward off her anger and disapproval and to win whatever affection he could. By the age of 10, Peter had adopted a pervasive personality style of moving toward people. His teachers described him as an especially responsible and well-behaved boy who was a pleasure to have in class. Now let's turn the clock forward 15 years and see how these developmental conflicts are expressed in Peter's early adulthood.

At the age of 25, Peter is a graduate student in psychology. Becoming a therapist felt like a perfect career choice to Peter. He prided himself on his sensitivity and concern for others and took pleasure in being able to help those in need. And now that he was carrying his own client caseload, it was great to find that he enjoyed being a therapist as much as he thought he would. Best of all, Peter was finding that he was often able to help his new clients. At least, all of his clients seemed to like him, and they kept returning to their therapy sessions each week. As his second semester practicum got underway, Peter felt that he was hitting his stride and on his way.

Later that semester, though, Peter had a setback. After presenting a video-taped recording of one of his therapy sessions in group supervision, Peter received some unexpected feedback. The practicum instructor told Peter that he was being "too nice" to his clients and that he seemed to need his clients' approval too much. The instructor went on to say that Peter seemed to be afraid of confrontations and was avoiding conflicts in the therapist-client relationship that should be addressed.

Peter was stunned. Although he was aware of his aversion to conflict, he did not truly understand what the instructor was talking about, and he felt hurt and confused by the criticism. It was important to Peter that his supervisor like him and approve of his clinical work. Peter tried carefully to explain that the instructor did not understand the close relationship that Peter was developing with his clients or recognize all of the important personal issues that his clients had been revealing. The instructor responded that, whether or not that was true, Peter was missing the point and repeated that Peter needed to address his aversion to conflict and its impact on his therapeutic style. To make matters worse, two students in the practicum group chimed in and agreed with the instructor's comments. With that, Peter's anxiety became so high that he could no longer defend or explain himself, let alone try to understand or learn from their comments. Peter finally stopped arguing with them and quietly nodded agreement throughout the rest of the supervision session.

The next few days, Peter felt sick. He was confused and dismayed by the criticism; his thoughts and feelings were racing. For a while, he thought that he should drop out of the practicum group, but then he decided that, if he only tried hard enough, he could make the instructor see that his critical observations were inaccurate. Peter's mind kept looking for a way to discount the feedback and stop the anxiety that was churning inside.

A week later, Peter found out that his girlfriend was having an affair with another student in the program. Although he tried to be understanding at first, he felt shocked and betrayed. He alternated between angrily telling her that their relationship was over and desperately trying to win her back. Peter felt shattered. He became so anxious that he was unable to eat or sleep, let alone study. It felt as if a motor were racing inside of him—accelerating out of control. Peter began hyperventilating and having anxiety attacks. By the end of the week, he was so anxious that he couldn't drive his car.

To make matters worse, Peter tried to keep all of this to himself. He thought he should remain calm and together, and he was afraid that his supervisors would not want him to see clients if he was "so messed up" that he was having anxiety attacks himself. But despite his attempts to cover up his distress, his individual supervisor soon asked him what was wrong.

Although he could not ask for it, Peter desperately wanted his individual supervisor's support, and he was greatly relieved to receive it. Peter explained how his practicum group was becoming one of the worst failure experiences of his life and how hard it was to accept what his girlfriend had done. The supervisor was supportive but did find a better way to say that the practicum instructor's comments fit with some of his own observations. Because he knew that the individual supervisor liked him, Peter was able to consider the feedback this time. The supervisor suggested that these were important issues for Peter to work with but that they were more than they could handle in supervision. They both agreed that Peter should begin seeing his own therapist at the student counseling center.

It helped that his supervisor knew about his anxiety attacks and still thought he could be a fine therapist, but Peter's anxiety was paralyzing as he began his own treatment. Fortunately, Peter was assigned to a skilled and experienced therapist, who soon formulated the predisposing vulnerability, interpersonal coping strategy, and situational stressors that precipitated his crisis. Peter's coping strategy of moving toward people in an accommodating and pleasing manner had worked well for him up to this point. However, both of the crises that Peter had just experienced ran headlong into the heart of his generic conflict and coping style. As Peter's ability to rise above his old unmet need by pleasing others failed, the anxiety associated with his developmental conflict broke through, and anxiety attacks resulted. Let's examine more closely why Peter developed these symptoms.

Precipitating Crises, Maladaptive Relational Templates, and Symptom Development

Many people would have coped with the two stressful events that Peter experienced without developing such significant symptoms. As we have seen, however, we must apply the concept of client response specificity. We will use Peter to illustrate how symptoms develop when situational stressors reenact old relational patterns, confirm pathogenic beliefs, or cause characterological interpersonal coping strategies to fail.

The first stressor for Peter was his practicum instructor's critical feedback. Criticism from a respected authority figure would be unsettling for most people, but generally they could cope with it. For someone like Peter, however, such disapproval carries far more weight. Because of Peter's unmet needs for approval, his history of receiving excessive criticism, and his lifestyle adaptation of trying to win approval by being helpful and responsive, the supervisor's criticism was devastating.

Second, the practicum instructor specifically challenged Peter's interpersonal coping style of moving toward others. This confrontation not only aroused the anxiety of receiving criticism, but also undermined Peter's primary means of warding off this anxiety (pleasing others and accommodating himself to their needs). Thus, the instructor's feedback intensified Peter's generic conflict and, simultaneously, weakened his defenses against it. In other words, Peter's developmental experiences made him acutely sensitive to criticism. He developed the attitude toward himself communicated by his early caretakers—that he was flawed and not worthy of approval. He also expected others to be critical of him. His attempt to defend against the anxiety evoked by this set of circumstances was to try and be perfect—to please others and meet their needs, so that this critical schema he had developed could be refuted by their appreciation, praise, and admiration. This coping style was, of course, a formula for failure: It was impossible for Peter to be perfect and loved and admired constantly by everyone he met. He experienced even constructive criticism, especially from an authority figure, as an attack on his sense of self, and he became flooded by feelings of shame for not being able to win everyone's approval and admiration.

If not for the subsequent crisis with his girlfriend, Peter probably could have recovered from the first setback without developing incapacitating symptoms. Most likely, he would have reconstituted his interpersonal coping style and been somewhat successful in winning the approval he needed from others in his life—from instructors, friends, and perhaps clients. As we will see, however, the subsequent stressor with his girlfriend also struck at the same generic conflict. At that point, his moving-toward coping style failed, and the strong anxiety associated with his insecure attachment history was evoked. More specifically, this was shame anxiety—anxiety over the threat of having his felt sense of himself as unlovable and unworthy revealed. This shame anxiety became too intense to be blocked; it broke through in anxiety attacks and symptom formation.

A partner's infidelity will be highly stressful for almost everyone. Here again, though, this particular stressor held far greater significance for Peter when viewed within the context of his developmental history and subjective worldview. Peter had to cope with far more than just the loss of trust with his girlfriend; he also suffered a blow to his identity and basic sense of self-worth. In Horney's neurotic pride system, Peter's coping style of moving toward entitled him to be special, so that others would love and prize him inordinately. Although Peter was largely unaware of these unrealistic claims, he expected his girlfriend to idealize him as the most sensitive and loving man there could possibly be. It was impossible for Peter to believe that his girlfriend could actually be interested in someone else. Peter's neurotic pride system was painfully shattered when he found that he was not the commanding center of his girlfriend's life that he felt entitled to be.

Multiple stressors occurred for Peter in a short period of time. In and of themselves, these would precipitate a crisis for many people and lead them to therapy. If these situational stressors do not tap into preexisting templates, faulty relational expectations, and pathogenic beliefs in the client's personality, however, they will not usually provoke such strong symptoms as anxiety attacks. The client will often be able to recover in a relatively short time with crisis intervention or short-term supportive therapy. In contrast, when life stresses tap squarely into a client's problematic internal working models, a client such as Peter has to cope with far more than just the demands of the current situational stressors.

Thus, when Peter's girlfriend became sexually involved with another man, he had to cope with much more than just the pain of this loss or betrayal. He also had to cope with his long-standing attachment conflicts, with relational templates of criticism and rejection, and with the wound to his self-esteem when his sense of being special was shattered. Clients such as Peter do not have a secure sense of self-esteem to fall back on in times of crisis. As unrealistic as it is, the sense of being special—yet acutely vigilant for slights—and the secondary gains that clients can earn by rising above their need is the only sense of self-worth that some clients have been able to achieve. It is understandable, then, that Peter was overwhelmed by anxiety attacks. Both of the situational stressors intensified Peter's old unmet need, and they both took away his lifestyle adaptation for defending against his generic conflict. Thus, we see that clients' presenting symptoms are not random, irrational, or unrealistic. They make sense once they are understood in the context of the client's developmental history and the templates, beliefs, and coping styles that result from this history.

The Course of Treatment

Fortunately, Peter's therapist was knowledgeable and kind. He was genuinely empathic to the pain that these situational crises brought on for Peter but also grasped the predisposing conflicts that made Peter so vulnerable to them. The therapist recognized that the current crisis provided an opportunity to resolve the more important developmental conflicts that left Peter prone to reexperience crises such as these whenever others rejected or disapproved of him. In the months that followed, Peter was able to resolve the precipitating crisis. More significantly, he was able, more than ever before, to come to terms with the impact of his developmental history and the pathogenic beliefs and maladaptive relational patterns that followed from it. As a result, Peter was able to step back somewhat from his rigid moving-toward coping style and adopt a more flexible and varied interpersonal repertoire. Here and in the next chapter, we will review the course of therapeutic events that allowed Peter to make these far-reaching changes.

Peter entered therapy in crisis, and the therapist was able to respond effectively to his intense anxiety and distress. From the first session, Peter felt understood and cared for by the therapist. As they developed a strong working alliance, Peter's anxiety attacks stopped. The intense anxiety and profound distress that had been so disorganizing for him steadily subsided as the therapist continued to provide an effective holding environment. As soon as Peter felt better, however, he began to be resistant to the feelings he had been sharing with the therapist and wondered about ending treatment.

The therapist observed that, as Peter felt understood and affirmed by him, Peter's unmet needs for emotional contact were assuaged and the attendant anxiety diminished. As a result, Peter felt safer, his symptoms abated, and he began to function better. At the same time, however, he also began to reconstitute his interpersonal coping style of moving toward others. Rather than terminating therapy at that point—which might need to occur in crisis intervention or time-limited treatment—the therapist was able to help Peter go further and address the developmental conflicts that made him so vulnerable to rejection and criticism. One way he did this was to help Peter recognize his habitual coping pattern; the therapist named it, so that they could talk about it together. Peter also learned to focus inward so as to identify what he was feeling and what was occurring with others at the instant when he felt compelled to employ this coping pattern. Usually, at such moments, he was expecting others to be disapproving of him or to disengage from him in some way.

Additionally, as he listened to Peter's narratives, the therapist would repeatedly highlight instances of Peter's key relational themes: (1) pleasing others; (2) avoiding anger and other interpersonal conflicts; and (3) expecting others to be critical and rejecting. Most importantly, the therapist used process comments to point out these three repetitive relational themes whenever he saw them occurring in their current interaction; this provided direction and structure to the treatment process, and served as the basis for developing mutually held treatment goals. For example, the therapist was able to encourage Peter to risk not pleasing him and took the opportunity to give Peter the experience of being cared about even when he was not employing his old coping style.

The therapist continued to focus Peter inward. In particular, they explored the feelings that were evoked for Peter when he felt the need to employ his usual coping style of pleasing or taking care of others but refrained from doing so and risked trying out other responses that he had rehearsed with the therapist. Gradually, by focusing on what he was experiencing just before he employed his usual coping patterns, Peter began to discover the emotional deprivation he had suffered as a child but had always been too ashamed to reveal to others or even acknowledge to himself. Relatively quickly, the therapist's warmth, acceptance, and calm presence enabled Peter to overcome the shame of having his unmet needs revealed. Thus, Peter, with all his strengths and flaws, could be fully seen and continue to be accepted and cared about. Peter began to understand how he had coped with his unmet emotional needs by taking care of and pleasing his mother, and then others as well. As this very significant progress occurred, however, Peter still continued to struggle with feelings of anxiety, shame, and sadness.

This triad of contradictory feelings continued to wax and wane for Peter over the next few months. Peter was increasingly recognizing how unloved, unwanted, and alone he had felt as a child and how ashamed of himself he had always felt. Accompanying these feelings, Peter realized that he had always believed that he was not loved because he was somehow unworthy and did not matter enough to be important to others. As Peter continued to feel emotionally held and contained by the therapist's caring presence, his long-withheld feelings of sadness over this deprivation and invalidation could be expressed for the first time and came through the old block. Peter received a far-reaching corrective emotional experience: The therapist's kind and affirming response was a powerful disconfirmation of his old relational expectations and was deeply comforting; it offered him something for which he had always longed. The therapist's response also provided him with a sense of being seen or known for the first time and accepted for who he really was (component 5 in Figure 7.1).

As Peter began to mourn his childhood loss, two reactive feelings followed closely behind. Peter found himself feeling furious that his father could value him so lightly and leave him so easily, and angry at his mother for making him feel responsible for trying to earn her love and ward off her disapproval. As soon as this anger emerged, however, the third feeling in Peter's affective constellation was aroused. Following his relational templates, Peter also became exceedingly anxious—afraid of being left helplessly on his own by the therapist, his girlfriend, his father, and anyone else with whom he would ever risk becoming close. Each of these three feelings in Peter's affective constellation—the shame surrounding his unmet emotional needs; his anger over being rejected and dismissed so readily; and the anxiety of being physically left alone again or emotionally cut off if he protested—was repeatedly aroused, and the therapist was gradually able to help Peter integrate or resolve these by experiencing, expressing, and containing them. This was not a smooth or simple process, however. As the primary feeling of sadness or deprivation emerged, Peter's own shame and internal blocking defenses were also activated (component 3 in Figure 7.1). Let's examine how the blocking side of Peter's compromise solution was expressed in therapy.

Peter's situational crises disabled his rising-above defense. He could no longer be special and win the approval he needed from either his practicum supervisor or his girlfriend. As his long-held interpersonal coping system failed, it broke open and exposed his shamefully unacceptable ("weak") unmet attachment needs and evoked intense anxiety. However, the other side of Peter's compromise solution was still operating: To defend against the painful feelings associated with the unmet needs, Peter turned the same type of block that was originally imposed by the environment against himself.

In treatment, and throughout his life, Peter had tried resisting each feeling in his affective constellation. He did this by blocking both the subjective experience and the interpersonal expression of his conflict; that is, his unmet attachment needs. For example, as the sadness of his childhood deprivation emerged, Peter thought the therapist would be critical and rejecting. He was afraid that the therapist would want him to "stop crying, grow up, and act like a man." This *transference reaction* again served to block his sadness until the therapist was able to draw it out more fully and clarify this misperception by talking it through with Peter. More importantly, Peter was able to experience the therapist's commitment and caring in the face of Peter's neediness. Peter learned in this relationship that he did not have to be perfect or take care of the therapist to be cared about by the therapist. Thus, although his relational templates led him to expect that the therapist would be irritated by his emotional needs and resent their burden, as his mother had been, Peter experienced a new and corrective response.

Next, as Peter's anger toward his girlfriend, and then both of his parents, emerged, Peter felt guilty: He felt that he should be understanding, accepting, and forgiving of them, as his coping style demanded. He also felt ashamed of his inability to sustain this response. This intrapsychic defense reflected his characterological moving-toward adaptation and served to block expression of his unmet need for secure relational ties again. With the therapist's skillful balance between affirming the validity of Peter's need and anger and acknowledging that Peter still loved his mother and had also received many good things from her, Peter finally felt safe enough to be able to risk experiencing both the need and the anger that had always been present but too anxiety-arousing for him. Why was it so threatening for Peter to experience his need and anger or share it with the therapist? While growing up in his family, Peter had learned that relationships were not resilient and his needs and his anger were unacceptable; he knew that, because of his father's unavailability and his mother's strong propensity toward love withdrawal, he could easily destroy the little emotional support that he had. Aware that expression of his needs and his anger threatened to break his already insecure emotional attachments, Peter convinced himself that his needs could never be met. Further, his anger aroused his separation anxieties; he feared that he would be further abandoned if he expressed it. Peter then tried to block these feelings by telling himself that he should be calm and accepting—especially if he were going to be a therapist and help other people. Fortunately, the therapist was able to approach each of these feelings and help Peter understand and contain them.

To sum up, the therapist provided a skillful balancing of the good news and the bad news in Peter's development. He was able to acknowledge the reality that Peter loved his mother and still felt loyal to her despite all that occurred and that some of the qualities he valued most about himself were qualities derived from her. However, these reality-based strengths were accompanied by the painful realities of Peter's emotional deprivation and rejection. By exploring with Peter and then affirming what was legitimately good in his relationship with his mother, rather than simply blaming or rejecting her, the therapist helped Peter preserve important aspects of this tie. This gave Peter the attachment security he needed to risk Bowlby's protest—his anger at both of his parents for what was seriously wrong in their relationship. Being able to protest what had been so unfair and hurtful, to say no to it rather than deny his pain, and to feel his anger and hurt fully for the first time led Peter to a stronger, more self-contained stance that he was able to generalize to other relationships. Peter became more confident and assertive; developed firmer boundaries and more clarity about who he was and who he wasn't; and began to disconfirm the core pathogenic belief that he was unlovable and too insignificant to matter to someone. Peter came to understand that his mother's rejection was unwarranted, had absolutely nothing to do with him, and was just the wounded projection of her own deep hopelessness about ever being loved. Significantly, Peter was able to feel some compassion toward her and appreciate the predicament she was in, but without denying the reality of his hurtful feelings and the significant consequences this had held for him.

As a result of these changes, Peter no longer needed to defend so strongly against his unmet attachment need and its attendant anxiety. He felt less urgent about making others like him, more comfortable with interpersonal conflict, and more confident. Finally, Peter was also able to keep the best parts of his moving-toward style: He remained a sensitive, caring, and responsive person, but this was no longer the compulsive, unidimensional response pattern that it had been in the past. Peter had grown through this crisis and become a more interesting and resilient person with more emotional depth. Therapy helped to immunize him against his predisposing vulnerability to rejection, criticism, and loss. By experiencing a more satisfying response to his old conflicts from the therapist and finding that relationships can be different from those he had experienced and expected, Peter expanded his internal working models of what can or must occur in relationships. In this way, the therapist helped Peter to stop carrying emotional baggage from his past into his present relationships and from reenacting the same maladaptive relational patterns and delimiting interpersonal coping strategies.

Peter will certainly have further crises in his life, especially when circumstances again tap into his old generic conflict, but the changes he has made will enable him to respond to future problems in a more adult, present-centered manner.

Two Case Summaries

In the case study of Peter, we have seen how our model for conceptualizing client dynamics may be utilized with a moving-toward client. Let's look at two additional case summaries that highlight salient features in the functioning and treatment of a moving-against client and a moving-away client.

Carlos: Moving against Others

Carlos had been riding high, on his way to becoming a real estate "king," when his business failed. Buoyed by his initial success in a booming real estate market, Carlos had been woefully overextended when interest rates rose unexpectedly. His business went bankrupt; his new home and sports car were repossessed; and friends deserted him as creditors and the IRS aggressively pursued him. Carlos entered therapy when his dreams of wealth and power were dashed; he felt desperate, frightened, and suicidally depressed.

After eight months of therapy, however, his suicidal preoccupation and intense distress had greatly improved. Carlos recognized how helpful the therapist had been to him and had recently told a friend that he might not be alive if not for the therapist's expertise and genuine concern. On this particular day, however, Carlos was ready to quit. He was frustrated that he had to come to therapy for such a long time, and he did not want to need the therapist anymore. In particular, he was fed up with the therapist's fees. As Carlos tended to do when he was upset, he began the session by bitterly taunting the therapist:

Carlos:
    Who do you think you are to charge so much? Do you really think you're worth all that?!

The therapist acknowledged the feeling in Carlos's provocation:

Therapist:
    You're angry about my fees. Let's talk about them.
Carlos:
    (*continuing his assault*) What are you doing—trying to prove that you can gouge as much money as the psychiatrist next door? Did your mother want you to be a "real" doctor and make a lot of money or what?

Fortunately, the therapist was able to remain nondefensive and stay emotionally connected to Carlos during this critical incident. The therapist did not retreat or counterattack, which is what Carlos's provocations usually elicited from others. Instead of acting on his own reflexive response to justify and explain, the therapist tried to find a more neutral or effective middle ground of involvement:

Therapist:

Yes, of course you're angry with me. I do charge you for my time, and it's hard for you. And there are still problems in your life that haven't changed yet.

The therapist's accepting response only infuriated Carlos further:

Carlos:

I hate your understanding, I hate coming here, I don't want to do this anymore! What do you do—be nice to people when they're falling apart so they become dependent on you? Then they can't leave you and you can keep taking all of their money. There's something wrong with you; you're sick!

Therapist:

It seems like I am trying to make you dependent on me so that I can take advantage of you financially?

Carlos:

(*livid*) Yes, you idiot, I want to leave and get out of here, but I can't. I'll probably just get depressed again if I leave. I need you and you know it, and you use that to take the little money I have left. I'm trapped by you. You've got me and I hate it. I hate you!

The intensity of Carlos's reactions told the therapist that, in addition to whatever reality-based issues needed to be addressed regarding fees, Carlos's basic developmental conflicts and maladaptive relational patterns were being activated by this issue. On the basis of other hypotheses he had already formulated, the therapist reasoned that, for Carlos, having a need of someone automatically meant being used or shamefully exploited by them. Trying to articulate and help to contain what Carlos was experiencing in their relationship right then, the therapist responded:

Therapist:

It's as if you don't experience the decision to remain in therapy as your own choice, Carlos. You feel trapped by your own need to be helped or understood. You have learned that to let yourself need somebody puts them in control of you and leaves you subject to their own selfish needs. Your only choice then is to resist by leaving the help you still want or to stay in treatment with me but feel exploited. You may have experienced something like this with others, Carlos, but I want you and I to have a different kind of relationship. I want to respond to your need for help, not to use your trust and vulnerability for my own gain.

Over the course of the session, Carlos gradually calmed down and was somewhat able to accept what the therapist was saying. Carlos's transference reaction—that the therapist had manipulated Carlos's dependency in order to meet his own financial need—encapsulated Carlos's developmental dilemma. Carlos had grown up in an orphanage and had been emotionally deprived throughout his early childhood. When he was placed with foster parents during second grade, they took advantage of his intense, unmet dependency needs. They manipulated Carlos to respond to their needs under the guise of their love for him.

By the time he was 9 years old, Carlos basically had become an exploited laborer on his foster parents' farm. In order to earn their approval and affection, he worked sunup to sundown on the days he wasn't in school. By the time he was 18 and ready to leave home, Carlos had long since realized that his foster parents were taking advantage of him and were not to be trusted. At some point during his early adolescence, Carlos's anxious efforts to please his foster parents had evolved into an angry defiance toward them and others. He had vowed to himself that he was never going to be dependent on anybody again. Attempting to disconfirm this shame, Carlos decided that he was going to be rich and powerful— strong enough that no one would ever be able to take advantage of him again. Carlos's feelings of exploitation and betrayal were further fueled by his racial and ethnic history. A Latino, Carlos was the child of undocumented migrant workers whose babysitter had been instructed to leave him at the orphanage if his parents were to be suddenly deported. Carlos was aware of his parents' history; he felt very angry at their exploitation by farmers but also angry at them for betraying him by never returning to claim him. These factors all contributed to Carlos's determination to never depend on or need anyone. He was going to be so financially successful that no one would ever use or betray him; he would simply pay money for whatever he wanted or needed.

The dramatic confrontation with his therapist proved to be a corrective emotional experience for Carlos. Although his old fear of being used was aroused in relation to the therapist, as it was whenever Carlos began to trust, need, or get close to anyone, his defensive coping strategy—his anger—did not succeed in pushing the other person away. The therapist was able to tolerate Carlos's provocative attacks and to remain engaged with him and responsive to his concerns. The therapist acknowledged and validated the negative impact of both Carlos's family history and experiences and the additional impact of his cultural background in the society in which he lived.

As the therapist remained available to Carlos and provided a different response to his moving-against style than Carlos expected, the next feeling in his affective constellation emerged: His primary feelings of hurt, deprivation, invalidation, and betrayal became the predominant issue in therapy. Just as the therapist was able to tolerate Carlos' anger, he was also able to provide a holding environment to contain these threatening feelings, and they, too, became less intense. Carlos's accompanying shame about not being strong and self-sufficient and about being vulnerable and his recurring fear of being used continued to be acted out in the therapeutic relationship. Thus, Carlos experienced the same conflicts and fears over and over again with the therapist. Each time these concerns were manifested in a conflict between them, however, Carlos received a genuinely empathic and understanding response from the therapist that was different from what he expected and had experienced in the past.

In addition, it was helpful to Carlos when the therapist validated his perception that he had indeed been taken advantage of as a child, that people of his culture were often devalued and exploited, and that elements of devaluation and exploitation were present in some of his current relationships. However, the therapist patiently but repeatedly questioned whether others were *always* trying to exploit him in his current relationships. Honoring his resistance or interpersonal coping strategy, the therapist also observed that, as a child, Carlos had made the best and only adaption that he could to his unsolvable predicament. The therapist pointed out that there was courage and strength in Carlos's defensive attempt to rise above his conflict by adopting an aggressive and defiant interpersonal style. Although this defensive attempt to rise above his conflict was no longer necessary or effective, it had indeed served an absolutely necessary purpose—to protect him from feelings of vulnerability and exploitation that he experienced as a child and was determined never to be exposed to again. Eventually, Carlos understood what the therapist meant when he suggested that it took just as much strength for Carlos to confront his old conflicts and feelings now, as an adult, as it originally did to defend against them as a child by compulsively striving to achieve success and power.

With the therapist's consistent support, Carlos was gradually able to reexperience his old feelings of desperate need and simultaneous outrage at being used by his foster parents and, later, at being left by his biological parents. As he gradually let go of his strong combative and competitive stance with the therapist, and let himself undergo those painful feelings of his childhood that had always seemed disgusting and weak, Carlos began to change. First, his combative, argumentative style gave way to a friendlier comradery with the therapist. Carlos began to disclose more easily and felt less competitive with the therapist. In particular, Carlos complained less about needing the therapist and being exploited by him.

These changes with the therapist began to carry over to other relationships as well. Carlos began to establish male friendships for the first time; his competitiveness and hostility had always precluded such friendships in the past. Likewise, Carlos's relationships with women had always been short-lived, superficial sexual contacts, which often were ingenuine and manipulative. Although not intending to be exploitative, Carlos felt less threatened by these short-term relationships, in which he often gave the women expensive gifts. As Carlos stopped defending against his shame and allowed himself to acknowledge to himself and the therapist how he had been used in his childhood, his patterns of using women faded. However, he was still unable to establish an intimate, egalitarian love relationship at this time in his life.

Other evidence of change came from Carlos's business associates, who told him he seemed less driven and less pressured than in the past. His attempts to rise above his conflicts through achieving wealth and power were modulated and gave way to an ambitious, achieving work ethic. When an established banker in town offered Carlos a good job as a loan officer and appraiser in his real estate division, Carlos felt ready to terminate therapy. All of his problems were certainly not solved, but Carlos now felt capable of successfully moving on in life on his own. It was important to Carlos to be reassured that he could return to therapy if he needed to in the future, and that doing this would not signify failure or weakness but rather strength in recognizing when he needed help or support. Assured of the therapist's continuing availability, Carlos terminated and did not recontact the therapist.

Maggie: Moving Away from Others

Four months after her adolescent daughter had been raped, Maggie was still in crisis. Nightmares disrupted her sleep; tension headaches tormented her by day. Helpless to comfort her daughter and enraged at the casual indifference of the police and courts, Maggie was afraid she was going out of control. At work, her supervisor evaluated her as irritable, sullen, and withdrawn and suggested that she seek counseling. Although she had always been reluctant to ask for help with anything, Maggie contacted a therapist when she realized that her job was in jeopardy.

The therapist was responsive to all Maggie's feelings of rage, guilt, and helplessness evoked by her daughter's tragedy. To her surprise, Maggie felt understood by the therapist and began looking forward to their meetings. She gradually came to trust the therapist and increasingly invested herself in their relationship. After several months in treatment, Maggie reported a dream from the previous night. In the dream, Maggie was alone in a vast desert night. No other people existed in this great, silent space. The desert night was black; no light shone from stars or moon. As she walked across the endless sand, a cool, dry wind began to move lightly across her face. Maggie laid down on the sand, closed her eyes, and silently slipped away into the darkness.

Maggie said that variations of this dream had recurred throughout her life. Accordingly, the therapist knew that the dream expressed Maggie's most central developmental conflicts. Hoping that the dream could also provide an avenue to reach her central feeling of aloneness, the therapist tried to bridge Maggie's moving-away orientation:

Therapist:

Can you close your eyes and see the dream again?

Maggie:

Yes. (*settles back and closes her eyes*)

Therapist:

Describe what you see to me.

Maggie:

I'm walking; it's quiet and dark. I'm alone. I can feel the breeze. Now I'm lying down on the sand and I close my eyes, like going to sleep.

Therapist:

I don't want you to be there alone. Will you let me join you?

Maggie:

(*pause, then slowly*) Yes, you can.

Therapist:

Keep your eyes closed, and hold that same, familiar image. But you're not alone this time. I am walking toward you, and I reach my right hand out toward yours. Will you take it?

Maggie:

OK.

Therapist:

I'm now in the dream with you. We're holding hands, and walking together through the desert night. Can you see that?

Maggie:

Yes, we're walking toward the lights of a distant city. (*opens her eyes, and looks kindly at the therapist*) I'm not alone. It's good to have you with me.

This joining experience was a turning point in therapy. Maggie had grown up in a silent void, much like the setting in her dream. She had never known her father, and her mother was often away. By 10 years of age, Maggie was regularly spending most of the weekend alone, fixing her own meals, and putting herself to bed while her mother was out. The dream reflected both the aloneness and emptiness of her childhood and her lifestyle adaptation of moving away from others. However, the crisis with her daughter overwhelmed her lifelong coping strategy to withdraw and be self-sufficient and aloof. In letting the therapist join her in her dark, empty space, Maggie took the enormous personal risk of accepting the human contact that she longed for but had long since learned to hold away. This single experience did not resolve her conflict or her interpersonal adaption to it, of course, but incidents of sharing continued in her relationship with the therapist. Maggie's sense of aloneness was heightened by her consciousness of being "different" because she was bicultural, as her therapist, herself a bicultural woman, was well aware. The therapist's warmth, support, and bold articulation of all the factors, familial and cultural, that contributed to Maggie's sense of isolation helped her begin to change in three ways.

First, Maggie's conflicted affective constellation emerged. The emptiness and longing resulting from her childhood abandonment and aloneness became accessible. To ensure that the lack of response that Maggie had experienced as a child was not recapitulated, the therapist was careful to let Maggie know that her feelings were being heard. The therapist was skilled in communicating to Maggie that she was not in a dark, silent void anymore, but in a caring and concerned relationship. Maggie became more comfortable sharing her deprivation and longing with the therapist. However, the intense anger at her mother for virtually abandoning her and the frightening separation anxieties aroused by her rage at her mother, to whom she was so insecurely attached, were worked through much more slowly.

Second, as the therapist continued to provide a corrective emotional experience in which all of Maggie's difficult feelings were accepted and affirmed, Maggie began to change in her relationship with the therapist. Although Maggie had become attached to the therapist because of her crisis state, she was still deeply reluctant to accept help from anyone; she held a part of herself back. Interestingly, the therapist's bicultural identity had given her an *ascribed* credibility, based on perceived similarities of race and gender, which helped Maggie stay in therapy in the beginning even though she was unsure of this process. Over time, however, the therapist's *achieved* credibility—her skillfulness and the effectiveness of her responses—took center stage. Thus, it was only at this later point in treatment that Maggie was able to further let go of her coping style and allow the therapist to help her. As a result, Maggie began to act in a less reserved manner toward the therapist and became more expressive than she had ever been with anyone else. She talked more freely about herself with the therapist and found herself being curious and interested in the therapist's own personal life; the therapist saw this as positive for this particular client and responded willingly at times. Maggie's nightmares and headaches had been alleviated months before, but now a sense of humor, interpersonal warmth, and graciousness were flowering as well.

Third, as Maggie talked about feeling "fuller" inside, the therapist observed that changes were occurring in her other relationships as well. At work, her supervisor was pleased to observe that Maggie was more involved and available than she had been, which made it easier for others in the office to work with her. Whenever interpersonal conflicts surfaced at work, Maggie's initial reaction was still to withdraw and go away, since this still felt safe to her, but she was also aware that it robbed her of the joy that relatedness can bring. As a result, Maggie became better able to discuss the problems and her reactions to them with the therapist. Together, they would find new ways for her to remain involved in the situation, rather than to withdraw or remain aloof as she had always done in the past. Maggie also became more accessible to her two adolescent children. For the first time, she talked more about her own feelings, interests, and personal history with them. The nearly grown children welcomed this contact and, in turn, they began to disclose more to her as well. In general, Maggie felt closer to her children than she had ever been and even felt somewhat more capable of helping her daughter cope with the painful aftereffects of her assault.

Finally, Maggie began to change how she responded to her boyfriend. This had been a superficial relationship, even though they had known each other for two years. Maggie now asked for more personal sharing and for more commitment from him; she initiated an agreement that they not date other people. She also asked him to talk about his feelings more and to spend more time with her, which he was able to do. It was exciting but anxiety-arousing to take each of these steps forward. Her old coping style of moving away was activated each time she took a step toward others, especially if the others responded inconsiderately, and she still reflexively defended against the hurt that relationships had originally held for her. As the therapist repeatedly helped her work through her concerns in each of these arenas and remained consistently available, supportive, and caring, Maggie tentatively suggested that it might be approaching time to end therapy.

Although things had been going well for the past few months, Maggie became depressed as soon as they talked about ending. She missed the next appointment and arrived late to another session. As before, the therapist continued to focus Maggie inward, and her old feelings of emptiness and abandonment emerged again. The therapist suggested that they put off setting a termination date for a while and work further with these conflicted emotions. With more vividness and detail than before, Maggie recalled her childhood depression at abandonment. Painful memories returned—for example, the memory of being 8 years old, sitting alone on the living room couch, and listening to the clock tick the empty afternoon away. Maggie sobbed as she recalled her desperate childhood wish for someone to come home to, which was only reenacted in her adulthood. Following her formative relational templates, Maggie had married a salesman whose job took him away from home for extended periods—and he had been emotionally unresponsive to her when he was home as well. Thus, in her first significant love relationship, Maggie continued her history of aloneness. However, touching the original pain so directly and sharing it so deeply with the therapist relieved the depression that the suggestion of terminating had precipitated. Before long, Maggie again felt ready to terminate. This time, with the therapist's reassurance that she could contact her again if necessary, Maggie successfully terminated.

Closing

In this chapter, we have studied a model for conceptualizing clients' dynamics. In an extended case study of Peter, a moving-toward client, we saw how this model can be applied to treatment. We also looked at two case summaries that illustrate its application to moving-against and moving-away clients. At this point, however, two issues remain incomplete.

First, we need further guidelines for conceptualizing clients' dynamics and using this conceptualization to shape clinical interventions. The information about clients' current interpersonal functioning provided in Chapter 8 will help therapists further conceptualize their clients. At this point, readers are prepared to utilize the structured guidelines for keeping process notes and writing case conceptualizations provided in Appendixes A and B. These guidelines will help therapists to conceptualize clients by recognizing the repetitive relational patterns that provide a focus for treatment, to intervene along the process dimension, and to formulate treatment plans and goals.

Second, we need further guidelines for negotiating the crucial stage of therapy in which the compromise solution fails and the client's generic conflict emerges. The relationship that the therapist provides the client at that stage determines whether the client will be able to resolve the conflict and change. In Chapters 9 and 10, we will discuss more fully how the therapist can best respond to the client's emerging conflict, provide a corrective emotional experience, and generalize this experience of change with the therapist to relationships outside of therapy.

Suggestions for Further Reading

1. Useful guidelines for conceptualizing client dynamics are found in Chapters 5 and 7 of *Psychotherapy in a New Key: A Guide to Time-Limited Dynamic Psychotherapy* by H. Strupp and J. Binder (New York: Basic Books, 1984). *Coping with Conflict: Supervising Counselors and Psychotherapists* by W. Mueller and B. Kell (New York: Appleton-Century-Crofts, 1972) is unsurpassed in helping trainees utilize their own reactions and the process dimension to conceptualize client dynamics and effect change.

2. Karen Horney's work is essential reading for interpersonally oriented clinicians; it has been widely incorporated by psychodynamic and cognitive behavioral theorists alike. Readers may be especially interested in *Neurosis and Human Growth* (New York: Norton, 1970) or *Our Inner Conflicts* (New York: Norton, 1966).

3. Therapists-in-training need case examples that illustrate how theory guides clinical practice. Unfortunately, such illustrations are not plentiful. One of the most effective examples of integrating theory with practice is James Masterson's trilogy on using a developmental, object relations approach for treating borderline disorders. Masterson's books will be illuminating even to readers who do not have a special interest in borderline functioning. See especially *Psychotherapy of the Borderline Adult: A Developmental Approach* (New York:

Brunner/Mazel, 1976) and *Treatment of the Borderline Adolescent: A Developmental Approach* (New York: Wiley, 1972). Extended case studies of late adolescents and clients of college counseling centers from an interpersonal process viewpoint are provided in *Child and Adolescent Therapy: A Multicultural-Relational Approach*, edited by F. McClure and E. Teyber (Fort Worth, TX: Harcourt Brace, 1996). Each case study presents the treatment of a client with a particular disorder or problem—such as sexual abuse, eating disorders, spiritual values clarification— and illustrates how multicultural factors in the case influenced treatment plans and intervention strategies; see especially Chapters 2, 4, 6, 8, 10, and 13.

4. *Cognitive Therapy of the Personality Disorders* by A. Beck and A. Freeman (New York: Guilford, 1990) masterfully links Horney's interpersonal coping styles to cognitively based interventions for clients with difficult-to-treat personality disorders.

CHAPTER EIGHT

Current Interpersonal Factors

Conceptual Overview

We now turn to the third component of our conceptual formulation. As we have seen, the first component comprises the three basic dimensions of family life introduced in Chapter 6. Most clients who enter therapy have experienced problems along each of these dimensions: They have usually come from families without a primary parental coalition, are often emotionally bound in cross-generational alliances and loyalty conflicts, and/or have not had their developmental needs for secure attachments and discipline met. When development goes awry along these dimensions, children are likely to develop enduring personality conflicts.

The second component of our conceptualization is the schematic model for understanding how client problems are expressed, maintained, and resolved (Chapter 7). Central to this conceptual model is clients' characteristic interpersonal style for coping with their unmet needs and relational problems. In the present chapter, we consider the third component of this conceptual formulation—clients' current interpersonal relations, which provide the most important information for conceptualizing clients' dynamics and understanding what is occurring in the therapeutic relationship. In particular, we will see how therapists' own feelings and reactions toward clients can be the best source of data for understanding clients' conflicts and interpersonal coping strategies. Thus, in this chapter, we examine how therapists can systematically utilize clients' current interpersonal functioning and the therapist's own experience of the clients to formulate treatment plans and guide their interventions.

Chapter Organization

In the first section, we examine three major ways in which client conflicts are expressed in the therapeutic relationship: eliciting maneuvers, testing behavior, and transference reactions. Next, we consider the degree of separateness versus relatedness in the therapist-client relationship. Using the process dimension, therapists must establish and maintain the optimal degree of interpersonal relatedness with clients if change is to occur. Finally, we look at the ambivalent or two-sided nature of clients' conflicts. Applying the guidelines for writing case formulations in Appendix B, therapists consider each aspect of clients' current interpersonal functioning in conceptualizing their clients' dynamics and developing interventions.

How Clients Bring Their Conflicts into the Therapeutic Relationship

The symptoms and problems that clients strive to resolve in treatment originated from, and are now being expressed in, close personal relationships. Before long, the relational conflicts that clients are experiencing with others also become a highly relevant aspect of the therapeutic relationship. As we have seen, clients do not merely talk with counselors in the abstract about their problems with others. In a far more personal and immediate way, they also recreate these problematic relational scenarios with the counselor and then try to work out a new, more satisfying solution to these problems in their real-life relationship with the counselor. Thus, in this section, we will examine three distinct but closely related ways in which clients bring their conflicts into the therapeutic relationship: eliciting maneuvers, testing behavior, and transference reactions.

Sometimes clients will employ certain interpersonal operations—such as *eliciting maneuvers*—(Sullivan, 1968)—with the therapist in order to avoid and defend against their conflicts. At other times, clients will employ different interpersonal mechanisms—such as *testing behavior* (Weiss, 1993; Weiss & Sampson, 1986)—to approach and try to resolve their conflicts. Testing behavior reflects a healthy, growth-oriented attempt to master conflicts, in which clients assess whether the therapist will respond in familiar, expected, problematic ways or will respond in the ways they actually need. Working with these complex and sometimes contradictory maneuvers is one of the most interesting aspects of psychotherapy.

The third way clients bring their conflicts into the therapeutic relationship is through *transference reactions*—the clients' systematic misperceptions of the therapist. As we will see, much confusion arises from two very different definitions and usages of the term *transference*; we will try to distinguish clearly between these definitions. These three interrelated concepts reflect the most fundamental aspects of interpersonal psychotherapy.

Eliciting Maneuvers

Interpersonal theorists have described how clients develop fixed interpersonal styles to avoid anxiety and defend against their conflicts (for example, Beck &

Freeman, 1990; Fromm, 1982; Horney, 1970; Kiesler, 1988; Sullivan, 1968). Most clients perceive others in rigid, inflexible ways, are restricted in the range of emotions that they can experience and express, and respond to others in stereotypic, limited ways, such as Horney's fixed interpersonal coping styles of moving toward, moving away, and moving against. In part, clients systematically employ these interpersonal styles to (1) elicit certain desired responses that will avoid their conflicts and diminish anxiety and (2) preclude responses from others that will arouse their generic conflicts. An *eliciting maneuver* is a behavior that brings about certain desired, safe responses. To illustrate how eliciting maneuvers serve to defend against conflicts, we will return to the case study of Peter, the moving-toward client in Chapter 7.

Peter's moving-toward style tended to elicit approval, acceptance, kindness, and trust from others. Peter was helpful; he sympathized, agreed, and cooperated with people. For example, Peter tried to understand his girlfriend's infidelity and to win his supervisor's approval. When Peter began to see a therapist, he employed the same interpersonal style: He tried to elicit his therapist's approval by being a good client who was quickly getting better—at least until the therapist began using process comments to question this compliance.

In contrast, Peter was never appropriately critical, forthright, assertive, angry, or skeptical. These responses, which a well-functioning person needs to employ at times, were not part of Peter's interpersonal repertoire. For example, he did not communicate how angry he was with his girlfriend or even experience how angry he was with both of his parents. Similarly, he did not assertively set limits with his practicum instructor or classmates regarding how much critical feedback he could incorporate at one time. His pleasing interpersonal style also discouraged angry, critical, and confrontive responses from others that would arouse anxiety for him.

If Peter's therapist had reflexively responded to what Peter's interpersonal style elicited, Peter would not have changed in therapy. If the therapist had merely supported Peter and met his need for approval, Peter would have only reenacted in therapy the same rising-above defense that he had used throughout his life. Shoring up these defensive coping styles is often the goal in some time-limited or supportive therapies, requires skill to accomplish, and is highly beneficial to many clients. The therapist did not wish to do this with Peter, however. Having the opportunity to seek broader and more enduring changes, and not merely symptom relief, the therapist used process comments to make Peter's moving-toward style overt and a focus for discussion in their relationship. Thus, instead of automatically responding to what Peter elicited, the therapist focused on it as part of Peter's problems that needed to be addressed in therapy.

Over the course of several months, the therapist patiently pointed out, in a supportive, noncritical way, each instance in which Peter's fixed interpersonal style of moving toward others was operating. Working as a team, Peter and his therapist considered both how this was serving to protect him from his anxieties and unmet needs and how Peter suffered as a result of this coping style. Once they named this pattern, Peter began observing in many aspects of his life how this fixed moving-toward style elicited from others the approval, support, and nurturance he had missed as a child. In addition, he began to recognize how this style discouraged the critical, angry, or rejecting responses from others that were so threatening for him. As Peter made progress in mastering his developmental conflicts in therapy, he gradually became less intent on winning others' approval and was able to expand his interpersonal repertoire. He began to be appropriately assertive, direct, and more confident for the first time in his life.

The eliciting maneuvers used by moving-toward clients like Peter are familiar, and may not be especially difficult for beginning therapists to work with. In contrast, moving-against clients employ eliciting maneuvers that are often challenging for beginning therapists to work with. In the first few minutes of the first session, these clients often do something to take command of the relationship. They seem to quickly find some way to intimidate the therapist and make him or her feel insecure. For example, clients may insist on sitting in the therapist's chair, question the therapist's adequacy, or criticize one of the therapist's early responses. This provocative presentation elicits anxiety and withdrawal in many beginning therapists, competition in some, and hostility in a few. If moving-against clients succeed in eliciting one of these responses from the therapist, therapy will not progress. The clients will have successfully neutralized the therapist's effectiveness and defended against their anxieties associated with having problems, asking for help, entering a new relationship, and relinquishing some control. In this way, clients' eliciting maneuvers protect them from conflicts, but at the price of change.

Unfortunately, therapists tend to respond reflexively or automatically to the client's eliciting maneuvers or *relational enactments* with their own countertransference propensities—with their own characteristic tendencies toward flight or fight. What can therapists do instead? First, they can try to retain legitimate control of themselves internally, by trying to understand what is occurring in their relationship with the client. They need to begin formulating working hypotheses about (1) the impact this behavior might have on others in the client's life and (2) what conflicts the client might be avoiding by these maneuvers. Next, on an interpersonal level, therapists can try to engage the client in some way in which the usual relational patterns of domination, intimidation, or competition are not enacted. There are several possibilities.

One is for the therapist simply to withstand the client's criticism, without avoiding the challenge.

Therapist:

> Your previous therapist was a more experienced psychiatrist, and you're not sure that a younger social worker such as myself can help you. Let's talk about that. Tell me more about your concern.

Another alternative is for the therapist to make a process comment about the current interaction.

Therapist:

> You're speaking in a loud voice right now. You sound angry. Are you aware of speaking loudly, and do you feel angry?

With some clients, the therapist may wish to take this approach one step further and begin to explore the effects of this interpersonal presentation on others.

Therapist:

> I am aware of how frequently you are critical of me; you're being critical right now. I wonder if criticizing others is something you do frequently, and I wonder how others usually respond when you do this.

Many of these clients will be unaware of the impact they have on others, and this interpersonal feedback can provide a productive new starting point. However, the purpose here is not to win the battle for control and get one up with the client. On a process level, this would only continue the same problematic interpersonal scenario that recurs with others. Rather, this approach is intended to help the therapist find some way to engage with the client other than by reflexively responding with anger or compliance to the eliciting defenses. It will also help the client develop increased awareness of his or her interpersonal style and its impact.

Still another possibility is for the therapist to focus on how the client's eliciting maneuvers affect the client.

Therapist:

> You have insisted on sitting in my chair, even though I asked you not to. Now that you are there, how does it feel?

Routinely, clients become aware of feeling alone, empty, or anxious at these moments, and the hollowness of their victory provides a new shared point of departure for the therapist and the client.

Particular types of responses will work well with one client and not at all with another. Therapists' goals in responding to such difficult clients are as follows:

1. To attend to their own subjective reactions and feeling states and observe what the client is eliciting in them
2. To formulate working hypotheses and try to conceptualize how others would typically respond and consider how this may serve to help the client avoid his or her conflicts

3. To try and find another way to respond (for example, via process comments) that does not reenact the same relational scenario that the client usually elicits

Although moving-against clients may intimidate beginning therapists, this usually does not last very long. With a little more experience and a supportive supervisor, beginning therapists learn not to charge so readily at a waving red flag. With more exposure to moving-against clients, therapists learn to appreciate that the intensity and rigidity of clients' eliciting maneuvers, no matter how alienating initially, are commensurate with clients' degree of anxiety and conflict. As therapists better understand the purpose of these distancing maneuvers, they may begin to find some compassion or feeling for the predicament such clients are living out, which is always the best way to step out of the old relational pattern being elicited. By trying out various tactics, therapists can usually find ways to engage with these clients that do not reenact the intimidated, competitive, or hostile responses that they expect.

As we have seen, the confrontational eliciting maneuvers of moving-against clients—such as Carlos in Chapter 7—place considerable demands on beginning therapists. However, the meeker, moving-away clients—such as Maggie in Chapter 7—pose a greater challenge in terms of treatment outcome. Most people who select counseling careers have strong needs for close, caring, and intimate relationships in their personal lives. Nevertheless, when moving-away clients continue to maintain emotional distance, many therapists—with very different relational needs—will also ultimately give up and disengage. When this occurs, the clients' eliciting behaviors have successfully defended them against relating to the therapist and thereby allowed them to avoid the threats entailed by closer involvement. Clients' conflicts will not be activated, but the therapist's only real vehicle for resolving these conflicts is closed as well. Rather than responding reflexively to these clients' aloofness by turning off, the therapist may use process comments to try and find other avenues for establishing a relationship. Beginning therapists must be patient with themselves and these clients, however, since it will be threatening initially for clients to relate in a different way.

To sum up, the therapist should attend to what clients systematically elicit in others and hypothesize how this interpersonal strategy protects them from their central conflicts. In doing this, the therapist must attend to the clients' developmental history and current life situation and to the cultural context in which these are embedded. More important, *the therapist must also attend to his or her own feelings and reactions that are elicited by clients.* This self-awareness is one of the therapist's most important sources of information about the client. If the therapist can observe what clients elicit, it will provide information about the nature of clients' conflicts, interpersonal defenses against their conflicts, and how the conflicts being played out in the therapeutic relationship may be impacting their lives outside the therapy setting also.

We have already seen that the therapist must generate working hypotheses to understand why clients elicit certain responses, such as reassurance, disinterest, competition, confusion, irritability, or nurturance. The therapist's task is to keep from responding reflexively to what clients elicit and, if possible, to try to respond in a way that does not reenact the clients' conflict. Over time, the therapist can begin to help clients *identify* their eliciting mechanisms, *recognize* the situations or types of interactions where they tend to employ these measures, and *understand* how these maneuvers have served to create safety in the past. It does take courage, however, for beginning therapists to tolerate their own bad or unacceptable feelings long enough to consider how they may be used to foster an understanding of clients' dynamics. The key element here, as in working with resistance, is for the therapist to remain nondefensive and accepting of his or her own internal reactions, no matter how unwanted, and reflect upon them—if necessary, in consultation with a supervisor—in order to glean the information that they contain.

Finally, the beginning therapist should be aware that significant problems arise when clients' eliciting behaviors tap into the therapist's own dynamics and immobilize the therapist. With this new development, we enter the arena of countertransference. For example, suppose that the client acts helpless and escalates her or his distress, and the therapist tries to help but fails at each attempt. Ideally, the therapist would remain involved and keep trying—for example, by making the process comment that whatever the therapist tries seems to disappoint the client. However, if the therapist's own personal history regarding performance demands, criticism, or adequacy concerns is activated, the therapist may lose his or her effectiveness. The therapist may feel overly responsible and inadequate, become angry or critical of the client, and emotionally withdraw, thereby recapitulating maladaptive relational patterns for the client.

Scenarios in which the clients' eliciting maneuvers tap into the counselor's countertransference propensities are commonplace. When this happens, clients often switch to the other side of their ambivalence and try to reengage the therapist, perhaps by reassuring the therapist about how helpful the sessions are. The clients do not really want to succeed in immobilizing the therapist or ending the relationship and will often try to get the disengaged or discouraged therapist involved in the relationship again. Premature and unnecessary terminations routinely occur when the therapist (and perhaps the supervisor) fails to hear clients' renewed bids for relationship because the disengaged therapist has given up on the relationship. Thus, therapists must keep clients' ambivalence in mind. Once clients' eliciting behaviors have succeeded in distancing or immobilizing the therapist, they subsequently try to reengage. Especially with more troubled clients, therapists must keep both of these engaging and disengaging maneuvers in mind.

It is essential for therapists to distinguish between their client's eliciting maneuvers and their own personal or countertransference reactions toward the client. This is sometimes referred to as *client-induced versus therapist-induced countertransference* (Springmann, 1986). How do counselors know if the client's eliciting maneuvers or their own personal issues are evoking their strong reactions? This is one of the most important questions therapists must answer; guidelines for making these critical distinctions will be given later in the chapter.

Testing Behavior

We have seen how clients employ their fixed interpersonal style to systematically elicit responses from others that often seem self-defeating but have served to effectively avoid anxiety in the past. These eliciting maneuvers—or security operations, as Sullivan (1968) sometimes called them—are interpersonal defenses.

Although interpersonal defenses are often used to avoid anxiety, at other times they are used to approach and try to resolve problems. While this may sound contradictory, it really reflects clients' ambivalence about treatment and what change might entail. Clients often carry out the approach side of their ambivalence about treatment by means of *testing behavior*. Clients systematically reenact their conflicts with the therapist, in some direct or metaphorical way, to test whether the therapist will respond in the familiar problematic way that others have in the past or whether the therapist can provide a new, more facilitating response that allows them to disconfirm pathogenic beliefs, expand their interpersonal range, and make progress on their problems. *Therapists should not underestimate how vitally interested clients are in assessing whether the therapist is going to reenact or resolve the old relational pattern.* Many of these clients have histories of betrayal, abandonment, and rejection; their trust of others has been severely damaged. Thus, their testing behaviors often reflect their intense wishes for healthier, safer relationships, as well as their fear of again being disappointed or hurt. As we will see, when therapists pass important tests, clients make progress in treatment, they have more confidence in themselves and in the therapist, their symptoms often decrease, and relevant new information is brought up. Most of the discussion that follows is based on aspects of control/mastery theory (Weiss, 1993; Weiss & Sampson, 1986).

In a healthy attempt to master conflicts, clients actively (but not consciously) elicit from the therapist the same conflicts they have had and are having with others in their lives. Although clients test to see if they can change old relational patterns and disconfirm the pathogenic beliefs that stem from these patterns, it is difficult for clients to find disconfirming evidence because repeated confirmation of the beliefs has made clients deaf to such evidence. It is only the immediate, in vivo relearning with the therapist that enables the client to relinquish deeply held core beliefs. The therapist can fail this test by responding in a way that reenacts old relational patterns, or the therapist can pass the test by behaviorally demonstrating that relationships can be different; this provides the client with a corrective emotional experience. Suppose that the client's pathogenic belief is: "If I become stronger and do what I want with my life, my parent will be hurt and withdraw, and I deserve to feel guilty." In that case, the therapist would fail the test and recapitulate the maladaptive relation pattern if she or he expressed a worry or concern about the result of a new venture that this overly inhibited client is considering. In contrast, the therapist would pass this test—and embolden the client—by expressing genuine interest in the venture and providing unambiguous support for the client's expansiveness.

Such *transference tests* pervade all counseling relationships—especially in the initial sessions, in situations where strong feelings are being evoked, and in circumstances where the current interaction between therapist and client closely parallels important relational conflicts that the client is having with others. Many of these tests are easy to recognize, and therapists can readily pass them and thereby allow clients to progress in treatment and move on to other issues. For example, suppose a compliant, dependent woman says to her male therapist, "Where should we start today?" The therapist will fail the test and reenact a familiar relational scenario if he says, "Tell me about _____." Here the therapist has confirmed the client's old pathogenic belief that she must assume the subservient role, that she ought not to assert herself or take the lead. This belief may reflect her own profound sense of inadequacy and incompetence— that she is incapable of knowing what is best for herself. The therapist will pass the test if he says, "I'd like to hear about what is most important to you. Let's begin with something that you would like to tell me about." This response demonstrates that he is capable of engaging with her in a more egalitarian way than the client has been able to do with other male figures in her life; it acknowl- edges the client as important and capable of deciding what she needs or wants to deal with. Repeated interactions of this sort will begin to bolster the client's self-esteem and sense of competence.

Tests that involve issues of power and control are more likely to prove challenging to beginning therapists. Clients with different developmental experi- ences will test the strength of the therapist. Before they feel safe expressing strong feelings or addressing threatening issues, clients usually need to assess whether the therapist can manage what they present. For example, when clients push limits with the therapist, they are testing whether they can control or manipulate the therapist—as they could their parent—or whether the therapist is stronger, can tolerate their disapproval and set limits, and can thereby provide the safety they did not receive before. Let's look at a case example.

Graduate students in Carl's clinical training program were supposed to videotape all therapy sessions. Early in treatment, however, Carl's client Lucy complained that the videotaping made her self-conscious. She said it was already hard for her to talk about what was really important and that the videotaping made an already difficult situation impossible. She continued to stress this concern for several weeks, and finally told Carl that he was not helping her because there were too many things that she could not talk about when the tape was on.

Carl was torn and did not know what to do. He knew the clinic rules, and that his supervisor would abide by them, but he also felt that he should be flexible enough to do what his client needed. Without consulting his supervisor, he reached a compromise with Lucy and agreed to turn the videotape off for the last five minutes of each session. Lucy was appreciative and did indeed disclose new material to Carl during those last five minutes. Carl felt badly about doing this without his supervisor's permission but also thought it was the only way to help his client and keep her in treatment.

Carl soon observed, however, that the therapy was not going well. Lucy became scattered and started jumping from topic to topic. Carl tried to pin her down and focus on her escalating feelings, but she slid away each time. Lucy communicated that she was very upset and needed more guidance from Carl, but whatever he did failed to help. Realizing that things were getting out of control, Carl decided to talk with his supervisor.

First, Carl and his supervisor spent some time honestly discussing what Carl's decision to turn off the tape meant for their relationship, and then his supervisor began helping Carl hypothesize what this test might mean about Lucy's dynamics. Assisted by some new ways of thinking about what was going on in the client-therapist relationship, Carl accepted his supervisor's firm insistence that the videotape remain on throughout every session. When Carl reinstated this limit at the beginning of the next session with Lucy, she protested strongly. She accused Carl of being untrustworthy because he broke their agreement, informed him that she could no longer talk about certain important things, and told him that she would have to think about whether she wanted to remain in treatment with him.

Carl felt badly about her threats, but he tolerated his discomfort and stuck to the rules. Although she continued to blame and complain, Lucy did return. In fact, despite her protests, she became calmer, more contained, and began to focus more productively in therapy. Over the next few weeks, Lucy went on to recall how disheartening it had been for her to be able to "boss my father around."

It is necessary and adaptive for clients like Lucy to express their conflicts and test whether the therapist will reenact or resolve them. Clients cannot progress to more conflicted, vulnerable material if the current therapeutic relationship is not safer than past relationships have been. There is limited value, however, in verbally reassuring clients. Moreover, most clients are not consciously aware of testing therapists in these ways, and it is not usually helpful to point out testing behavior to them. Instead, as in all human relations, it is what the therapist does that counts, rather than what the therapist says. Therapists pass tests and provide a corrective emotional experience when they consistently respond in ways that behaviorally disconfirm problematic relational patterns. Fortunately, even though all therapists will fail tests at times, as Carl did, they will have many opportunities to recover from failed tests or mistakes if they stay emotionally present in the relationship, attend closely to the process, and assess whether therapy is progressing or not. Therapists can respond more appropriately the next time this issue or theme comes up—often later in the same session or in the next session.

When the counselor passes a client's transference test and responds more effectively than certain others have in the past, the client experiences a new degree of safety in the therapeutic relationship, and begins to remember, feel, and share the developmental experiences that originally caused his or her pain.

How can counselors conceptualize the specific interpersonal response that will allow them to pass a particular client's tests and provide the client with a reparative experience? To determine what kind of response the client needs, therapists must (1) consider client response specificity and (2) learn how to assess the client's positive or negative reactions to different therapist responses and interventions. It is important to keep in mind that the client's cultural background will impact client presentation and response; accordingly, therapists need to clarify with the client the meaning of certain behaviors and responses.

Client response specificity   As discussed in Chapter 1, therapists must respond differently to different clients. The same therapeutic response that passes a test for one client may fail the test and reenact conflicts for a client with a different developmental background. For example, in response to pressure from her client, Beverly agreed to reduce her fee substantially. Almost immediately, her client began wearing expensive jewelry, delighted in telling Beverly about expensive shopping trips, and described how wonderful it was to fly first class on her vacation. Realizing that the client was taking advantage of her, just as the client had done with her doting mother and indulgent husband, Beverly renegotiated the fee. Later in the session, the client brought up important new material about feeling lonely and empty when she felt she had too much control over others, and entered a productive new phase of treatment. Beverly concluded that, by being firm, she had passed the client's test and provided her with the corrective response she needed. The client could only feel safe enough to move closer to these difficult feelings when she was sure that she could not manipulate Beverly as she could her mother, her husband, and others.

Now let's look at an example where lowering the fee was a helpful response. A client was coping with a diminished income following a recent divorce and was having trouble making ends meet. Knowing that this client grew up with demanding, withholding parents, the therapist realized that offering a more flexible fee arrangement to this client would be a significant overture. The client could not ask the therapist for a reduced fee directly and was initially reluctant to accept any of the alternative fee arrangements that the therapist suggested. When the therapist asked about her reluctance to even consider his offer, the client burst into tears. As their discussion soon revealed, the therapist's generous response disconfirmed the client's belief that she didn't matter and evoked the longing she had felt for her parents to reach out and respond to her. Changing the payment schedule with this client disconfirmed certain pathogenic beliefs about herself and relationships and moved the treatment into more intensive work about her emotional deprivation and her accompanying belief that she did not matter much to others. As these two examples illustrate, *the same response by the therapist will often have very different effects on different clients.*

Clients' ethnic backgrounds will further complicate response specificity. For example, people of color who have experienced the feeling that they don't matter in a broader political context will have an additional layer of issues to sort through. Moreover, racial, gender, religious, or other differences between therapist and client will affect how the client responds to the therapist's offer.

Client response specificity puts substantial demands on the therapist. It takes away the security of having a rule-based approach to therapy. It requires the therapist to decenter, enter the subjective worldview of the client, and remain close to the unique and personal experience of each client. However, within this highly idiographic context, some guidelines are available to help the therapist respond more accurately to the specific needs of differing clients. The therapist can learn to assess the specific relational experiences that each individual client needs by (1) identifying the problematic relational patterns that repeat in their current interaction; (2) observing the maladaptive relational scenarios and interpersonal themes that are presently recurring with significant others in the client's life; and (3) listening for and clarifying the repetitive transactional patterns in the client's family of origin that engendered these generic conflicts. In addition, the therapist also learns how to pass tests and provide clients with the specific relational experiences they need in order to change by assessing client reactions to therapeutic interventions.

Assessing client reactions   How do therapists know if they have passed or failed the client's test? Are there guidelines to assess whether the therapists' response has provided a corrective emotional experience or reenacted some aspect of the client's conflict? If clinical trainees rely only on their supervisor's advice, they will remain dependent and insecure. Further, they will have a false sense of security if they foreclose on a particular theoretical approach to avoid the anxieties and ambiguities inherent in finding their own identity as a therapist, but then they run the risk of becoming rigid and dogmatic. What better alternatives exist? Although supervisors and theories are of great help, beginning therapists achieve some legitimate autonomy by learning how to directly assess client responses in order to determine the effectiveness of particular therapeutic interventions.

Clients inform the observant therapist of the responses they need in treatment. They do not usually verbalize this, but they predictably guide the therapist by their behavioral responses to the therapist's interventions. Let's examine closely how this happens.

When the therapist responds effectively to a client's test and disconfirms problematic relational patterns, the client usually feels safer and begins to behave in new ways. For example, in the minutes after the therapist passes a test, the client may stop talking about other people and begin talking more openly or directly about his or her own personal experience. Perhaps the client will relate a vignette that is richer, clearer, or more meaningful than its predecessors. In general, the therapeutic process may acquire more immediacy or relevancy and become less intellectualized or repetitive. The client will often initiate a stronger, more direct interpersonal contact with the therapist, by expressing warm feelings toward the therapist or bringing up a problem he or she is having with the therapist. Further, the client's affect will often emerge or intensify. The client's feelings about what has happened—happy, sad, calm, or angry—become clearer and more accessible or are shared more fully with the therapist.

Therapists also know that they have responded correctly and passed the test when clients bring up new material and enter new conflict areas. In such moments, the client may also make useful connections between the current behavior and past relationships, without any therapist input, interpretation, or guidance. The insights that they generate in these circumstances are enlivening and hold enduring meaning for them. As a general rule of thumb, it is not usually productive to lead clients back to historical connections; however, significant behavioral changes result when clients spontaneously make these links on their own.

To put it simply, the therapist has passed the client's test when the client responds to the therapist's intervention with stronger behavior, more direct engagement in the work, or plans to try out a new behavior. For example, when Carl turned the videotape on for the full session, Lucy complained and even threatened to terminate. But by maintaining appropriate professional boundaries, Carl demonstrated that he was different from Lucy's father and was therefore safer. Carl could tolerate Lucy's disapproval and do what she actually needed rather than what she demanded. As a result, Lucy became less demanding, began to work more productively, and introduced relevant new material into her treatment.

If they identify and remember what they have done to pass (or fail) clients' tests, therapists will be able to successfully formulate case conceptualizations. These conceptualizations can then be used to guide subsequent interventions and to continue providing clients with the corrective emotional experiences they need in order to change. For example, at other anxiety-arousing points in treatment, Lucy will again push the limits with Carl. If Carl has learned from their previous interaction, however, he will be better prepared to provide the response Lucy needs.

In their behavioral responses to therapeutic interventions, clients also tell therapists when they have failed. On having failed a test, the therapist will often observe that clients become weaker—more anxious, dependent, confused, or distant. These reactions are often evident almost immediately (within the next minute or two) and may continue into subsequent sessions. Therapists should formulate working hypotheses about how their recent interaction may have reenacted the clients' conflict when (1) clients begin talking about others rather than themselves; (2) the therapeutic process becomes repetitive or intellectualized; or (3) clients become compliant, lose their initiative, or cannot find meaningful material to talk about.

We have just seen that when therapists pass tests (for example, take pleasure in the client's success rather than feeling threatened, envious, or competitive, as significant others have been), clients usually acknowledge the therapist's achievement by acting stronger and progressing in treatment. In addition, when a therapist repeatedly fails a specific test (for example, by brushing aside the potential validity of a client's criticism of the therapist, just as the client's parent always needed to be right), the client may inform the therapist that something is awry by recalling other relationships in which the same conflict was enacted. Thus, the client may begin to tell the therapist stories about other people the client has known or characters in books or movies who are enacting the same problematic relational pattern. *The therapist should always consider the possibility that, whenever the client is talking about another relationship, the client may be using this as a metaphor or encoded reference to describe what is going on between the therapist and the client* (Kahn, 1997; Kell & Mueller, 1966).

The therapist must listen for the relational themes that characterize the vignettes clients choose to relay. If clients repetitively tell stories in which, for example, trust is betrayed, someone is left, or control battles are being enacted, they are often expressing how these same conflicts are currently being enacted in the therapist-client relationship. When this happens, the therapist can use a process comment to make this relational theme overt and bring the conflict back into the immediacy of the therapeutic relationship—where something can be done to resolve it.

Therapist:
   Both of these people you have been telling me about failed to hear something important that others were trying to tell them. I'm wondering if something like that could be going on between you and me right now. Maybe you told me something earlier that was important to you that I didn't hear very well. Do any possibilities come to mind?

To sum up, just as clients use eliciting behaviors to avoid and defend against their conflicts, they also use transference tests to try and resolve conflicts in the therapeutic relationship. By tracking the therapeutic process in the ways suggested here, therapists can assess whether they are providing clients with the corrective relational experiences they need. This approach refines therapists' conceptualizations of clients' dynamics and gives therapists the flexibility they need to respond to the varying conflicts that clients present.

Transference Reactions

The third interpersonal factor for conceptualizing clients is *transference reactions*. Transference is one of the most important but misunderstood concepts in psychotherapy. Therapists use this term to convey very different meanings, and, over the years, competing definitions have evolved. Traditionally, transference has referred to the thoughts, feelings, and perceptions that the client has toward the therapist. This psychoanalytic model of transference is based on the notion of the therapist as a blank screen. Since the therapist is blank, whatever emotional reactions the client has toward the therapist are *distortions* that have been inappropriately transferred to the therapist and may represent old relationships and relational patterns or wished-for relationships and relational patterns. Especially during vulnerable or affect-laden moments, most clients do have repetitive or characteristic predispositions to distort or misperceive the therapist and others in certain systematic ways. These persistent distortions of the other (as wonderful, withholding, contemptuous, and so forth) are an enduring characteristic of the clients' inner life, repeatedly disrupt their close relationships, and are almost certain to be evoked toward the therapist at times. When these types of transference reactions occur, they systematically distort clients' perceptions of, and emotional responses to, the therapist.

A more useful definition of transference refers to *all* of the feelings, perceptions, and responses that the client has toward the therapist. These reactions, which may be positive or negative, range on a continuum from fully accurate and reality-based perceptions to highly distorted responses that are based on past relationships and/or wishes. They do not represent pathology but are clients' attempts to make sense of important life experiences. Thus, although traditional transference distortions certainly occur, the contemporary definition emphasizes that many of the client's reactions toward the therapist are reality-based, and almost all hold at least a kernel of truth . For example, the client's attraction to the counselor may be a genuine response to the very real and appropriate caring and trustworthiness the client feels in this relationship. Similarly, the client's anger or competitiveness may be based on the reality that the counselor is often late for sessions, preoccupied, emotionally unavailable, uncomprehending of what is most important to the client, or invested in the client making certain decisions regarding marriage, divorce, abortion, career, or child-rearing practices. Counselors make a significant error when they too readily attribute the client's positive or negative feelings toward them to traditional transference distortions. Before they do so, they should seriously examine the potential or partial accuracy of those feelings and consider what reasonable conjecture or understandable misperception might have arisen at the interface between the client's subjective worldview and the therapist's behavior.

As we will see, therapists can use transference reactions both to conceptualize client dynamics and to help clients resolve their conflicts within the therapeutic relationship. More specifically, the short-term dynamic therapies have encouraged an important change in how therapists work with transference reactions. Whereas transference reactions were previously used as a springboard to discuss historical relationships and the genesis of conflicts with parents and early caregivers, they are now used to understand and change the current interaction between the therapist and the client (Mills, Bauer, & Miar, 1989). Although some clients do indeed make meaningful connections to the past, it will usually be more fruitful for clients to first address and resolve these distortions or interpersonal conflicts in the real-life relationship with the therapist. In this approach, working with transference in the immediacy of the current relationship rather than through the safer and more distant approach of trying to interpret or lead the client back to historical relationships, the goal is to: (1) help the client recognize or become aware of repeated misperceptions of the counselor; (2) clarify the self-defeating aspect of the relational pattern and how it diminishes or disrupts the relationship; and (3) work together to replace this repetitive pattern with a more flexible and reality-based relationship.

In many cases, however, client perceptions, feelings, and responses toward the therapist accurately capture aspects of the interaction with the therapist. It is essential for the therapist to acknowledge the potential validity of the client's perceptions and to enter into a genuine dialogue with the client to clarify what each was thinking, feeling, and intending during any misunderstanding or disruption. If the therapist is willing to explore nondefensively what he or she may have done that activated the client's relational templates, the client feels empowered and is able to be more forthright and authentic in the relationship.

In the interpersonal process approach, the therapist recognizes the validity of many client reactions and tries to resolve actual distortions in the current relationship rather than focusing on the historical genesis of conflicts. Ironically, after the therapist and client have addressed and resolved a significant misperception or misunderstanding between them, the client will often go on to make meaningful historical connections between formative and current relationship patterns. Before going further with this clinical focus, let's become better acquainted with the everyday expressions of transference.

Transference in everyday life   Transference reactions are commonplace. We all systematically misperceive others as a result of overgeneralizing previous learning to the present situation. For example, imagine a male university professor who stands in front of a large lecture hall about to convene the first class meeting of Psychology 100. Before he has even begun to lecture, many students will be transferring expectations from significant male authority figures in their own lives onto the professor. Thus, in the first row of the lecture hall, 18-year-old Mary looks up to the professor admiringly. Just as she still somewhat idealizes her father, Mary respects the middle-aged professor, even though she actually knows little about him.

In contrast, Joe looks down on the professor disparagingly from high up in the last row of seats. Joe anticipates that the professor does not know as much as he thinks he does and that he will not be very well prepared for class. Just like his own father, who always had to be right and on top, this professor will probably try to get away with a lot in class. Joe has already decided that this know-it-all professor is not going to get away with much if Joe can help it.

Let's hope that the professor is aware that people tend to transfer expectations onto authority figures who hold power and onto those who have high public visibility (movie stars, sports heroes, and political leaders). If the professor is not sensitized to transference reactions, his self-confidence as a teacher will take a roller-coaster ride. One day, Mary will follow him out of class and tell him what an interesting lecture he gave; the next day, Joe will drop his books loudly when the professor starts to speak and demand that he defend half of what he says.

Transference reactions such as Mary's and Joe's are everyday occurrences. The more conflicted or disturbed one is, however, the more one distorts the current reality by inaccurately displacing reactions from the past onto the present. For example, Joe and Mary have strong initial reactions to the professor that are based on little real-life experience with him. If Joe and Mary function fairly well psychologically, their internal working models will be flexible and varied and they will become more realistic in their assessment of the professor as they have more exposure to his teaching. In contrast, the more thoroughly Mary idealizes her father and the more pervasively Joe disparages his father, the more likely it is that they will not be able to accurately perceive the strengths and limitations of the professor. Instead, their relational templates will be more fixed and self-fulfilling, and they will maintain their initial perceptions and expectations despite the fact that they are inaccurate or overdetermined.

Transference reactions are also more likely in emotion-laden situations. For example, transference reactions are especially pronounced when two people are falling in love. Many married couples recall their courtship as the happiest time in their lives; during this time, each member systematically misperceives the partner as someone who can fulfill unmet developmental needs and make one feel lovable, capable, or worthy when one has not felt this way before. This romantic period is usually short-lived, however. The transference projections soon break down as they both realize that the other person cannot fulfill all of their unmet needs. If the unmet needs and resulting transference distortions are especially strong, the result can be dashed dreams and angry feelings of betrayal. In contrast, if the transference distortions of the partner have not been extreme, they can readily be shed, and a more realistic and enduring relationship can develop. Relinquishing these transference distortions of the new spouse is one of the major psychological tasks of early marriage.

In therapy, the therapist can systematically utilize the client's transference reactions to better understand the client's conflicts. One of the most effective ways to discern the client's internal working models—and their interpersonal expression in maladaptive relational patterns—is to note how the client perceives and reacts to the therapist. Almost without exception, aspects of the client's maladaptive relational patterns that cause problems with others will be reenacted in the therapeutic relationship. If the therapist tracks these transference reactions, they provide an invaluable tool for conceptualizing the client's conflicts, and changing the problematic interpersonal process that is being reenacted.

Using transference reactions to conceptualize clients   Some therapists are uninterested in the client's transference reactions because they do not see these reactions occurring in therapy. The therapist will often miss the client's transference reactions, and the important information they provide, unless the therapist actively attempts to draw them out. How can the therapist highlight the client's transference reactions? Let's look at two ways of making the client's transference distortions overt in such a manner that the client can comfortably recognize them and join with the therapist in a collaborative effort to understand them.

First, the therapist should find tactful but direct ways to inquire about the client's feelings and reactions toward the therapist. As we saw when discussing resistance, it is important for both therapist and client to learn about the client's reactions to being a client and seeking help. Further along in treatment, it is even more important to learn about the client's perceptions of, and feelings toward, the therapist—especially in the event of strong emotions or conflicts between the therapist and the client. The sample questions that follow can provide a wealth of new information if they are well-timed and integrated into the ongoing discussion:

> As you were driving to our session today, how did you feel about coming to see me?
> When you find yourself thinking about therapy, what kind of thoughts do you have about me?
> What are you feeling toward me right now as we talk about this sensitive issue?

Second, the therapist should also explore what the client projects onto the therapist—what the client believes the therapist is thinking, expecting, or feeling toward the client at that moment:

What do you think I am feeling toward you as you tell me that?
How do you think I am going to respond to you if you do that?
What do you think I expect you to do in that situation?

It is often uncomfortable for the beginning therapist to break ingrained familial and cultural norms and invite the client's feelings and perceptions so forthrightly. The therapist must find ways to be direct, yet sensitive and respectful, without being demanding or intrusive. If therapists take this risk, clients' transference reactions will be made overt. The issues that are revealed often surprise both the therapist and the client:

I think you are feeling disappointed in me.
I think you'll act nice, but you really wish I would stop coming.
I think that you believe I should get divorced.

Once these misperceptions have been made overt, therapists can correct them and clarify their own responses to the client; this often provides the client with a corrective emotional experience. When transference reactions are revealed in this way, therapists have the opportunity to respond to the client with an immediacy and authenticity that is not often found in other relationships and is powerful and highly engaging for both.

Most beginning therapists find it difficult to believe that their clients have significant feelings toward them that are not based on reality. Seasoned therapists have become familiar with clients' distortions and comfortably expect them, but beginning counselors usually find it hard to accept that their clients have unspoken fears, concerns, and wishes toward them. If unaddressed, however, such client reactions will influence the therapeutic relationship in significant but covert ways.

Exploring clients' subjective reactions toward the therapist reveals three important types of information. First, just as adopting an internal focus will uncover new conflicts that would not have surfaced otherwise (Chapter 4), exploring clients' reactions toward the therapist will also reveal new issues—central to the clients' concerns—that need to be addressed in treatment. Detailed examples of this important intervention will be given in Chapters 9 and 10. Second, exploring transference will also show how aspects of the same problems that clients are having with others—and that they originally experienced in formative family relationships—are being reenacted with the therapist. Unless the clients' perceptions of, and reactions toward, the therapist are directly solicited, these reenactments are far less likely to be revealed, addressed, and resolved. Thus, transference reactions reveal clients' generic conflicts and, as we will explore further in the next chapter, also provide an opportunity to resolve them in the therapeutic relationship.

Third, transference reactions provide evidence that the clients' feelings are trustworthy. Because clients' emotional reactions may not fit the situation in which they are aroused, they may not make sense at first to the clients or others. As a result, many clients experience their emotional responses toward spouses, children, employers, and others as irrational. Consequently, they do not trust their own internal processes or the validity of their own experience, and their self-efficacy is significantly diminished. This also occurs in the therapeutic relationship when clients have emotional reactions toward the therapist that seem uncalled for, exaggerated, inappropriate, or unrealistic. However, clients' emotional reactions are not irrational. Though they may not fully fit the present circumstances, they do make sense when they can be placed within their original context; real-life events have legitimately caused clients to feel the way they do. If the therapist helps clients trace the feelings back to their source (by asking, for example, "When is the first time you can remember feeling this way? Who were you with and what was happening?"), they will prove to be an *appropriate* reaction to someone else in another time and place. Once the therapist and client have been able to clarify the transference distortion in their current interaction, this developmental link can often be made readily and easily. In contrast, the developmental context will not come alive or be meaningful to the client if it is not first resolved in the current interaction with the therapist. Thus, even though this may seem paradoxical at first, transference reactions ultimately reveal that the client's feelings always make sense.

To sum up, clients transfer emotional reactions from past formative relationships onto the current relationship with the therapist. First, therapists must clarify and resolve these misperceptions in the current relationship with the therapist:

Therapist:
   No, I'm not thinking you're selfish and demanding if you ask him—or me, for that matter—for what you want.

                                        OR

Therapist:
   No, I'm not tearing because I'm burdened by your needs. I'm sad about how much you've been hurt.

This gives clients a real-life experience of change; they learn that their inaccurate expectations or problematic working models do not fit in this relationship. As they experience another way to be in a relationship, clients are often able to make meaningful connections to past experiences. By themselves, without historical interpretations offered by the therapist, clients often begin to understand the source of their emotional reactions and find that their emotional lives do make sense and have validity. This powerful experiential sequence leaves clients feeling affirmed and in control of themselves and of their emotional reactions—perhaps for the first time in their lives. When clients find that their feelings do make sense, even though in another time and place, they feel validated and empowered. By resolving the transference distortion in the real-life relationship with the therapist, and then making meaningful connections between the past and the present, clients are able to leave the past behind, and new ways of responding can emerge.

Optimum Interpersonal Balance

In understanding clients' interpersonal functioning in the therapeutic relationship, we need to apply the separateness-relatedness dialectic (Chapter 6). In order to establish a relationship that is capable of producing change, therapists must have an effective degree of involvement with their clients. If the therapist is inappropriately close, it will not be safe for clients to take the risk of letting the therapist influence them. Similarly, if the therapist is too distant or removed, a corrective emotional experience cannot occur, because the relationship is too insignificant to effect change. Thus, throughout treatment, therapists must monitor the balance of separateness versus relatedness in the therapeutic relationship. At times, all therapists will have difficulty maintaining an optimum interpersonal balance with some clients. The therapist must be both a genuine participant in the relationship and an objective observer of it; Sullivan (1968) captures this double role in his enduring term *participant-observer*, while more recent discussions speak of *empathy* and therapist-client *boundaries* (Jordan et al., 1991). Combining these emotional and cognitive components is no easy task for beginning therapists.

Enmeshment

At times, all therapists will become inappropriately overidentified—enmeshed—with certain clients who tap into the therapists' own issues and dynamics. Moreover, certain therapists, because of their own personality dynamics, will consistently become overinvolved with clients. How can therapists recognize when this traditional countertransference is occurring? Whenever therapists find that they are angry or critical of a client for not changing, they are overly invested in the client. Usually, therapists are also enmeshed when they have dreams about a client, often think about a client outside of the therapy session, feel depressed when a client is not changing, see the client as being just like themselves, or become envious or elated over positive changes in the client's life. This overinvolvement is different from genuine care and concern for the client, which is appropriate and necessary. When enmeshment occurs, therapists usually need the client to change in order to meet their own needs—for example, to shore up their own feelings of adequacy or to manage conflicts of their own that resemble the client's.

Substantial problems occur when therapists become overinvolved with a client. When they do so, they lose sight of the fact that the client has his or her own subjective worldview, shaped by many factors that actually differ from those of the therapist. They tend to become controlling of the client and to inappropriately project their own personal conflicts and solutions onto the client. This is likely to recapitulate the client's conflict in the therapeutic relationship. For example, many clients have grown up in families with overly controlling or intrusive parents. When the therapist's own experiential range is too limited to encourage clients' own autonomy, the therapist is responding in the same problematic way as the parents. This reenactment will prevent clients from being able to resolve their conflicts around intimacy and control. Thus, the basic reason for paying attention to countertransference is to avoid treating clients in the same ways that originally caused their problems.

Further, when therapists become too close or overidentified, they lose their ability to think objectively about the client. Therapists' heightened emotional reactivity to the client will cloud their understanding of the client. Instead of accurate empathy, therapists will offer a global, undifferentiated sympathy. Most importantly, therapists will fail to discriminate the differences between themselves and the client and will begin to see the client as just like themselves. Although this closeness or identification may be reassuring to clients at first, little enduring change occurs in an enmeshed relationship.

Disengagement

All therapists will hold themselves too far away from certain clients at times—usually when the client arouses the therapists' own personality conflicts. Also, some therapists will consistently be too removed from their clients, because of their own enduring personality conflicts and interpersonal coping style. This distance renders therapy ineffective. When therapists are not emotionally available to the client, therapy loses its intensity and becomes intellectualized. Therapists lose their intuition, empathy, and creativity. When this occurs, the client does not have the supportive relationship or holding environment necessary to explore conflicted emotions. Psychotherapy is more than an intellectual discourse, and enduring personality change will not occur in an emotionally disengaged relationship. Without the impetus provided by genuine personal involvement, a corrective emotional experience will not occur.

A therapeutic relationship in which the therapist is too distant or uninvolved will also recapitulate many clients' generic conflicts. For example, for clients whose aloof, self-centered, or authoritarian parents could not meet their childhood needs for emotional contact and support, a distant therapist might be frustrating but ultimately safe, because their anxieties over their insecure attachments—and their conflicts associated with being close to others—will not be activated in the therapeutic relationship. Unfortunately, just as the client's conflicts are not aroused in a disengaged relationship, the opportunity to resolve these conflicts is lost as well. Thus, when therapists defend themselves against their anxieties that are aroused by an authentic relationship and genuine emotional involvement with clients, many clients will not be able to change. Novice clinicians who are consistently unwilling to risk genuine involvement with clients are much less likely to become effective therapists than are counselors who are willing to risk a real relationship but tend to become overinvolved at times.

Effective Involvement

In order to be effective, the therapist must find an optimum middle ground of involvement. Maintaining this effective middle ground of involvement throughout the course of therapy is one of the most challenging aspects of psychotherapy, for two basic reasons. First, therapists are repeatedly confronted with anxiety-arousing issues in their own lives by the affect-laden material that clients present. All therapists must become aware of how they tend to respond when clients arouse their own anxieties. For example, some therapists will become passive or withdraw from the client; some will distance themselves by intellectualizing the relationship—for example, by interpreting the client's affect before the client fully experiences it; others become controlling and overly directive and will tell clients what they should do in a particular situation. Therapists must learn to hold these countertransference propensities in check.

A useful exercise for beginning therapists is to write down or role-play their initial reaction propensities in different anxiety-arousing situations—for instance, situations in which an angry client is critical of the therapist's abilities, a dependent client asks to sit on the therapist's lap, or a competitive client questions whether the therapist is really bright enough to be of help. In small discussion groups of three or four, trainees can provide each other with feedback and share their observations of the other therapists' responses in these challenging situations.

Second, the balance between overinvolvement and underinvolvement is difficult because clients often try to move the therapist along this continuum of involvement, as an interpersonal strategy to defend against their conflicts. When the therapist has been responding effectively, clients may become threatened by the anxiety-arousing material that begins to emerge for them. To defend against the difficult emotions that are being aroused by the therapist's effectiveness, clients may try to create either distance or enmeshment in the therapeutic relationship. Thus, clients may try to bore the therapist or push the therapist away by talking about issues they are not really concerned about, or they may try to elicit too much involvement from the therapist, by exaggerating their distress and escalating their demands for help. Thus, forces from within the therapist's own dynamics and from the client's defensive eliciting behaviors may conspire to pressure the therapist away from an effective middle ground of involvement.

If they can maintain this middle ground of involvement, therapists will be more dependable and provide a more even therapy, with no stormy ups and downs or lengthy impasses of emotional disengagement. It is also safer for the client to be vulnerable and trust when the therapist is consistently concerned and available, without swinging to either side—without becoming controlling or overreactive, on the one hand, or unresponsive, on the other. If the therapist maintains an optimum interpersonal balance, the client does not have to be concerned about disappointing an òverinvolved therapist and does not have to elicit the attention or concern of an uninvolved therapist. It seems that only the attachment researchers have appreciated the importance of consistency of caregiver response and the development of security and trust.

The essential balance is for therapists to maintain their own personal boundaries while still being available and responsive to the client. They need to fully enter the client's subjective worldview—to be fully present and connected both emotionally and cognitively—but still maintain appropriate self-boundaries and self/other differentiation. Regardless of the therapist's theoretical orientation, therapy is usually successful when the therapist can maintain this optimum interpersonal involvement. Conversely, therapy usually reaches an impasse—and ends before the client's work is complete—when the therapist and client become enmeshed or do not emotionally connect with each other.

To help therapists recognize the process they are enacting with their clients and assess their balance of involvement with clients, therapists should consider the following questions after every session:

1. What are my feelings and reactions toward this client?
2. How do my feelings and reactions parallel those of significant others in the client's life?
3. What is the client doing to arouse such feelings and reactions in me, and how do these behaviors tie into my conceptualization of the client's generic conflict and interpersonal coping strategies?
4. Do my feelings typify my reactions to significant others in my life and/or to other clients, or are they more confined to this particular client?

By writing answers to these questions after each session, therapists ensure that they are attending to the process dimension. Asking themselves such questions will also help therapists to distinguish their own countertransference reactions toward the client—therapist-induced countertransference reactions—from the client's eliciting behaviors and interpersonal coping style—client-induced countertransference reactions. Beyond these four questions, beginning therapists are again encouraged to utilize the guidelines for writing process notes in Appendix A after each session.

As we have already noted, all therapists will sometimes become overinvolved with a client. Indeed many theorists have described the therapeutic process as the therapist's ability to become immersed in clients' modes of relating and then to work their way out (Gill & Muslin, 1976; Levenson, 1982; Mitchell, 1988). However, beginning therapists are especially prone to become overidentified with their clients and to perceive their clients' problems and lives as fundamentally the same as their own. This is the most common way in which clients' conflicts come to be reenacted in the therapeutic process. Beginning therapists often feel discouraged when their supervisors point out repeatedly that they are metaphorically reenacting the client's conflict in the therapeutic relationship. Beginning therapists must be patient with themselves and accept that it is difficult to maintain the effective middle ground of involvement that safeguards against recapitulating the client's conflict. By gaining more clinical experience, examining their own experiential range, and tracking the separateness-relatedness dialectic in the therapeutic relationship, beginning therapists will become increasingly capable of maintaining an optimum interpersonal balance.

When therapists realize that they have become overinvolved or underinvolved with a client and find that they cannot realign the relationship on their own, they must consult with a colleague or supervisor. In fact, the most productive use of supervision is to help therapists regain control of their own emotional reactions so that they can reestablish an effective degree of involvement. Maintaining this optimum interpersonal balance is the best way to keep from recapitulating the client's conflict in the therapeutic relationship.

The Ambivalent Nature of Conflict

The Two Sides of Clients' Conflicts

In conceptualizing client dynamics, therapists must recognize the ambivalent nature of conflict. There are two sides to the client's generic conflict, and therapists must respond to both sides before problems can be resolved. Too often, however, therapists fail to recognize both sides of the conflict that underlies clients' problems and only respond to one side of the conflict. Therapists can better understand their clients' problems when they can identify the two conflicting needs, wishes, or impulses that comprise the conflict.

Colloquial expressions convey the two-sided nature of conflict. We often say that someone is "stuck between a rock and a hard place," that someone has grabbed "both horns of the dilemma," or that a difficult relationship is "like a double-edged sword." These expressions capture the push-pull nature of conflict—that you are "damned if you do and damned if you don't." For example, therapists often fail to appreciate that most clients have concerns both about being hurt by the therapist and others in the ways they were hurt originally and about hurting others. The field has well-developed concepts regarding clients' concerns about being hurt, ignored, rejected, shamed, abandoned, or controlled by the therapist. However, less attention has been paid to clients' concerns about hurting others by getting stronger, succeeding, finding their own voice, acting on their own interests, advocating on their own behalf, or relinquishing problematic familial roles (such as parentified, perfect, bad hero or irresponsible child). When clients improve in therapy or successfully pursue their own interests in life (by graduating from college, receiving a promotion, or having a satisfying love relationship), many will retreat from these success experiences by sabotaging themselves or simply not be able to sustain their progress. Often, they will become depressed as a result of separation guilt and/or survivor guilt or become anxious as success threatens their object ties and/or family loyalty.

To highlight the two-sided or double-binding nature of many clients' enduring conflicts, let's examine these elements in two previous case studies. As we will see, once the therapist is able to recognize the two-sided structure of the conflict and articulate the ambivalent feelings that result from the client's bind, the client's seemingly irrational and self-defeating behavior often becomes more understandable and resolvable.

Recall Anna, the young adult client who was discussing her dream with her therapist at the end of Chapter 4. Anna wanted to end her chronic depression, become more outgoing and involved with friends, and find a career path that she could pursue. Although she had expressed these wishes for many years, she could not follow through and realize them. She rarely initiated activities with peers and often declined the invitations she received from others. She repeatedly enrolled in college but would eventually lose interest and drop out after one or two semesters. Without recognizing the two-sided conflict that immobilized Anna, her behavior seems irrational.

Anna was immobilized between two conflicting needs. On the one hand, she wanted to grow up and have her own life, friends, career, independence, and identity. But, at the same time, she felt a need to be close to her mother and to feel her mother's continuing support and availability. Anna's repetitive relational scenario is that her mother will withdraw, feel lonely and hurt, and induce guilt when she starts to pursue her own goals and succeed with friends, in school, or with life in general. Unable to break this tie that binds, Anna repeatedly fails and has to return home, thereby continuing the lifelong, repetitive pattern in her attachment history. Only upon her return home does her mother again become available to her. Having to sacrifice their own independence to secure the support they need, clients like Anna become depressed over their inability to do what they want and experience internalized anger over such binding control.

After a few months, the same cycle will be reenacted again. Anna will start to act on her own need for autonomy, seek peer contacts, and reenroll in school. As before, however, her mother will withdraw, and Anna will not be able to individuate without her mother's support. In a few months, she will return home feeling hopeless and a failure. Completing the cycle, the mother will again become comforting and supportive of Anna in her defeat. This type of cyclic transactional pattern can continue for decades, until both sides of the client's conflict are made overt and worked through in therapy.

Without understanding these family dynamics (discussed in Haley, 1980), the therapist and others might conclude that Anna is lazy, sick, masochistic, unmotivated, or receiving too many secondary gains from being home and depressed. As we see, however, such pejorative labels are inappropriate. Clients' contradictory behavior does make sense once the two-sided, push-pull, or double-binding nature of their conflict is recognized.

Now recall Jean, the client in Chapter 5 who began to change when she could experience her conflicted feelings with the therapist. Until we recognize her underlying conflict, Jean's behavior also seems irrational and self-defeating. Jean's primary symptom was her continuous involvement with selfish, dominating, and demeaning men. Jean complained that she felt unhappy, used, and afraid in these unsupportive relationships, but she was unable to disengage from them. Following her relational templates, she remained in the relationship no matter how bad things got for her. Paradoxically, when the man decided to leave her, she would feel panicked, rather than relieved, and would quickly become reinvolved in a similar relationship. This pattern had continued throughout her 20s and 30s, and finally led her to seek help for her low self-esteem, chronic anxiety and depression, and budding agoraphobia.

When Jean entered therapy, she was describing herself as masochistic and stupid. Jean's friends also described her as crazy and hopeless for continually reinvolving herself in the same type of destructive relationship. As with Anna, however, Jean's seemingly irrational and self-destructive behavior makes sense once the two-sided nature of her conflict is recognized.

For years, Jean had said she wanted to be able to meet her own needs, set limits, say no sometimes, and leave relationships that were not supportive. But, year after year, she continued to meet a man's needs at the expense of her own well-being. She could not say no and set self-preservative limits, even though she had learned the skills to do so in assertion training classes. Why? A countervailing force covertly opposed all of Jean's attempts to change. As soon as Jean started to act on her own needs, set limits, or try to leave a relationship, she felt cold, small, and alone. In other words, she was gripped by intense separation anxieties whenever she made any movement toward her own individuation. Her only means of abating these separation anxieties and feeling related to others again was to retreat to the familiar dependent/victim role. Thus, her symptoms preserved attachment ties, and improving in therapy evoked the pain and distress of intensely insecure attachments.

This paralyzing conflict developed in Jean's family of origin. Growing up, Jean was an extremely parentified child. She was expected to respond only to her parents' needs, and she was harshly rebuffed or painfully shamed for expressing her own "selfish will." Throughout her childhood, she received so much criticism from both of her parents, and so little emotional support, that she desperately needed whatever thin threads of approval she could earn by devoting herself to them. Afraid of receiving further rejection from her parents, she was eager to serve them and comply with whatever they asked of her. Thus, as such an insecurely attached child, Jean became a captive of her own need for relatedness, which she could neither relinquish nor fulfill. By meeting the caregivers' needs, Jean and other parentified children take responsibility for the parent-child relationship; this interpersonal coping strategy is aimed at fashioning a more secure attachment. Such offspring, laboring under the myth of being able to control the attachment tie, often struggle with control conflicts in adulthood; they commonly experience eating disorders.

Further, to maintain some minimal sense of relatedness, Jean *identified* with her parents' critical rejection of her. She internalized her parents' contempt for her and began hating herself for being bad and selfish. (See component 3 in Figure 7.1.) As an adult, she recapitulated the feelings and experiences of her childhood, by selecting punitive, selfish men who confirmed and fostered her own internalized feelings of self-loathing and worthlessness.

With this developmental background in mind, we can appreciate how Jean's seemingly irrational behavior does in fact make sense. Her only sense of herself as a person—and her only means of being in a relationship—is to meet another person's needs. When she finally becomes depleted by this role and begins to respond to her own needs (set limits, ask for what she wants, leave a hurtful relationship), intense anxiety results. As in her childhood, she is overwhelmed by separation anxieties, fear of being shamed when her attachment disruption is exposed, and self-hatred. Lacking any better working models, she tries to diminish these threats by retreating to frustrating relationships and reenacting the same pattern. She again responds to her boyfriend's demeaning demands and suffocates her own needs, feelings, and life. The same cycle will soon start again, however, as her frustration and resentment over having her own life denied begin to reemerge. She will start to respond to herself and her own needs, only to be immobilized again by the same separation anxieties and guilt.

A cyclic dynamic such as this is difficult to change. As Jean illustrates, however, change is possible if two things occur. First, Jean experienced a relationship with the therapist in which her own needs were listened to, responded to, and valued for the first time. This was difficult to provide, however, because Jean was adept at rejecting the care and concern that she desperately wanted from the therapist. In particular, she tended to elicit the same feelings of helplessness in the therapist that her parents had originally engendered in her. Second, the therapist was also able to help Jean understand the two-sided structure of the

the conflict that had bewildered and immobilized her throughout her life. Understanding the conflict helped Jean change by allowing her to feel more compassionate toward herself and empathic toward her own predicament and hence to feel more accepting of the choices and compromises that she had been forced to make throughout her life in order to survive. This acceptance, in turn, helped her to disengage from her self-hatred and value herself and her own needs more than she had ever been able to do before. These more accepting responses from the therapist led to important internal changes for Jean and, after several months, she became involved with a supportive and responsive man for the first time.

Ambivalent Feelings

As we have seen, the client's conflict includes two opposing forces that the client has not been able to reconcile. Therapists cannot conceptualize the client's dynamics until they recognize its two-sided structure. In addition to framing the client's problem as a conflict, therapists must be able to respond affirmingly to the opposing feelings that accompany both sides of the client's conflict. Therapists are ineffective if they only respond to one half of the client's ambivalent feelings, as the following case vignette illustrates.

Marie, a depressed, 25-year-old graduate student, entered time-limited (12-session) treatment at the student counseling center. During the first few sessions, Marie discovered how angry she was at her mother. Her mother always expected Marie to be perfect, insisted that Marie never had any problems, and always needed Marie to be happy. In therapy, Marie began to realize how much she resented her mother for denying so many of her true feelings and for having so many expectations of how she should be. Marie's therapist resonated with her new awareness and actively supported her anger.

At first, it was liberating for Marie to be able to express her long- suppressed anger. She was both excited and relieved to realize that she no longer needed to fulfill her mother's unrealistic expectations. This good feeling was short-lived, however. In the next few sessions, Marie's new-found freedom gave way to a growing sense of futility and pessimism. The therapist thought that Marie must be feeling guilty about defying her mother's expectations and depressed because she was defensively turning the anger she felt toward her mother back on herself.

While repeatedly offering this interpretation, the therapist continued to draw out and encourage Marie's anger. Moving in a different direction than the therapist, however, Marie began to share fond memories about her mother and recalled with longing special times they had spent together making cookies after school. The therapist was not very responsive to Marie's positive feelings toward her mother, however. The therapist thought that Marie's positive feelings were part of denying her anger toward her mother, avoiding the conflicts in their relationship, and continuing to idealize her mother and thereby remain committed to meeting her mother's demanding expectations.

In contrast to Peter's therapist in Chapter 7, Marie's therapist only recognized one side of her ambivalence and failed to respond to the feelings on both sides of her conflict. Yes, Marie was angry at her mother, and these suppressed feelings needed to be explored, expressed, and validated. At the same time, Marie knew that her mother had offered her many fine things as a parent and she did not want to forgo the good things they had shared. Marie's conflict was that she wanted to keep the good things she had experienced with her mother without having to internalize the problematic aspects of their relationship as well. Thus, on the one hand, Marie wanted to reject the unrealistic demands to be perfect and happy that her mother placed on her. But, on the other, Marie became depressed when it seemed that the only way to do this was to give up her identification with her mother altogether. Marie did not want to risk losing her most important female role model and the good parts of this relationship as well.

Realizing that time was short and therapy was stagnating, her therapist sought consultation from a colleague. The colleague quickly recognized both sides of Marie's conflict and encouraged the therapist to allow Marie to simultaneously have contradictory feelings of appreciation and anger toward her mother. As a result of this helpful consultation, the therapist began to directly invite and actively affirm the other side of Marie's feelings as well.

Therapist:
You feel two different ways toward your mother at the same time. You're angry at her for expecting you to be perfect, but you also treasure many of the fond times you've had together. Tell me about both your angry and your loving feelings toward her.

As the therapist encouraged Marie to explore and express both sides of her feelings, her depression improved. During their final sessions, Marie began to clarify what aspects of her mother she wanted to keep and make a part of herself and what aspects of her mother and her mother's expectations she wanted to discard. This process of differentiation also allowed Marie to clarify which issues she wanted to confront and try to change in her current relationship with her mother and which issues she preferred to let go of and simply accommodate herself to. Marie progressed in therapy as soon as the therapist allowed her to come to terms with both sides of her ambivalent feelings. Therapy was success-fully terminated at the 12th session as Marie better integrated aspects of her own identity as a woman and initiated a rapprochement with her mother.

To summarize, clients change when the therapist can conceptualize and articulate the two-sided structure of their conflicts and help clients explore and integrate both sides of their ambivalent feelings. Trained to look for pathology, therapists tend to miss positive aspects. However, clients will make far more progress when the therapist can allow clients to keep the good parts of the parental or spousal relationship while they are complaining about the problematic parts. The therapist can tell clients explicitly that they are not being disloyal if they talk about problems with a parent or spouse and that the therapist knows that their complaints do not reflect all of their feelings about this person.

Just as therapists in individual practice must affirm both sides of the client's conflict, couple therapists must affirm both sides in an interpersonal dispute. Their primary task is to resist the couple's eliciting maneuvers and keep from *taking sides* by participating in the good guy/bad guy, overadequate/inadequate, healthy/neurotic, or other polarizations that the couple presents. When couple counseling fails, one member of the couple usually stops treatment because she or he accurately perceives that the therapist has taken sides in the marital conflict.

Closing

In this chapter, we examined three dimensions of clients' current interpersonal functioning: how clients' conflicts are brought into the therapeutic relationship through eliciting maneuvers, testing behaviors, and transference reactions; how therapists can establish an optimum degree of interpersonal relatedness and enact a corrective balance of separateness-relatedness; and the two-sided structure of clients' conflicts. This discussion may help therapists better understand clients' self-defeating or contradictory behavior and their seemingly irrational feelings. In turn, this greater understanding may enable therapists to be more compassionate toward the conflicts with which clients struggle. These interpersonal factors, together with the developmental/familial perspective and interpersonal schema presented in the previous two chapters, will enable therapists to better understand and intervene with the problems clients present.

With these conceptual guidelines in mind, we now return to intervention issues. The next chapter examines how therapists can use the interpersonal process they enact with clients to begin resolving clients' problems in the therapeutic relationship. If they are to utilize process-oriented interventions or other aspects of the therapeutic relationship to effect change, therapists must be able to talk directly with clients about their current interaction or relationship. Working with the process dimension—sometimes known as metacommunicating, or "talking about you and me"—is essential to all forms of interpersonal or relationally oriented psychotherapy but often feels uncomfortable for beginning therapists initially. Thus, in Chapter 9, we explore therapists' concerns about working with the process dimension, and look at guidelines to help counselors intervene more effectively within the therapeutic relationship.

Suggestions for Further Reading

1. Sullivan originally emphasized how clients' anxiety served to inform the therapist about their conflicts. His classic primer on interviewing, *The Psychiatric Interview* (New York: Norton, 1970), is still an illuminating clinical guide. Sullivan's interpersonal theory is presented in *The Interpersonal Theory of Psychiatry* (New York: Norton, 1968). Both of these works are recommended to interpersonally oriented therapists.

2. Alice Miller's *Prisoners of Childhood* (New York: Basic Books, 1981) helps sensitize therapists to the developmental arrests and internalized conflicts that interfere with adult functioning.

3. Chapter 2 of Jay Haley's *Leaving Home* (New York: McGraw-Hill, 1980) elucidates the family dynamics and structural family relationships that impede individuation and emancipation for many young adult clients.

4. Readers are encouraged to learn more about clients' testing behavior— and the interesting theory of control/mastery—by reading Chapters 4–6 of *The Psychoanalytic Process,* by J. Weiss and H. Sampson (New York: Guilford Press, 1986). A popular account of control/mastery theory and the often unrecognized role of unconscious guilt in many clients' symptoms and problems may be found in *Hidden Guilt* by L. Engel and T. Ferguson (New York: Pocket Books, 1990), which will be helpful for many guilt-ridden clients.

5. Aspects of Margaret Mahler's developmental theory—in particular, her model of parent-toddler interaction—can help therapists establish an optimum degree of interpersonal relatedness with their clients. Readers are encouraged to read Chapter 6, "Rapprochement," in *The Psychological Birth of the Human Infant: Symbiosis and Individuation* by M. Mahler, F. Pine, and A. Bergman (New York: Basic Books, 1975).

6. Family systems concepts of cohesion and adaptability can also be used to describe an optimal degree of interpersonal relatedness. Interested readers may examine *Families: What Makes Them Work* by D. H. Olsen et al. (Newbury Park, CA: Sage, 1983).

PART FOUR

Resolution and Change

CHAPTER NINE
An Interpersonal Solution

CHAPTER TEN
Working Through and Termination

An Interpersonal Solution

Conceptual Overview

Once the therapist has conceptualized the client's dynamics, how does change occur? Are the client's problems resolved through an emotional catharsis of childhood traumas, from having unmet developmental needs fulfilled, or from delving into illuminating insights about the past? No. Therapists do obtain essential information for case conceptualizations and treatment plans from understanding familial relationships and developmental experiences. Also, clients need to develop their own narrative or framework that allows them to make sense of the influences and events that shaped their development and who they have become. Unless they do so, it will be harder for them to guide where they are going in the future and to maintain changes after treatment has stopped, and they may start to stray from what they have learned. However, a consistent focus on the past will not produce change and may even serve to defend clients against the anxiety of confronting current conflicts. Certainly, information on the problematic relational patterns that clients experienced in the past does help both the therapist and clients to understand how faulty relational templates have developed, and why current relationships are being shaped in a particular way. Resolution must be found in the current relationship with the therapist, however. The primary reason for exploring developmental and familial relationships is to throw light on clients' coping strategies and characteristic ways of responding in current relationships. In other words, understanding formative relationships is a necessary component of change but is in itself insufficient to produce enduring change. Enduring change will only come about when clients enact a resolution of their conflicts in their current interaction with the therapist and then are helped to generalize this experience of change to others.

If the primary focus of therapy and fulcrum of change is the current interaction between the therapist and the client, what must the therapist do to help clients change? Does the therapist need to provide clients with the love they never received? Is the therapist's role to reparent the client? No. The therapist cannot meet clients' unmet developmental needs, although many clients may wish that were possible. Instead, the therapist has more modest but attainable goals. The therapist can help clients enormously by responding empathically and affirmingly to their unmet developmental needs. This allows these unacceptable, shame-based or anxiety-arousing parts of the client to be acknowledged, understood, and integrated rather than to remain split-off and disavowed. The therapist must also provide clients with a response to their old relational patterns that is different from what they have received in the past and have come to expect from others. Enduring change results as the new and more satisfying relationship with the therapist expands clients' internal working models, alters their beliefs about themselves and expectations of others, and increases their interpersonal range. This experience creates greater interpersonal safety than the client has ever experienced before. It is this experience of safety that frees clients to become more authentically connected to themselves and fosters greater choice and flexibility in their views of themselves and construction of current relationships.

The corrective relationship with the therapist evokes the conflicted emotions accompanying clients' problematic relational patterns, but it also allows clients to undergo as an adult—to remember, experience, and share—those feelings, needs, and experiences that were too threatening or painful for clients to contain on their own. In turn, the holding environment provided by the therapist's understanding and care allows clients to integrate these conflicted feelings that they have not been able to resolve on their own. In the safety of the holding environment, clients resolve their conflicts by facing honestly and grieving openly for the emotional support they missed or for what was hurtful while they grew. Clients also develop the capacity to feel compassion and empathy for themselves, which was not possible before. As a result, clients feel empowered to take more responsibility for their own choices, to act on behalf of their own needs and wishes, and to relate more authentically to themselves and others. Thus, the issue is not just reexperiencing what happened in the past. Instead, as clients experience a corrective relationship with the therapist, this new interpersonal safety allows clients to come to terms with the strong feelings— shame, guilt, sadness, anxiety and so on—that have both resulted from and contributed to their problems. In doing this, clients develop greater intrapersonal and interpersonal authenticity, which leads to a more fulfilling and meaningful life.

We have already seen in Chapter 8 that the client's presenting problems and conflicts are not just talked about abstractly in therapy; rather, they are brought into the therapeutic relationship and reenacted in their interpersonal process. In order for the client to change, the therapist and client must not recapitulate the client's conflict in their interpersonal process, as so commonly occurs. Instead, they must mutually work out a resolution to them. As the attachment researchers put it, you do not get an empathic child by teaching or admonishing the child to be empathic; you get an empathic child by being empathic with the child (Karen, 1994).

As we will see in Chapter 10, once this corrective emotional experience becomes an ongoing pattern of the therapeutic interaction, it will be relatively easy for the therapist to help the client *generalize* this emotional relearning to other arenas in the client's life where the same themes and patterns are being enacted. Conversely, if the therapist and client talk about issues and dynamics, but their interpersonal process does not enact a resolution of the conflicts they are discussing, enduring change will not occur. Thus, the purpose of this chapter is to further clarify the process dimension and illustrate how the therapeutic relationship is the best avenue available to resolve clients' conflicts.

Chapter Organization

This chapter is divided into two sections. In the first section, we examine how the interpersonal process that therapists enact with clients can be used to resolve clients' conflicts. First, we will review the course of therapy up to this point in treatment and summarize how clients' conflicts are brought into the treatment setting and reenacted with the therapist. Next, we look at five case vignettes that demonstrate more concretely how therapists can use the process dimension to help clients resolve these conflicts that have been activated within the therapeutic relationship.

In the second section, we explore why beginning therapists may be reluctant to use the process dimension and to address how aspects of the clients' conflicts may be occurring in the therapeutic relationship. We examine six reasons why therapists-in-training may have difficulty utilizing interpersonal interventions that focus on the current interaction between the therapist and the client, along with guidelines to help therapists begin working in this way.

Enacting a Resolution of Clients' Conflicts in the Interpersonal Process
Bringing Clients' Conflicts into the Therapeutic Relationship

Before we examine how therapists can use the process dimension to resolve conflicts, let's briefly review the course of therapy. This review serves to introduce the next stage of therapy and to place it within the overall context of the developmentally unfolding course of therapy.

Problematic patterns of repetitive interaction with caregivers give rise to conflicts that include feelings so anxiety-arousing that the client feels unable to integrate or resolve them. Symptoms and problems emerge as the client tries to gain control of these hurtful relational patterns and conflicted emotions by adopting particular interpersonal coping strategies. Although, in childhood, these coping patterns were necessary and effective, they are maladaptive and self-defeating in most current relationships. The therapist's first task is to establish a collaborative alliance with the client; they then work together to identify and change the recurrent relational patterns that occur in the therapeutic relationship and with others, as well as the core conflicted emotions and pathogenic beliefs that accompany them. As this work proceeds, the therapist must not comply with the client's strong pull to reenact these relational themes in the therapeutic relationship. Instead, the therapist's goal is to provide a different type of relationship that resolves, rather than reenacts, the client's recurrent relational themes. By responding in more empathic and affirming ways, the therapist gives the client the opportunity to experience a new way of relating and to evaluate more openly the usefulness of his or her previously fixed and inflexible coping style. In this process, the client becomes empowered to choose how to cope with, and construct, current relationships and to develop a more realistic and affirming self-concept. In other words, by enacting this corrective relational experience, the therapist provides the client with the interpersonal safety he or she needs in order to experience, explore, and share the painful or unacceptable emotions that have accompanied these relational problems. Previously, those emotions have been too threatening for the client to contain and integrate. This experience then frees the client to change, grow, and develop a stronger sense of self.

The therapist's primary role is not to give advice, explain, reassure, self-disclose, or focus on the behavior or motives of others, although these and other responses will all be effective at times. Instead, the therapist encourages the client to set the direction of therapy. What issues does the client feel are most important? While following the client's lead, the therapist participates in shaping the course of treatment and providing a treatment focus by listening for (1) the relational patterns, (2) pathogenic beliefs and faulty expectations about self and others, and (3) affective themes that *recur* throughout the material the client presents. The therapist continues to focus the client inward, responds to the client's affect, looks for opportunities to link the client's problematic interpersonal patterns with others to their own current interaction, and works to maintain a collaborative alliance by ensuring that the client is an active participant who feels ownership of the change process. With a growing understanding of each client and continuing refinement of the working hypotheses about this particular client, the therapist will be able to conceptualize the generic conflicts and interpersonal coping styles that are expressed in the different issues and concerns that each client presents.

This conceptual formulation enables the therapist to provide a treatment focus and direction for the ongoing course of therapy. For example, the therapist will be able to do the following:

- To point out when and how clients employ their interpersonal coping strategy to rise above their conflicts
- To help clients become aware of how they block their own needs and feelings by responding to themselves in the same way that others originally responded to them
- To help clients explore why they become anxious at a particular time, what types of responses they tend to elicit from the therapist and others to manage this anxiety, and how they systematically avoid certain interpersonal modes or affects

In these and other ways, the therapist helps clients become aware of their generic conflicts and the maladaptive ways they have learned to cope with them. Although this conceptual awareness will be an important part of helping clients change, something more is needed.

At the same time as the therapist and client explore the content of the client's dynamics, they will also play out these same dynamics in their interpersonal process. As we have seen, the client and therapist do not just talk about conflicts in therapy; they actually relive them in the therapeutic relationship. This occurs in three ways.

First, transference reactions bring the client's generic conflict into the current relationship with the therapist. As the client's conflicts begin to emerge in therapy, the client will become concerned that the therapist has been responding—or is going to respond—in the same frustrating or unwanted way that significant others have done in the past. These fears and misperceptions of the therapist follow from the client's relational templates, and are most likely to occur when strong feelings have been evoked and the client feels distressed or vulnerable. In some cases, therapy will reach an impasse until the therapist can help the client differentiate the therapist's response from the problematic responses that the client expects and has in fact received in most other relationships.

Second, clients will systematically elicit responses from the therapist that metaphorically reenact their relational expectations. When therapy stalls or ends prematurely, clients have usually recapitulated their generic conflict in the therapeutic relationship, by eliciting responses from the therapist that are thematically similar to those they have received from others in the past—responses perceived as rejecting, idealizing, sexualizing, competitive, controlling, abandoning, critical, hopeless, and so forth. Clients elicit these old problematic responses from the therapist for different reasons, sometimes defensively—to avoid their internal conflicts and maintain object ties—and sometimes more adaptively—in an attempt to master their conflicts by testing whether they can obtain a new and more satisfying response from the therapist than they have received before.

Third, the manner in which the therapist responds to clients—the therapeutic process—may unwittingly reenact their conflicts. Clients' relationship with the therapist must provide a more satisfying response to their conflict than they have received in the past. It is not always easy to provide this corrective emotional experience, however. Clients will expect, and often successfully elicit, the same hurtful responses from the therapist that they have received in the past. The task for the therapist, then, is to identify how the interpersonal process with clients may be reenacting their conflict and, when this is occurring, to use a process comment to make this interaction a topic for discussion. The therapist then works together with clients (and with the help of a supervisor when the clients' dynamics have evoked the therapist's own matching or reciprocal conflicts) to establish a different pattern of interaction that does not recapitulate the old relational pattern. As we have seen, however, this idea is far easier to understand than to put into practice.

Thus, if it is to lead to change, the therapeutic process must enact a resolution of clients' conflict rather than a repetition of it. When this occurs, clients have received far more than just an explanation for their problems. They have experienced a meaningful relationship, in which their old conflicts have been aroused, but they have found a relational outcome more satisfying than the relational pattern they have come to expect. They have been able to reveal themselves, to act more boldly, to express a need, and so forth, without being ignored, rejected, exploited, abandoned, dominated, or diminished as they have come to expect.

The experience of an interaction with the therapist that is incompatible with old maladaptive templates and patterns does not make up for clients' childhood deprivations or deficiencies. However, it does behaviorally demonstrate that change can occur—that at least some relationships can play out in a different way, at least sometimes. This corrective emotional experience broadens clients' internal working models and interpersonal modes of responding with others. At this critical juncture in treatment, when clients are having a meaningful experience of change with the therapist and the therapeutic process is providing a resolution of the clients' conflicts, all other intervention techniques become more effective. Cognitive, interpretive, self-monitoring, educational, skill development, and other interventions can all be utilized more productively by clients, *because the useful new content in these interventions is congruent with the corrective process they are enacting*. In addition, as clients have repeated experiences of change with the therapist, the therapist can help clients generalize this experience of change to other arenas in their lives where similar conflicts are being enacted. This next period of therapy—generalizing this emotional relearning that has occurred with the therapist to other relationships—is called the *working-through* phase of treatment and will be discussed in the next chapter.

With this overview of the change process in mind, let's now examine more closely how therapists can use the interpersonal process they enact with their clients to effect change.

Using the Process Dimension to Resolve Conflicts

Clients must experience a new and more satisfying response to their conflicts in their relationship with the therapist. To provide this, it is not enough for the therapist to listen empathically, suggest relevant explanations and interpretations, or facilitate client insights about the historical roots of problems. Nor is it enough to train new skills and direct clients to try out new, more adaptive behavioral responses in their environment. Although these and many other interventions are useful, they are usually insufficient to produce enduring change.

In addition, clients must experience a relationship with the therapist in which their old conflicts are aroused but the therapist does not respond in the problematic way that clients expect. This is deceptively difficult to do. Therapists commonly fail to recognize how they are reenacting clients' conflicts in their interpersonal process. Clients do not just talk about their conflicts in the abstract; they reenact the same conflicts with the therapist, in the three ways described earlier. When therapy fails, the therapeutic process has usually reenacted the same conflict that the client has been struggling with in other relationships and that originally brought the client to therapy. Further, this usually occurs because the client's eliciting maneuvers, "inductions," or interpersonal coping strategies have tapped into the therapist's own personal dynamics (perhaps the client's need to control others taps into the therapist's strong concerns about being controlled, for example), and the therapist and client have not been able to work themselves out of their mutual conflict.

Thus, through transference reactions, testing behavior, and eliciting phenomenon, there is a strong propensity for the therapeutic process to recapitulate the client's core conflictual relational patterns. As we will see, it challenges the therapist's own personhood to utilize the process dimension and enact a solution to the client's conflict rather than a repetition of it. We now look at five different case examples to understand how the process dimension can be utilized to effect change.

Example 1: An inability to recognize the process   Whereas therapists find it easy to understand the process dimension conceptually and relatively easy to identify when a client's conflict is being reenacted with another therapist, it is far more difficult for therapists to recognize how the interpersonal process they enact with their own clients may be recapitulating the conflict.

An alcoholic client began therapy by expressing how much pain he was in as a result of his drinking problem. His wife was threatening to leave him, and he was on probation at work. His 8-year-old son had recently seen him in an embarrassing situation while intoxicated and had asked him not to drink anymore. The client was distraught that, even though the structure of his life was collapsing, he could not stop drinking. During their first session, the client sobbed, "Nothing ever works out; relationships all go bad; everything I try to do falls apart in the end."

The therapist was moved by the client's plight but became overinvested in helping him change. The therapist found it difficult to listen to how hopeless he sounded, and she kept reassuring him that things would get better. The therapist disclosed that she had had a serious drinking problem of her own many years ago and that she knew what he needed to do in order to stop drinking. The therapist and client quickly established a friendly relationship, in which the client successfully elicited a great deal of nurturance from the responsive therapist. The client enjoyed this support immensely and sincerely tried to follow the therapist's advice and suggestions.

Things went better for about a month, but then the client fell off the wagon and began drinking heavily again. When he started arriving late for his therapy sessions and then missed several of them, the therapist began to feel let down. It was the final straw for the therapist when the client arrived drunk for therapy. The therapist felt angry and betrayed, because she had extended herself to him more than she usually did with other clients. She felt that he was letting them both down and told him so. In this way, the therapist became punitive toward the client, and she also induced guilt over the impact of his "irresponsible" drinking on his family. The client was contrite and tried, unsuccessfully, to elicit her sympathy and support again. The client did not show up for his next appointment, however, and did not return to therapy.

This interaction recapitulated the client's conflict in two ways. First, the client initially elicited a promise of love and support from the therapist, but the therapist eventually turned to criticism and control. This confirmed the client's disillusionment that "nothing ever works out; relationships all go bad"—the specific relational pattern about which he had forewarned the therapist during their initial session. Second, the therapist benevolently tried to rescue this client, who was behaving as a victim. The caretaking, advice-giving response from the therapist inadvertently defined the client as incapable of managing his own life. This response paralleled a similarly diminishing attitude that the client had received from his indulgent but belittling parent. The rescuer-victim relationship that was enacted in therapy prevented the client from addressing his real conflicts over adequacy and achievement, which he defended against, in part, through drinking.

In reviewing this case several weeks later, the therapist reported that "the client just wasn't ready to stop drinking yet. He's going to have to get worse and really hit bottom before he'll be able to admit there's a problem and do something about it." The therapist did not recognize that the interpersonal process enacted in therapy unwittingly recapitulated the client's conflict and prevented change from occurring. As is often the case when therapy fails, the therapist attended only to the content of what they talked about and not to how their process might be recapitulating the client's conflicts.

From an emotionally neutral vantage point—that of a therapist's supervisor or colleague looking at the session on videotape—it is relatively easy to see how the client's conflicts have been reenacted in the interpersonal process. In humbling contrast, it can be exasperatingly difficult for therapists to see their own interpersonal process while engaged in an intense, affect-laden relationship. This is true even for the well-seasoned practitioner but especially for the novice clinician. Beginning therapists commonly will reenact the client's conflict in their interpersonal process without realizing what is happening. It is difficult to evaluate and simultaneously perceive at both the content level and the process level. To be effective, therapists must be able to attend both to the content of what is being discussed and to the process that is being enacted as the discussion proceeds. Once they begin observing the process level, however, trainees frequently have difficulty *using* the responses that they now see occurring with their clients. Therapists-in-training must be patient with themselves and know that the ability first to recognize and then to respond to the process level usually is acquired gradually over a period of years. In their attempts to adopt a process-oriented approach, therapists will benefit from further reading on this topic (see the suggestions at the end of the chapter) and the assistance of a supportive supervisor.

Example 2: Process comments keep therapists from reenacting conflicts   In the previous example, the therapeutic process recapitulated the client's conflict. In the next example, a similar dynamic begins, but the therapist recognizes their interpersonal process and effectively realigns this maladaptive pattern of interaction by means of a process comment.

Many clients enter therapy because of depression. They feel sad, believe they are bad, and see themselves as powerless to change their circumstances. In many cases, these clients communicate their very real suffering to the therapist with emotional pleas for help. However, whatever the therapist tries does not work. Often, the therapist's efforts to meet the client's request for help are met with "Yes, but . . . " A problematic cycle begins: The client feels increasingly desperate and intensifies his or her emotional pleas for help. For example, the client might even exclaim, "I can't go on living if things don't change soon!" In response, the concerned therapist becomes more active and tries even harder to respond in some helpful way. However, nothing the therapist does has any impact or provides any relief for the client.

As the client rejects the help he or she elicits, the therapist's own need to help is frustrated. The therapist's own sense of adequacy as a helping person may be threatened in cases where the client's eliciting behavior has tapped into the therapist's own personal issues and dynamics. In response to this, some therapists will become punitive and critical toward the client; others will withdraw and emotionally disengage. When this occurs, the therapeutic process thematically repeats the client's core conflicted interactions.

When the therapist remains critical toward, or disengaged from, the client, therapy is unsuccessful. The depressed client's conflict is recapitulated, as the client again feels dependent on a relationship with someone who is critical, is inconsistently available, or threatens to terminate the relationship. What is the alternative? In successful therapy, the therapist is able to make a process comment about what is going on in the present interaction with the client. Rather than focus solely on the content of what they are talking about (for example, depression), the therapist can point out how they are responding to each other. This is a very effective way to see and change the maladaptive, repetitive pattern of interaction.

Therapist:
Let's talk about what is going on between us right now and see if we can understand what's happening in our relationship. It seems to me that you keep asking me for help, but keep saying "Yes, but. . . ." I think that I get frustrated with you then, feel that you don't ever allow me to help you, and then become more demanding of you. What do you see going on between us?

With a process comment of this type, the therapist is joining with the client in trying to understand their mutual interaction. The therapist is responding to the client's need by helping the client look at the depression and how it is expressed to, and experienced by, others. As we will see, the process comment also allows the therapist to break the cycle of escalating client need and therapist frustration, at least temporarily.

How does the process comment help the client to change? In this example, the client is experiencing a relationship with the therapist that provides a new and different solution to an old and problematic pattern of interaction. The client is not rejected or abandoned, as has happened in the past, but remains engaged in a relationship with someone who consistently responds to the client's concerns. The client has not experienced this consistency, availability, or acceptance in past relationships. If this corrective emotional experience is repeated in many other ways—large and small—throughout therapy, the client is experiencing a resolution of the conflict in the relationship with the therapist. When clients have the experience of change, rather than passively hearing advice, reassurance, or interpretations, they are better prepared to make enduring changes.

A key concept in the interpersonal process approach is that the capacity for change increases with increase in the personal significance that the therapeutic relationship holds for the client. When an interpersonal solution is enacted in a meaningful relationship with the therapist, change is facilitated in two ways. First, in the safer context of this new type of relationship, clients can now come to terms more directly with the conflicted feelings, unmet needs, and pathogenic beliefs that accompany their problematic relational patterns. In other words, clients have the supportive relationship necessary to undergo (remember, experience, share, contain, and integrate) those experiences that were too threatening to be dealt with in childhood, rather than continuing to split off or disavow them. Second, the experience of change with the therapist demonstrates to clients, rather than merely tells them, that current relationships can be different and more satisfying than they have learned to expect. It does this by expanding their internal working models of relationships—by altering their basic beliefs, expectations, or schemas about the possibilities that relationships hold for them. With the therapist's assistance, the client can then begin to establish more satisfying relationships with others along these new, more flexible, and self-affirming lines.

As an example, let's consider the depressed client's response to the therapist's process comment:

Client:

Yeah, I do keep saying "Yes, but . . . " to you—like my wife always says I say to her. This is ridiculous. If I'm doing this with you, too, then there's no way I'm going to get better. Maybe we should just forget it and stop now?

Therapist:

Oh, no—that's not what I'm suggesting at all! It makes sense that you are having the same problems in here with me that you have with other people. In a way, it is a problem, but in another way, it gives us the opportunity to resolve your problem right here in our relationship.

Client:

How?

Therapist:

If you and I can work out a more satisfying relationship than you have had in the past—one that doesn't repeat this same old pattern—I think it will go a long way toward helping you change.

Client:

Do you think we can do that now if we haven't been able to up to this point?

Therapist:

Yes, I think it's very possible. In fact, I think we are breaking that pattern right now, just by talking about the way we interact together. Tell me, how is it to be talking with me about our relationship and the way we respond to each other?

Client:

I like it. It's different.

Therapist:

I like it, too. I feel like I'm working with you, rather than being pushed away as I was feeling before.

In this example, the therapist has used a process comment to effectively alter the current interaction with the client. For the moment, the process comment has kept their interaction from recapitulating the client's conflict. Some variation on the old "Yes, but . . . " pattern will probably soon reappear, and the therapist will have to make another process comment and work through a similar cycle again. With the process comment, however, the therapist is temporarily providing the client with a different response that does not recapitulate the client's conflict. If the therapist continues to find ways to provide this response throughout treatment, the client will experience a corrective relationship, and change will occur. The next example illustrates a macroperspective that tracks the process dimension over the course of therapy.

Example 3: How the process dimension bridges differing theoretical orientations
Therapists with different personalities and therapists working from different theoretical orientations can all help clients change. Every theoretical approach clarifies some important aspects of client problems, and every theoretical approach has limitations. The primary component of change is the nature of the relationship that the therapist provides for the client (Bergin & Garfield, 1994; Walborn, 1996). Regardless of the therapy, techniques, and interventions employed, change is likely to occur whenever the way in which the therapist and client interact provides a resolution of the client's conflicts rather than a repetition of them.

After two years in analytically oriented therapy, Rachel still could not take charge of her life. She was always complying with her husband's demands, and it was impossible for her to tell her children what to do and have them obey her. She also complained of her stultifying daily routine as a homemaker, yet she was never able to do anything to improve it. At her husband's behest, she would periodically enroll in a class or interview for an office job, but she was never able to follow through on any of his suggestions.

In therapy, Rachel had spent many hours reviewing her childhood. Her therapist had an uncanny ability to see unifying themes in many of the memories and recollections she shared with him. For example, her therapist repeatedly observed how she was subtly discouraged from initiating activities on her own as a child. She was also subtly not allowed to feel good about her success experiences: Anytime she expressed an enthusiastic interest, tried out a new venture, or achieved something she was proud of as a child, her parents did not notice it—or, if they did notice, they didn't seem very happy about it.

Rachel's therapist once explained to her, in a sensitive and informative way, that she had a passive-dependent personality style. Rather than feeling labeled or put down, Rachel was impressed that her therapist seemed to understand her so well. He knew so many things about her without her even having to tell him. Although her problems had not changed much yet, she still believed her therapist would cure her. He was so bright and insightful, and he seemed genuinely to care about her. He didn't like to tell her what to do but, when things became too much, he could usually help by explaining what the problem really meant. It was so comforting to be with him; Rachel did not know what she would do without him.

It was Rachel's husband, Frank, who finally ran out of patience with the slow course of treatment. After two years, he was fed up with the big therapy bills and his wife's unremitting discontent. His wife's helpless dissatisfaction felt like a subtle but unending demand for him to love her more, give her more, or somehow fill up her life. He was tired of these subtle, nagging demands, and he wanted a change.

A man of action, Frank was quickly able to obtain the name of a behaviorally oriented therapist on the faculty of a nearby college. A friend told Frank that this therapist was a "problem-solving realist" who could make things happen quickly. This sounded like just the right approach to him. Frank insisted that Rachel stop treatment with her present therapist and begin immediately with the new therapist. Rachel was panicked at the thought of leaving her therapist, but she sensed that Frank was truly at the end of his rope. Although she still believed in her therapist and felt very close to him, she was afraid that Frank might actually leave her if she did not do what he said.

After her first few sessions with the new therapist, Rachel was surprised to think that perhaps Frank had been right after all. The new therapist wasted no time in taking charge of the situation. It was exciting to have the therapist outline specific steps for her to follow, and it seemed sensible to have a specific plan for change laid out. The therapist discussed a list of treatment goals with her, and they planned a set of graduated steps for meeting these goals on a scheduled timetable.

In their first hour together, the therapist had Rachel role-play how she responded to her children when they disobeyed her. Then, with the therapist serving as a model and coach, they rehearsed more assertive responses that Rachel could try with her children. The therapist also had Rachel enroll in an assertiveness training class that the therapist was running for some other female clients. Each week Rachel was also to complete a homework assignment. For the first week, she was to call one new person she might like to get to know and ask her to lunch. She was supposed to report back to the therapist on this assignment at the beginning of their next appointment.

Frank was encouraged by this practical, problem-solving approach to his wife's problems. He began to think that something might change after all. Rachel was hopeful, too. She felt reassured by her new therapist's goal-oriented, problem-solving approach. In fact, Rachel became determined to make this therapy work, even though she didn't want it initially, and promised herself that she was going to try and do everything the therapist asked of her.

Therapy progressed well for the first few weeks, but then things started to slow down again. Although she did not really know why, Rachel began to find it hard to muster the energy to attend the assertiveness class. She knew her therapist would be disappointed in her, but she just could not help it. Although she felt guilty and confused about it, Rachel began to come late to her therapy sessions and then to miss them altogether; she faded out of treatment over the next month.

Both of these therapists failed to have a significant impact on Rachel's problems, even though the treatment approaches they used were seemingly very different. Theoretically, the analytically oriented therapist might attribute the unsuccessful outcome to the great difficulty in restructuring the passive dependency needs of a basic oral character. The behaviorist might note that Rachel was not sufficiently motivated to change, because she was receiving too many secondary gains from her help-seeking behavior. If we look at the interpersonal process that transpired between both therapists and the client, however, a different picture emerges. These two therapists actually responded to Rachel in a very similar—and problematic—way.

We must look carefully at what Rachel *experienced* in therapy. With both therapists, Rachel reenacted the same maladaptive relational pattern that she had with her children and with her husband. Her presenting problem was that her children ran over her and she could not make them listen to her or obey her. Her disturbing degree of compliance with her take-charge husband was a profound example of the same problem. Therapy failed in this case because her pattern of being passive and compliant recurred with both therapists. Unwittingly, both therapists provided a hierarchical relationship in which she remained the passive helpee led by an all-knowing helper. This interpersonal process reinforced her basic belief that the source of strength and problem resolution did not reside in her but in the therapist or others.

In order for Rachel to change, she needs to experience a therapeutic relationship that affirms her own efficacy—a relationship in which she is able to initiate what she wants, set personal limits with others, and make her own decisions and choices. The client's symptoms will gradually improve in the context of this new and reparative interpersonal process, where the therapist encourages the client to act more independently in their relationship and compassionately helps her explore and understand the significant anxiety and guilt that arises each time she tries to act more assertively with the therapist. This experience of change with the therapist shows her that her own strengths and resources are valuable and can be utilized, along with the therapist's skill and understanding, in a more productive and collaborative relationship.

Unfortunately, both of her therapists were comfortable with the helper-helpee relational process and did not bring it up as a focus for treatment. Neither therapist made the process comment that, in certain ways, their interaction repeated the interpersonal style that was problematic for Rachel in other relationships. Such a process comment, given in a tactful and caring way, would have allowed Rachel to begin looking at this problematic behavior in a supportive environment. More importantly, it would have given her permission to change her passive and compliant style and to begin finding—and then acting on—her own feelings, interests, and authentic voice, first in her relationship with the therapist and later in her relationships with others.

What do these changes in the relationship with the therapist have to do with solving the problems that brought her to treatment? As long as the therapists were telling her what to do, Rachel could not set limits with her children, take a more assertive stance with her husband, or follow through and do what she wanted to do for herself. That is, the highly relevant content of what she discussed with both therapists (issues about autonomy, independence, assertiveness) was not matched by the interpersonal process they enacted. *Unless Rachel has the actual experience of behaving as an active, strong participant in her relationship with the therapist, she will not be able to adopt this healthy, stronger stance in other areas of her life.* In other words, the process must be congruent with the content.

To enact a more egalitarian, collaborative relationship with someone like Rachel is not a simple task for the therapist, however. It requires therapeutic skill and careful monitoring of the process. In response to her life experiences, Rachel has become accomplished at getting her therapist, her husband, her children, and others to lead, direct, and control her. Both the behaviorally and analytically oriented therapists could have successfully used their differing theories and techniques to help Rachel if they had enacted a different process in their relationship. If they had encouraged her to initiate and lead, Rachel's central conflict would have emerged more overtly in the therapeutic relationship. As soon as either of the therapists encouraged her to initiate, respond to her own interests, disagree with the therapist, or behave more assertively or competently in the therapeutic relationship, Rachel would have become anxious. Either therapist could then have focused her inward on this anxiety, so that they could begin examining together why it was threatening, guilt-evoking, or unacceptable for her to step out of her compliant and help-seeking mode.

Simultaneously, either therapist could have given her permission to act more assertively and competently within the therapeutic relationship and worked with her in a way that facilitated this new behavior in the sessions. The therapist could do this in two ways: first, by carefully attending to, and affirming, this effective new stance whenever it emerged in their relationship or with others; second, by responding to instances in which Rachel "undid" herself by retreating to the safer, unassertive way after she had risked acting in a stronger way with the therapist. Given Rachel's life experiences and interpersonal style, she was likely to become apologetic or confused, or to act dependently and ask for direction, soon after disagreeing with the therapist, making an insightful connection on her own, risking a confrontation with the therapist, or effectively redirecting the session. If therapists have formulated working hypotheses about this and track this potential reenactment in their interpersonal process, they will be prepared to help her explore the threat or danger she feels when she has just retreated from her new, stronger stance with the therapist.

Therapist:
> What do you think I might be feeling, and how are you afraid I might respond, after you have just acted more strongly by disagreeing with me like that?

With both of her therapists, however, Rachel only continued to defend against her anxiety over being more assertive and effective by retreating to her dependent, help-seeking role and, in their interpersonal process, merely reenacting this maladaptive relational pattern.

In most therapeutic relationships, the therapist and client will temporarily reenact the client's generic conflict in their interpersonal process. In successful therapy of every theoretical orientation, however, the therapist and client do not continue to reenact the old relational pattern. Instead, they are able to identify and label this familiar but problematic process and work out a relationship that resolves, rather than reenacts, the client's conflict. Once clients find that their old conflicts can be activated with the therapist but do not have to result in the same hurtful or frustrating outcome, they are able to generalize this experiential or emotional relearning to other relationships as well.

Example 4: Resolving conflicts by working with the process dimension   Therapists-in-training often feel pressured to do something to help their clients. This urgency may lead to a premature emphasis on intervention techniques. Unless the therapist has conceptualized what has gone awry for the client in other current and formative relationships and considered how these problematic relational scenarios could be reenacted in the therapeutic process, intervention techniques will often fail. In contrast, once the therapist has several good working hypotheses about the interaction between the therapist and client—and the specific relational experiences that the client needs in therapy—it is usually relatively easy to find effective ways to intervene. The therapist can employ a wide range of techniques from different theoretical modalities (Wachtel, 1982, 1993), evaluate their effectiveness in terms of the client's progress in treatment, and flexibly modify the interventions to best fit this particular client. Thus, counselors should repeatedly ask themselves "What does this mean?" rather than "What should I do?" As we will see, the second question is usually answered by the first.

As an illustration, we examine two critical incidents in the treatment of an incest survivor. In these incidents, the therapist uses intervention techniques that have a highly significant impact on the client: validating the resistance and role-playing. It is not the intervention techniques in themselves that facilitate change, however. Both interventions are effective because they follow from the therapist's understanding of the client's dynamics in the current interaction and because the therapist is able to work with these dynamics in the immediacy of the client-therapist relationship.

On the basis of her presenting symptoms, Sandy's male therapist hypothesized that she may be an incest victim and anticipated that trust would be a central issue in their relationship. Therapy had gotten off to a good start but, several months into treatment, progress began to slow down; the material that Sandy presented was becoming repetitious. About this time, Sandy recounted two different stories in which the relational theme was feeling unsafe with men. The therapist hypothesized that, without being aware of it, Sandy was using these stories to address, indirectly or metaphorically, the topic of trust in their relationship and that her profound concerns about safety, betrayal, and vulnerability were now being activated with him.

The therapist began to talk with Sandy about trust in their relationship and asked about the different feelings she was having toward him. Sandy genuinely liked the therapist and was finding him very helpful. However, when the therapist pursued the trust issue further and asked specifically whether she felt safe with him, her affirmative response seemed distant and unconnected. It soon became clear to both of them that Sandy was emotionally removing herself from the therapist as they talked more directly about safety in their relationship.

The therapist responded to Sandy's concern by accepting both sides of her ambivalent feelings and tying her concerns directly to their relationship.

Therapist:
I know that one part of you likes and trusts me, but it makes sense to me that another part of you doesn't feel safe at all. I think that both sides of your feelings toward me are valid and are important for us to work with.

Sandy sheepishly nodded and gestured vaguely to indicate that this was true.

Rather than trying to talk her out of these concerns or convince her of his trustworthiness, however, the therapist validated her distrust and went on to characterize it more fully:

Therapist:
If I violated your trust in some way after you had taken the risk to ask me to help you, it would be very bad for you. If I tried to approach you sexually or foster any other kind of relationship between us, it would hurt you very much. In fact, I think it would hurt you so much that you might not be able to risk trusting or asking for help again.

Sandy began tearing, looked at the therapist, and slowly nodded in agreement. The therapist continued to elaborate Sandy's concern in terms of their relationship:

Therapist:
If I took advantage of our relationship, you would be without hope. I think you would again get very depressed, enter into other relationships that

would be destructive for you, and start thinking seriously again that you do not want to be alive.

As the therapist spoke and made Sandy's concern overt in terms of their relationship, her whole demeanor changed. She remained tearful but became alert and present with him; she agreed with what he was saying. As they continued to talk about this, the therapist went on to describe his own attitude:

Therapist:

> I understand how much it would hurt you if I betrayed your trust, and no part of me wants you to have that terrible experience again. In fact, I *like* the part of you that doesn't trust me. That distrustful, removed part of you is your friend and ally. We need her—because she is committed to not letting you get hurt again. That's why it's so important that we go at your speed in here, that you have control over what we talk about, and that you can say no to me and know that I will honor your limits.

After the therapist responded affirmingly to Sandy's resistance and worked with this conflict in terms of their relationship, important changes occurred. Whereas in the past Sandy had only alluded vaguely to her childhood abuse, she now began to share it more explicitly. Over the next six weeks, she recalled in vivid detail, and with strong emotions, how she had been repeatedly sexually abused over a period of years by her stepbrother, who was 12 years older. When she went to her mother for help, her mother was not supportive, denied that the abuse was occurring, and told Sandy never to talk about anything like that again. Sandy did not even consider seeking protection from her stepfather, who had always been distant and unresponsive. After failing to receive help, Sandy recalled sitting alone on the floor of her closet for long periods with the door closed, feeling afraid, ashamed, and deeply alone. In response to this deep sharing, the therapist was comforting, validated her experience, and began to help her with the many significant connections she began to make between this familial tragedy and her current life.

During one of these sessions, the therapist used a role-playing technique to provide a different response to the trauma she had suffered.

Therapist:

> I wish someone could have been there to stop him and protect you. No one was there for you but, if I had been there, I would have walked into your bedroom when he was there, turned on the light, and in a loud voice commanded: "Stop it! Get away from her and leave her alone right now! I see what you're doing and it's not fair. You're hurting her and I won't allow it."

After speaking this loudly and forcefully, as if he were actually saying it to the perpetrator, the therapist stood up and rolled Sandy's coat into a ball. Using it as a little Sandy doll, he held it close and spoke reassuringly:

Therapist:

> You're safe now, and he's gone for good. I'm going to call the police now, and keep protecting you so that you'll never have to worry about him hurting you again.

As the therapist symbolically gave Sandy the response that she desperately longed for—but did not get—as a little girl, the full intensity of her pain and rage was evoked. When these powerful feelings ran their course, Sandy became more composed and said, "I'm going to be all right." The therapist, still holding the rolled-up jacket tenderly, carefully tucked it in Sandy's arms.

Therapist:

> This is the little girl you were. She needs you to open up your heart to her and give her a home. You need to take care of her—and not push her away anymore like they did and you have done. You need to hold her, talk to her, and listen to what she tells you. I want you to join me in taking care of this part of you, so this little girl is no longer hurt and alone behind the closet door.

Sandy readily accepted this responsibility, and later bought a doll that she used to represent the part of her that needed to be cared for but had not been protected. Although all of Sandy's problems did not go away, long-standing symptoms disappeared and did not return.

Now in her late 30s, Sandy had become increasingly anxious, dependent, and almost incapable of leaving home alone. However, two weeks after this session, Sandy got her driver's license renewed and began driving for the first time in several years. Soon afterward, she got a job as a waitress—her first paid employment in six years. Although she had always foiled any type of success for herself, she enrolled in a local college and began receiving predominantly A's in her classes. Whereas Sandy had felt helpless, had characteristically acted as a victim, and had been repeatedly taken advantage of by others, she gradually became more assertive, forthright, and independent.

What allowed Sandy to get stronger in these very significant ways? Several factors made this therapeutic intervention effective. By using a role-playing technique, the therapist brought Sandy's conflict into the therapeutic relationship and provided her with the corrective interpersonal responses she needed. In sharp contrast to what occurred in her family, this time Sandy experienced protection, validation, appropriate boundaries, and a supportive holding environment. The therapist's compassionate and protective response also helped Sandy shift from an identification with her parent's rejection of her—and her own resulting shame and self-hatred—to an identification with the therapist and his compassion for her. Eventually, this healthy new identification led her to be able to care better for herself and believe that she did matter for the first time in her life. These far-reaching consequences were set in motion by the therapist's affirming and comforting response to Sandy's vulnerability. This response, even though it was just enacted in fantasy, provided Sandy with a corrective emotional experience that played an important part in allowing her to become stronger and capable of protecting herself for the first time.

Example 5: Recognizing the process dimension in all interpersonal relations   As we have seen, the most likely reason for therapy to fail is that the therapist and client unwittingly reenact aspects of the client's conflict in their interpersonal process. Therapists-in-training often find it challenging at first to track the process dimension and to distinguish the process that is enacted from the content that is discussed. For our final example, let's look at how the process dimension operates outside the psychotherapy setting—at a sporting event.

In a championship basketball game, the defending champions, mature and confident, face an underdog team made up of talented but younger players who are far less experienced. The defending champions are heavily favored to win.

It is late in the fourth quarter, and the game is tied. The fans, on their feet, are cheering at the prospect of an upset. The coach of the young team can feel that the game is about to slip away, however. His players are becoming rattled by the screaming fans, the pressure of time, and the unrelenting play of the opponents. To help compose his less experienced players, the coach calls a 2-minute time-out. His team has played well up to this point, but the seasoned coach knows that, in these stressful closing minutes, the other team's experience is likely to make the difference.

Crouched on the sideline, the coach positions his five players in a semicircle in front of him. With the roaring crowd only a few feet away, the coach quietly makes eye contact with the first player on his left. Without speaking, he calmly holds the first player's gaze for 12–15 seconds (a very long time amid such pandemonium). Having compelled the first player's attention, the coach turns to the second player and maintains eye contact with him for another 12–15 seconds, and so on with each of the five players. Each time, as he turns to engage the next player, the previous player's attention remains riveted on the coach. Finally, after making contact with each player in this way, the coach holds up a chalkboard with one word written on it: POISE.

The players and coach sit together silently for the remaining seconds of the time-out. The referee blows his whistle, the players return to the court, and the underdogs defeat the defending champions. The players did not make mental or emotional errors in the closing minutes of the game: They did not turn the ball over to the other team; they made clutch-free throws in the final seconds of the game; and they smoothly executed the plays they had practiced all year. In a word, they played with poise.

This coach astutely identified the central issue that could defeat his team in the final minutes: being intimidated by a more confident and experienced team, losing their composure, and making mental mistakes that would cost them the game. Having recognized the central issue, however, the coach did not just *tell* his team that they had to remain poised. Instead, he gave them the *experience* of composure during the time-out. The players were able to finish the game with such composure because the coach used the relationship he had developed with each of his players to give them the experience of composure during the final tense minutes. During the time-out, the players were able to generalize this experience to their performance on the court and play with poise.

Similarly, it is far more effective if the therapist-client interaction provides an experience of change than if the therapist merely tells the client what to do or explains what something means. It is often difficult to provide a relationship that enacts a resolution of clients' conflicts, however, and there are no easy formulas that tell the therapist how to do so. Every client is unique; the therapist must take the personal risk of entering into a significant emotional relationship with each client and must have the flexibility to provide the new relational experiences that this particular client needs in order to change.

Working with Clients' Conflicts in the Therapeutic Relationship

As we saw in Chapter 8, an essential element of interpersonal psychotherapy is to recognize how clients' conflicts are expressed in the therapist-client interaction. Working together, therapist and client then find new ways to relate that resolve, rather than reenact, the client's problematic relational patterns. For example, cultural, ethnic, or racial issues that have caused problems for the client in relationships can be made overt in the therapeutic process and resolved differently with the therapist.

Therapist:
> How do you think your cultural context is influencing this issue in our relationship?

Thus, the best way to begin this work is by asking the client in an open-ended way how the problem that he or she is talking about in relation to others may be occurring in the relationship with the therapist:

Therapist:
> How does this battle for control you have been describing with your partner go on between us? Where do you see the control issue in our relationship?

By bringing the client's conflicts directly into the therapeutic relationship in this way, the therapist creates the opportunity for a corrective emotional experience. In order to change, this client must find that the therapist is neither trying to control, nor willing to be controlled by, the client, which occurs for the client in other relationships. Only if the client has this real-life experience of change can the client resolve his or her conflict, find that control can be shared in close relationships, and learn that some relationships can be different from those in the past. Although this immediate, real-life experience of change is a powerful way to intervene, most beginning therapists find it uncomfortable to make this observation, query, or process comment. Thus, the purpose of this section is to help therapists take the risk of judiciously using their real-life relationships with clients to help clients change.

In the first part of this section, we review how therapists can bring clients' problems into the therapeutic relationship and work with them there. In the second section, we examine why process work may be anxiety-arousing for therapists at first and look at helpful guidelines.

Intervening within the Therapeutic Relationship

To some extent, most of the interventions discussed so far have asked the therapist to work on the clients' issues within the context of the therapist-client relationship. At the beginning of treatment, therapists need to speak directly to clients about their current interaction when they are trying to establish a collaborative relationship (Chapter 2):

Therapist:
  Am I understanding what is most important to you here?

Talking about what is going on between the therapist and client is even more important in working with resistances to entering treatment (Chapter 3):

Therapist:
  What's it been like for you to talk with me today?

Working directly with the current interaction is also helpful when establishing an internal focus (Chapter 4):

Therapist:
  You keep talking about others, and I keep asking about you. What do you see going on between us?

Therapists need to work with the client's affect in a way that brings the full intensity of whatever feelings the client is experiencing into the immediacy of the therapist-client relationship (Chapter 5):

Therapist:
  I can see how much this has hurt you and how sad you are feeling right now.

Finally, this focus on the interaction between the therapist and the client is central to working with transference reactions (Chapter 8):

Therapist:
  How do you think I am going to react if you do that?

Thus, the unifying theme in all of these interventions is to bring the client's issues and concerns with others into the therapeutic relationship, where they can be dealt with in terms of what is currently going on between the therapist and the client. Thus, the client's conflicts are not just talked about in abstract discussions about others; they are reexperienced, and resolved, in the real-life relationship with the therapist. Let's return to the client whose basic relational theme is a control battle. The best way to resolve the client's conflict is to make it overt and address it directly in terms of the client-therapist relationship.

Therapist:
Does that control battle ever get going between us, too? Does it ever feel like you and I are fighting over control in our relationship?

Making such a comment may seem a dreadful prospect to the beginning therapist. However, by taking this risk, the therapist can bring real-life immediacy to the therapist-client interaction: Problems with others are not just being talked about intellectually; instead, for the first time perhaps, they are being addressed directly and constructively with someone as they occur. If the therapist can use good timing and respond nondefensively and respectfully, the potential for change is at hand.

Let's consider a range of different responses to the therapist's query. Least challenging to the therapist is a disconfirming response.

Client:
No, I don't feel like we are in a control battle.

The therapist can use this response to begin exploring what is different about their relationship.

Therapist:
Good, I'm glad that's not a problem for us. Any ideas about what makes our relationship different?

The client's answer may provide useful information:

Client:
You treat me with respect; that's what's different.

The therapist can follow up on this comment by exploring the client's concerns about not being respected in other relationships, and can build on it to establish that other kinds of relationships can exist.

Second, the client may avoid the immediacy of his or her conflicts by discounting the meaning or relevance of the therapist:

Client:
We're not in a control battle because this isn't a real relationship. You're just my therapist.

Clients cannot make significant gains in therapy until this distancing defense is resolved. At this point, it is essential to clarify that the therapist and the client really are two people who have been having a relationship for some time, even though there are constraints on their relationship. The therapist can then begin to explore the threats aroused for the client by more meaningful involvement— such as being controlled or not being respected. The therapist and the client can then agree to watch for these concerns and address them if or when they arise in their relationship. It is usually most effective if therapists do not bring up these relational patterns in the abstract, however, but wait until they think the client is experiencing them in their current interaction before addressing them.

Finally, the client may respond to the therapist's question affirmatively.

Client:
> Of course you're in control. You insist that we stop at ten to the hour, whether I'm finished or not. And you're subtle about it, but you often take control by making things go where you want.

Most beginning therapists fear this response. They may believe that they have done something wrong if the client disapproves of them in this way. This possibility must certainly be considered but, in most cases, the client's reactions have far more to do with the client's own internal working models than with the reality of how the therapist has responded. Thus, rather than posing a problem, this affirmative response is one of the most important windows of opportunity that will occur in treatment. Openly acknowledging that the same conflict the client has with others also exists with the therapist creates the opportunity to discuss this issue and try to resolve it in their real-life relationship. The therapist can do this in the following ways:

1. By remaining nondefensive and accepting the validity of the client's concerns whenever possible

Therapist:
> Yes, we do have to stop at ten to the hour, and that is an unnatural ending for you. I can see how you would feel controlled by those time constraints.

2. By exploring the client's perceptions further and working with the client to understand them better

Therapist:
> What do you think is going on for me when I try to take things somewhere? What does it seem I am trying to do at those times?

3. By differentiating the therapist from other figures in the client's life

Therapist:
> Yes, I have had ideas about where we should go, but I have also been interested in following your lead, too. In fact, I may be different from some other people in your life because I genuinely liked it when you disagreed

with me last week and told me what you thought instead. I like it when neither of us feel controlled and we can both say what we want. It makes our relationship feel more alive to me.

4. By offering to be sensitive to this concern in their relationship, expressing a willingness to handle the issue differently in the future, and inviting the client to tell the therapist whenever this conflict occurs between them

Therapist:
I don't want you to be controlled in our relationship as you have been in others. That's no good for you or anybody else. Let's try to do something about it. From now on, I will watch the clock and let you know when it's five minutes before we have to stop. And any time you feel like I am directing you away from where you want to go, or being controlling in any way, tell me, and we'll stop right then and try to change it. How does that sound to you?

Adopting these approaches will not always be easy, as many clients will insist repeatedly on reestablishing the same constricted relational patterns. However, if the therapist is patient, willing to remain nondefensive, and able to tolerate personal discomfort and sustain this effort, clients often become far more engaged in their relationship with the therapist and highly motivated to explore this and other conflicts more deeply. As clients gradually find that the familiar, expected, but unwanted relational patterns do not recur with the therapist, they experience greater interpersonal safety than they have known before. As a result, clients often can allow important new issues and previously threatening feelings to emerge. This provides new material for exploration and a clearer treatment focus. As we will see in the next chapter, clients also begin to relate to others in new and more effective ways.

The real-life experience of change within the therapeutic relationship is powerful for clients, but working in this way can be uncomfortable for clinicians. Next, we explore several reasons for such discomfort, and we also examine some practical guidelines to help counselors work with the process dimension more effectively.

Therapists' Initial Reluctance to Address the Process

Although the best vehicle for effecting change is the therapeutic relationship, it is often anxiety-arousing for beginning therapists to take the plunge and say, "How does that problem go on between us here in therapy?" This direct intervention breaks the cultural norms and family rules that most therapists grew up with, and seemingly places great performance demands on the therapist. Because of this, beginning therapists are often reluctant initially to make process comments and explore the client's conflicts in terms of the therapeutic relationship.

A few beginning therapists find it easy to work with clients using a process approach. They have always been able to talk forthrightly with others about problems in their relationships, and it makes sense to them that what goes on for clients in other relationships will also come into play with the therapist at times. For these therapists, it is enlivening to make potential problems or misunderstandings overt and to talk about them openly but tactfully in the therapeutic relationship. The process approach simply gives them the permission they may need to work with clients in this direct and immediate way.

For many practicum students and interns, however, the process approach is more problematic at first. Although bringing clients' conflicts into the therapeutic relationship makes sense to them intellectually, they find it hard to imagine putting it into practice. First- and second-year graduate students often have little confidence, because they have had little direct clinical experience. Understandably, they are not eager to have what little confidence they may possess shaken by stepping outside of familiar bounds. Further, even if they think that this type of intervention might be useful with a particular client, they are often uncertain whether they are really observing the client's conflict or whether their perceptions simply reflect their own dynamics. Even if they feel confident that it is indeed a problematic relational pattern for the client, they are unsure of how best to address this reenactment and find a safe way to make it overt.

A few therapists feel concerned that to be forthright or direct with clients would be hurtful, because it would expose clients' pain, or they have seen directness used to induce shame or guilt. Counselors never want to intrude on clients, and impairing clients' sense of personal safety will not help them to progress in treatment. However, with appropriate care, therapists are able to address the process dimension in sensitive and respectful ways, without hurting or intimidating clients. We now examine six of beginning therapists' most common concerns about working with clients' conflicts in terms of the therapeutic relationship and suggest some possible solutions.

1. **Uncertainty about when to intervene**   Beginning therapists are often unsure when or if the client's conflicts are being reenacted in the therapeutic relationship, because they have not had much experience in conceptualizing clients' dynamics or systematically tracking the process dimension. As beginning therapists become more skilled and more comfortable with the tremendous amount of new information they are assimilating, they will become more confident and better able to attempt process interventions. It is a personal risk for the counselor to suggest, "I'm noticing something that may be going on between us right now, and wondering if. . . ." All counselors will probably be accurate with some of their observations; the client will not resonate with others. Therapists have not failed if they are inaccurate; they are simply trying to understand. This sincere effort is often an important aspect of the corrective emotional experience that they are

trying to provide for clients. Nevertheless, novice clinicians should not be in a hurry to make these interventions until they feel ready to do so. Responding out of a sense of pressure does not feel good and may cause the process the therapist is trying to create with the client to go awry: When therapists are disempowered because they did not feel in control of choosing when or how they intervene with the client, they usually will be far less capable of empowering the client.

How can therapists know when the client's conflicts are being enacted between them? One way to find out is to consult with a supportive supervisor. Once the supervisor has conceptualized how some aspect of the client's conflict is likely being replayed in the therapist-client interaction, the supervisee can explore his or her own personal concerns about trying to make this interaction overt with the client and role-play or rehearse alternative responses with the supervisor. It will often take two or three years before these responses feel natural and come effortlessly, because these new ways of responding do tend to break cultural norms and familial rules against forthright communication and bringing such immediacy to the relationship. Again, good advice for the novice therapist is to be patient.

2. Fear of offending the client   Many beginning therapists are concerned that clients will feel the therapist is being too blunt or personal if he or she directly addresses what is going on in the relationship. Therapists worry that the client will feel hurt, not like them, or—worse yet—just stare blankly back at them! Of course, process interventions, like any other interventions, can be made in blunt, insensitive, or otherwise ineffective ways. Warmth, tact, respect, a sense of humor, and good social graces will go a long way toward making every intervention more effective.

Counselors may also be worried about trespassing on social norms and expectations. Speaking directly about what may be going on between two people often breaks unspoken social rules, and most trainees are concerned that clients may be surprised or taken back by this. To help ease the transition toward this more open and genuine dialogue, therapists can offer contextual remarks that facilitate the bid for more open communication. For example, the first few times the therapist addresses the process dimension with a client, the therapist can provide an introductory remark that acknowledges the shift to another level of discourse:

> Can we be forthright with each other and speak directly about something?
> Let me break the social rules for a minute and ask about something that might be going on between us.
> I know that people don't usually talk together this way, but I think it would help if we could talk about . . .

Therapists will not be too blunt or direct for clients if they respond respectfully and provide transition comments such as these to help clients shift from the socially polite to this more honest, direct approach. Rather than being threatened by this invitation for more forthright communication, the vast majority of clients are hungry for a more authentic exchange, enthusiastically welcome it from the beginning, and are reassured by the counselor's willingness to address their problems and move closer to their real experience and problems.

3. Therapists' insecurity and countertransference issues   Being so genuine with the client may not feel safe for the therapists. When they begin to talk with the client about their relationship, therapists are becoming more emotionally present. In these circumstances, sessions become more intense and productive. At the same time, therapists' greater emotional availability (which is different from self-disclosure) makes them more open to being personally affected by the client, which can be anxiety-provoking. However, it's important to remember that a corrective emotional experience cannot occur unless the relationship is significant to both the client and the therapist. By talking honestly about their relationship, client and therapist create greater meaning in that relationship. Of course, this increasing *mutuality* (Jordan et al., 1991) may activate therapists' own conflicts at times. For example, in order to become more authentically responsive and closer, therapists must relinquish some control over the relationship, which may feel uncomfortable at times. Therapists may also be concerned that this genuine emotional contact with the client will lead to a loss of appropriate therapeutic boundaries and result in overinvolvement or acting-out on the part of the client or therapist. Clearly, every therapist will have moments of over-involvement. However, by applying the guidelines on optimum interpersonal relatedness in Chapter 6, therapists will be able to recognize when their own countertransference issues are prompting them to become too close to—or too distant from—the client.

Further, in making process comments, how do therapists know whether what is going on is due to their own personal issues or to the client's dynamics—to client-induced or therapist-induced countertransference? Therapists have two very useful methods of distinguishing whether the client's dynamics—eliciting maneuvers, testing behavior, projective identifications, transference distortions, and so forth—or the therapist's own dynamics are operating. In order to help sort out their issues from the client's dynamics, supervisees must have a supportive supervisor who closely tracks the therapeutic process. In the beginning, it is unrealistic to expect trainees to be a participant-observer who can maintain a balance of emotional relatedness and objectivity without assistance.

Another way beginning therapists can safeguard against confusing their own and clients' issues is by not jumping in and making observations about the therapist-client relationship as soon as they see something significant happening. Instead, it is more effective for the therapist to generate working hypotheses first, and then to wait and see whether the observation also applies to subsequent exchanges between therapist and client. If the issue, theme, or relational pattern is relevant for the client, it will systematically recur. If it does not, then it is likely that the therapist's own issues are involved, or that it is not an important concern for the client, and this hypothesis should be discarded. Thus, the therapist should not be cavalier about making process comments. When in doubt about whether an issue is the therapist's or the client's, it is best for the counselor to wait, gather additional information about the therapeutic process, and/or consult with a supervisor before raising the issue with the client. Remember, too, that all process comments are simply observations; they are offered to the client rather tentatively, as possibilities for joint exploration and mutual clarification, rather than as dictums, judgments, or established fact.

In psychotherapy the therapist's relationship with the client is the most significant means of effecting change. At the same time, the most likely reason for therapy to terminate prematurely is that the client's interpersonal coping strategies have activated reciprocal conflicts in the therapist, who has not been able to recognize or resolve his or her own issues. Therapists need to be aware of countertransference issues and to monitor the interpersonal process for signs of their emergence; to this end, therapists should generate working hypotheses about how each client's central conflict may be recapitulated in treatment (see Appendix B, part 6). The complex nature of the therapeutic venture requires that beginning and experienced therapists alike must:

- Make a lifelong commitment to being open to their own countertransference propensities
- Consult with supervisors and colleagues as a regular ongoing career activity
- Seek their own personal therapy when countertransference issues persist

Countertransference issues are most likely to create problems when therapists disregard them. Counselors who are aware that they are susceptible to countertransference—both client-induced and therapist-induced—and who discuss possible instances of countertransference with supervisors as they arise should feel unfettered about working with clients in a process-oriented manner.

4. Concern about appearing confrontational   As seen, some therapists fear that, if they address issues directly, the client will construe their behavior as a hostile or intimidating confrontation. Such a confrontive stance is not the intent here and never needs to occur. Process interventions can be presented in an aggressive or confrontational manner, as can any intervention, but there is nothing insensitive about simply communicating forthrightly. If therapists think these or any other interventions may be perceived by the client as hostile, judgmental, intrusive, or nonrespectful in any way, they should not use them. Therapists can either wait for what feels like a good time to try this particular intervention with this client or simply respond in other ways.

Similarly, some therapists fear that, by initiating more directness in therapy, they will "hurt" the client. These expectations of client vulnerability are usually inaccurate; most clients will welcome the opportunity for a more straightforward dialogue. As clients find it safe and productive to speak with the therapist about what goes on between them, novice therapists will feel more confident that it is possible to be sensitive and direct at the same time.

Fear about hurting the client takes on more personal significance for a few therapists who grew up with highly authoritarian parents (Chapter 6). Such therapists may have been exposed to double-binding family communications, in which there was a great discrepancy between what was being done (for example, the child may have routinely been threatened, humiliated, or hit) and what was being said (for example, "We are a close and loving family. There are no problems here, and everyone is happy"). The essential element in the double-bind is the clearly understood but unspoken family rule that the child cannot acknowledge the incongruency in any way. For example, the child in a double-bind cannot make these contradictory messages overt by saying, "You're telling me to clean my plate and making fun of me for being fat. Stop it, you're driving me crazy!" Children who have grown up with these double-binds are very afraid of making them overt because, among other reasons, further rejection and abandonment have been routinely threatened or implied.

Such double-binding communications occur in many dysfunctional families and are generic in emotionally, physically, and sexually abusive families. Therapists who have suffered such mistreatment may find it threatening (or at other times liberating) to use process comments and to *metacommunicate* about what is occurring. These therapists sometimes fear that, if they speak directly about what is going on, the client will be damaged by them, and their pathogenic belief in their own inherent destructiveness or badness will be confirmed. Thus, old threats can be evoked—but also resolved—by breaking double-binds and violating unfair family rules about how people should communicate. These dynamics are most appropriately resolved in the therapist's own therapy, rather than in the supervisory relationship.

5. Therapists' fear of revealing their own inadequacies   Therapists may fear that, if they bring out how the client's conflict is being replayed in the therapeutic relationship, their own inadequacies will be revealed. For example, the counselor may think: "Yes, there it is, the same conflict that he has with his wife is going on between us right now. And it sure is getting in our way, but I don't have any idea what to do about it." In that case, making the conflict overt may seem like the last thing in the world the therapist would want to do.

Whether the therapist chooses to address them or not, such reenactments will usually occur. By making them overt, however, therapists do not assume responsibility for *causing* the client's conflicts or pain because their effective responses have revealed them (a common pathogenic belief for many who enter the helping professions) or assume sole responsibility for changing them. Rather, the therapist is simply clarifying what is happening in the relationship and proposing that therapist and client work together to try to change the problematic pattern.

Therapists will feel more adequate if they follow the steps suggested earlier to address reenactments with the therapist. Let's review these again briefly. First, the therapist must *clarify the repetitive sequence of responses* that usually goes on in problematic interactions with others. Next, therapist and client must work together to clarify if and when these problematic patterns have been occurring in the therapeutic relationship. If they have, therapists can acknowledge their own participation in this shared conflict and express their willingness to change and to try to make this relationship better for the client. Working collaboratively, the therapist and the client find a new way to relate that does not continue the old relational pattern. By doing this, the client no longer feels invalidated, controlled, or rejected and is empowered to construct more rewarding relationships.

Therapists' concerns regarding inadequacy make sense here because when they make these reenactments overt, clients may feel discouraged or even blame-worthy. Initially, the client may feel hopeless because this reenactment confirms the client's pathogenic belief that relationships cannot be different from or better than they have been in the past. *The client may recognize intellectually that other ways of relating are possible, but this possibility holds little meaning experientially when the current enactment with the therapist confirms the client's relational templates and follows the problematic interpersonal scenario that he or she has experienced repeatedly.* However, therapists' feelings of inadequacy will be resolved when they realize they are already responding adequately to the client by recognizing that this is how the client's relationships have gone in the past, empathically understanding that this has been an unsolvable and painful problem, and accepting the historical validity of the client's hopeless feelings, based on past experiences. By doing this, the therapist is behaviorally demonstrating that the relationship between the therapist and client is no longer following the same well-worn, problematic patterns and is giving the client the corrective relationship he or she needs for change to occur.

By thoughtful preparation—mentally formulating and revising working hypotheses, keeping written process notes, and writing case conceptualizations—beginning therapists will be able to resolve legitimate concerns about their own adequacy and performance. In the beginning, therapists cannot successfully track the process dimension just by thinking on their feet. Counselors must be prepared by considering various hypotheses about what may be occurring along the process dimension if they are to effectively bring the client's conflicts into the therapeutic relationship. As we have already noted, therapists should not make a process comment until they have seen an issue occur several times and until they have some tentative understanding of what it may mean. Therapists must also wait when the client's coping strategy elicits anger or some other strong emotion, and should not respond until they understand their reaction and can reestablish their neutrality. Counselors who are well prepared with working hypotheses and case conceptualizations will also be better able to hear the client's disagreement with their ideas and to comfortably accept it.

As developing therapists become more confident, it will be easier for them to explore more open-endedly what may be occurring in the client-therapist relationship. For example, an experienced therapist, when feeling confused, may be able to inquire simply, "How did we get into this anyway?" In the beginning, however, therapists should wait until they understand better what is transpiring before addressing it with the client.

6. Concerns about owning personal power   A common but generally unacknowledged issue is counselors' conflicts about owning their own personal power. Many new therapists feel uncomfortable about allowing themselves to become so important to clients and having such a significant impact on clients' lives. Especially in the beginning, it is often anxiety-arousing to make strong interventions that can influence clients profoundly and to accept the responsibility that comes with such power. Thus, many beginning therapists overqualify their comments and dilute the emotional impact of their interventions.

There are many reasons why the therapist's own effectiveness can be anxiety-arousing. As discussed in Chapter 6, some therapists were parentified or aggrandized as children. For such therapists, legitimate competencies can be readily exaggerated into unrealistic all-powerful or all-responsible grandiosities. Initially, it may be exciting to be specially powerful in this way, but it soon becomes lonely, intimidating, or burdensome. As a result, these therapists may undo, or quickly retreat from, strong interventions they make or may avoid them altogether.

A more common reason for beginning therapists' reluctance to have a strong influence on clients is therapists' own separation anxiety or separation guilt. These therapists' experience, in their family of origin, was that competent or independent functioning threatened their emotional ties to parental caregivers. As the young child began to explore and to move toward greater autonomy, certain caregivers may have looked sad, acted indifferent, demanded more, ridiculed attempts at mastery, or paired anxiety with competence strivings. Therapists with these developmental experiences also tend to retreat from strong and effective interventions, and can often be heard saying, "I don't know what to do," "I'm afraid of hurting the client," or "I'm so screwed up myself that I have no right to try and help somebody else." All therapists certainly have their own personal problems and limitations. When such comments persist, however, it may indicate that the therapist wants to avoid the responsibility of clearly and effectively addressing the client's conflicts.

What can help trainees with these concerns? An affirming supervisory relationship is crucial if therapists are to embrace their own personal power and be as effective as they can be. The supervisor-supervisee relationship must be such that the supervisee

- Feels personally supported by the supervisor
- Receives conceptual information and practical guidelines when needed
- Obtains caring and nonjudgmental assistance in sorting out his or her own conflicted reactions toward the client that have been triggered by the client's eliciting maneuvers or by the counselor's own personal issues
- Is able to address and resolve the conflicts that are likely to arise in the supervisor-supervisee relationship

If the supervisee feels that the supervisor is solely directing the case, the interpersonal process between the supervisor and supervisee is problematic. Even though the supervisor is ultimately responsible for the client, supervisor and supervisee should be consulting together in a way that allows the supervisee to feel ownership of the treatment process. The supervisee must have primary responsibility for what occurs in sessions and must be free to act on his or her own ideas, while taking into account the supervisor's. Otherwise, the supervisee will not be able to experience his or her own successes and failures and thus will not learn from them.

Ideally, the supervisor will help the supervisee evaluate the effectiveness of interventions and consider alternatives without taking away the supervisee's own initiative. When the supervisor and supervisee become stuck on a conflict in their relationship, the same issues often carry over to the supervisee-client relationship and will be reenacted there. Often, the therapeutic relationship will not progress until the conflicts in the supervisory relationship are resolved. In contrast, when the supervisor and supervisee can maintain a collaborative relationship, the supervisee will be able to make more effective interventions that have a significant impact on the client.

Finally, same-gender supervisor-supervisee relationships may be preferable at times, although they are not essential for supervisees to develop their own professional identities and sense of efficacy. Similarly, people of color may feel more personal safety and fewer performance demands with a supervisor of the same race, especially in early training or in the event that their experiences with individuals of other races have been disempowering. However, positive interpersonal processes are possible with supervisors of any gender, race, or theoretical orientation. Issues of similarity or difference relate only to supervisors' initial ascribed credibility; supervisors of all cultural backgrounds can achieve credibility by their sensitivity to the familial and cultural backgrounds of both the supervisee and the supervisee's clients and to the ways in which these may impact the therapeutic process. As we have seen, it may take several years before therapists can allow themselves to have as significant an impact on clients as possible and to fully possess their own personal power; this is a gradual developmental process. If beginning therapists do not progress along this feeling-of-adequacy dimension as they move through their training, they should discuss this issue with their supervisors and/or seek treatment for themselves.

Closing

Therapists will find that, as in individual therapy, working with the process dimension is also the central focus in marital counseling, group therapy, and family therapy; the interpersonal patterns and relational conflicts between couples, family members, and group members similarly begin to involve the counselor. As in individual therapy, process comments that describe the current interaction with the therapist, or what is transpiring between group or family members, can be used to identify and change these maladaptive patterns. In couple counseling, for example, the therapist may repeatedly make explicit observations about the way in which the couple is interacting or talking together.

Therapist:
> You two are arguing about who takes out the garbage, but it seems like the issue is really about who has the power to make decisions or the right to tell the other what to do. I'm wondering if it would be helpful for us to talk about that issue more directly?

Finally, the underlying assumption in all interpersonal dynamic therapies is that the therapeutic relationship will come to resemble other prototypic relationships in the client's life. Clients' conflicts will emerge in the therapeutic relationship, especially if the therapist tracks the process dimension and attends to what reactions clients elicit from the therapist and others. However, therapists should not artificially construct situations to recreate clients' conflicts. In other words, therapists should respond to clients in genuine ways. Therapists should not contrive or manipulate events in the therapeutic relationship in order to strategically recreate clients' conflicts and so create "therapeutic experiences" for clients. This undermines the authenticity and integrity of the therapeutic relationship and diminishes the client's efficacy by taking control from the client; such treatment strategies arise from the therapist's agenda rather than from a collaborative working alliance. Commonly, this process merely recapitulates the client's developmental history and the client is again forced to comply with others or decipher incongruent metacommunications. In contrast, if therapists are willing to work with the natural expression of the client's conflicts in the therapeutic relationship, they can offer the client a meaningful, real-life experience of trust, empowerment, and change.

Suggestions for Further Reading

1. Practical guidelines to help therapists learn how to metacommunicate and when to intervene with the process dimension are found in Chapter 15 of *Handbook of Interpersonal Psychotherapy*, edited by J. Anchin and D. Kiesler (New York: Pergamon, 1982); see especially pages 285–294.

2. S. Cashdan's *Object Relations Therapy* (New York: Norton, 1988) provides superb clinical illustrations of the interpersonal process, helps therapists recognize and respond to metacommunicative messages, and demonstrates how the therapeutic relationship can be used to effect change; see especially Chapters 3 and 5.

3. The film *Ordinary People* (directed by Robert Redford, 1980) provides an effective illustration of the corrective emotional experience presented here and illustrates how clients can resolve their conflicts with others only after they have been able to confront and resolve them in their real-life relationship with the therapist.

4. Useful illustrations of how therapists can integrate other theoretical modalities with the interpersonal process approach are provided by P. Wachtel in Chapter 3 of *Handbook of Interpersonal Psychotherapy*, edited by J. Anchin and D. Kiesler(New York: Pergamon, 1982). Readers are also encouraged to examine two other texts by Wachtel that provide a practical integration of psychodynamic and cognitive-behavioral approaches, with effective case illustrations: *Psychoanalysis and Behavior Therapy: Toward an Integration* (New York: Basic Books, 1982) and especially *Therapeutic Communication: Principles and Effective Practice.* (New York: Guilford, 1993).

CHAPTER TEN

Working Through and Termination

Conceptual Overview

Working through and *termination* are two distinct phases of therapy. As clients continue to have a corrective emotional experience with the therapist, the task is to generalize this experience of change beyond the therapy setting and to achieve broad-based personality and behavioral change; the central conflicts that are being resolved in the therapeutic relationship must now be worked through in other relationships as well. Although this may be the longest phase of therapy for some clients, it is an exciting period of growth and change. During this phase, clients are assimilating and applying the emotional relearning that has occurred with the therapist. As the therapist helps clients address and resolve with others the conflicts that they have already been able to resolve with the therapist, therapy evolves to a natural close. The termination phase provides one last opportunity to reexperience and master old conflicts, to internalize the helping relationship with the therapist, and to successfully leave therapy.

Chapter Organization

This chapter has two sections. In the first section, we consider the working-through process. Following an overview of client change that places the working-through phase within the context of the overall course of therapy, we examine the details of the working-through process. Sample therapist-client dialogues illustrate how the emotional relearning that has occurred with the therapist can be generalized to the conflicts that the client is experiencing with others. We conclude this section by highlighting the client's transition from concerns about problems and conflicts to future plans that reflect life-enhancing aspirations and goals. This transition from conflict- to growth-oriented concerns is often marked by the emergence of the client's Dream.

In the second section, we discuss the critically important phase of termination. We examine specific guidelines for effecting successful terminations and common pitfalls that occur around termination.

Working Through
The Course of Client Change: An Overview

Beginning therapists usually lack a conceptual overview of how client change comes about. Because they have not worked with many clients or may not have experienced their own effective therapy, most beginning therapists do not have a sense of the ordered sequence in which change often occurs. In the previous chapter, we reviewed the sequence of therapist activities over the course of therapy. We begin this section by extending this framework for conceptualizing the course of therapy. To get an overview of the change process, we look at when and how clients resolve their conflicts and adopt new, more adaptive responses. Clients usually change in a predictable sequence of steps, and this unfolding course of change is reviewed from the beginning of therapy to the final working-through and termination phases.

For some clients, change begins as soon as they decide to enter therapy. That decision involves acknowledging that there really is a problem—that they are no longer going to deny or avoid the problem. It is also an acknowledgment that clients cannot resolve the problem alone and need help. Some clients recognize that their resolution to seek help is a healthy step forward, and they feel good about making this decision. In contrast, others may experience it as a self-condemning confirmation of their inherent failure, weakness, or inadequacy or as disloyalty to their family or a violation of cultural beliefs. When clients can allow themselves to feel good about the decision to seek help, or the therapist can help the client accept or reframe their need with more self-empathy or compassion, some initial relief of symptoms—such as anxiety or depression—may result. This internal commitment to seek help is a crucial first step.

Some small changes in feeling and behavior may also occur during clients' initial sessions with the therapist. Clients are reassured when the therapist:

- Invites them to express their concerns directly and fully
- Listens intently with respect and genuine concern for their distress
- Is able to enter their subjective worldview and grasp the central or emotional meaning that these concerns hold for them

As clients find that the therapist can understand their experience and respond compassionately to them, a supportive holding environment is provided, and this often reduces anxiety and depression. Such validation and support does not resolve most clients' conflicts, but it does engender hope and helps to relieve their distress. Taken together, these initial facilitative responses—which in fact are complex and sophisticated interventions that require skillful application—comprise the essence of crisis intervention treatment. They also form the bedrock of therapeutic effectiveness for clinicians working within every theoretical orientation and treatment modality.

More substantial changes begin when the therapist succeeds in focusing clients inward on their own thoughts and feelings and away from their preoccupation with the problematic behavior of others. When clients examine their own internal and interpersonal reactions, they often recognize how their behavior contributes to an interpersonal conflict and may be able to change their participation in it. Adopting an internal focus for change is also an important way to reframe how clients think about their problems. Redefining the conflict with the other person as, in part, an internal problem usually reveals a wider array of new and more adaptive alternatives that clients can adopt.

Next, as the therapist focuses clients inward on their own experience, their conflicted emotions will emerge. These conflicted emotions are central to most clients' problems. The pace of change accelerates as clients begin to experience and express these emotions and as the therapist is able to respond to each sequential feeling in clients' affective constellations. Clients have the opportunity to resolve their problems when they stop defending against their recurring conflicted emotions, feel the safety to begin experiencing and expressing them, and receive a more affirming and understanding response from the therapist than they have come to expect in other relationships. Far-reaching changes are set in motion as clients begin to understand, integrate, and resolve their affective constellations in this way.

As clients' conflicted emotions emerge, the therapist is also better able to conceptualize clients' relational templates and prototypic relational scenarios—and the pathogenic beliefs and core conflicted emotions that accompany them. The therapist then focuses treatment along these conceptual lines and clarifies how clients may be reenacting the same relational patterns, core beliefs, and affective themes throughout the various issues and concerns they present. Some further change may occur as the therapist helps clients to identify these patterns in their current interaction and to begin to recognize when and how they are occurring with others. In and of itself, this conceptual awareness may or may not lead to significant behavioral change, but it does facilitate the next and most significant point of change.

The same conflicts that originally led clients to seek therapy, and that they have been discussing with the therapist, will usually be reenacted in the therapeutic relationship, especially along the process dimension—that is, the way in which the therapist and client interact. This reenactment in the therapeutic relationship provides both an opportunity for therapy to fail—by recapitulating the client's conflict—and an opportunity for therapy to succeed—by resolving the conflict that has emerged in the therapeutic relationship. Therapy reaches a critical point when clients feel the therapist is responding in the same problematic manner as significant others have done in the past. This usually occurs in one of three ways.

1. *Transference:* At a point when clients are distressed or feeling vulnerable, they are likely to systematically misperceive what the therapist has said or done, on the basis of their maladaptive relational templates.
2. *Client-induced countertransference:* The client's eliciting maneuvers, interpersonal coping style, or testing behavior elicits the same problematic response from the therapist that the client has received in the past.
3. *Therapist-induced countertransference:* The client activates the therapist's own historical issues or personal conflicts that recapitulate the client's conflict.

These types of reenactments predictably occur in most therapeutic relationships. The client cannot resolve his or her conflicts with others unless this reenactment is resolved with the therapist, however. The therapist and client do this by jointly working out a different and more satisfying solution to the client's conflict in their real-life relationship. Only when the therapist and client can resolve this conflict in their relationship does the client have a real-life *experience* of change. This is the pivotal step that determines whether enduring change will occur. Note that, in most cases, the client will not change until corrective emotional experiences with the therapist have occurred repeatedly; one episode of relearning will not usually be sufficient.

At this point, the therapist begins in earnest to start generalizing the client's emotional relearning to other relationships. Let's now consider the working-through phase of therapy, in which client change with the therapist is integrated and generalized to others.

The Working-Through Process

When clients change, it begins in the therapy setting. With just the therapist's encouragement and advice, some clients who function well can successfully adopt new behavior. When the new behavior is integrally linked to their generic conflict and interpersonal coping style, however, most clients will need to practice this new response in the therapeutic relationship. Thus, therapists must give clients permission to try out new ways of responding in the therapy setting and must reassure clients that they intend to greet clients' new responses in affirming ways, rather than in the problematic ways that clients have grown to expect. If therapists give this permission verbally *and* enact it behaviorally, clients will soon test the therapist and try out anxiety-arousing new responses (for example, by being more assertive with the therapist or asking the therapist to respond to a need or personal request). If the therapist passes the clients' test—responds positively and adaptively to the new behavior, rather than metaphorically recapitulating unsatisfying but familiar responses—clients can resolve their conflicts and change.

This corrective emotional experience is a powerful agent for change but, as we have noted, one trial learning is not sufficient to effect change. Most clients will need to reenact this and other dynamic themes over and over again in the therapeutic relationship. As a rule of thumb, the more they have been hurt or traumatized, the longer or more often they will need to reexperience a new and safer response that doesn't fit the old template. In addition, the therapist must also help clients generalize this new experience of change to other relationships beyond the therapy setting. Typically—but by no means always—clients try out new ways of responding with others in the following progression.

First, change occurs in the client's relationship with the therapist. Clients then change with acquaintances they do not know well or with others who are not especially important to them. This is often followed by change with supportive others who are important to the client, such as caring friends, teachers, or mentors. Next, clients often change old response patterns with the historical figures with whom the conflicts originally arose, such as caregivers and important family members. Finally, clients change their behavior with primary others with whom the conflict is currently being lived out, such as spouses or children. This final arena is the most challenging, because the consequences have so much greater import.

During the working-through stage of therapy, some clients will rapidly assimilate the new ways of responding and readily apply them throughout their lives. Other clients, who have more pervasive conflicts or who have been traumatized or deprived more severely, will work through their conflicts more slowly. These clients must confront the same fears, expectations, and habitual patterns of response over and over again. Each time, the therapist must patiently point out how the same relational patterns, affective themes, or faulty beliefs are being played out in this situation and help clients master this particular manifestation of their generic conflict. In this regard, the generic conflict will be repeatedly expressed in four areas of client functioning:

1. In the current interaction with the therapist
2. When reviewing crisis events in the client's life or the crisis events that originally prompted the client to seek treatment
3. In current relationships with friends and significant others in which the client's emotional problems are being aroused
4. In historical relationships with family members

For many therapists, working through is the most rewarding phase of therapy. The therapist often serves as a cheering squad who takes pleasure in clients' newly obtained mastery and encourages them as they successfully adopt new responses in progressively more challenging situations. At this late stage of therapy, the therapist can also become more actively involved in providing information, suggestions, and guidelines to help clients enact new behavior. Some therapists find that cognitive interventions and behavioral procedures for rehearsing alternative responses are especially helpful during this phase. Once the interpersonal process is enacting a corrective experience, the client also responds more readily to other interventions, such as assertiveness training and parenting education; self-monitoring and self-instructional training; role-playing and other modeling techniques; and educational inputs or readings. The nature of the therapeutic relationship also changes during this period. As clients improve through the working-through period, they will gradually come to perceive the therapist in more realistic terms. As this occurs—that is, once the transference projections and eliciting maneuvers have been jointly identified and are being resolved—the therapist can disclose more personal information to the client, and there can be a gradual restructuring of the relationship on even more mutual or collaborative terms. However, the therapist must still be prepared to respond to all of the client's reenactments, relational patterns, and resistances.

At this point in treatment, clients are living out a new and different relationship with the therapist—a relationship that is resolving rather than reenacting the old relational patterns. As clients experience this new level of interpersonal safety and find that new ways of relating with others are possible, they will begin—on their own—to try out these more adaptive responses with others in their lives. This is a critical juncture in treatment, and therapists must actively help clients to anticipate and negotiate the successes and failures that are likely to follow. Clients need to be aware that, although some individuals will respond positively to the clients' new changes, others will not. It is exciting when others accept or respond affirmingly to clients' changes, but it can also be deeply discouraging when clients receive the same unwanted but familiar responses. Most clients will have experiences that affirm the new ways of being and other experiences that painfully reenact the old relational patterns or expectations. Thus, we must examine (1) how these setbacks are likely to occur; (2) what therapists can do to help clients anticipate these disappointments, which are an inevitable part of the change process; and (3) how therapists can prepare clients to respond more effectively to such disappointments than they have in the past.

As clients find that they can successfully and safely respond in new ways with the therapist and then with certain individuals in their current lives, their expectations of change in some other relationships may become unrealistically high. For example, suppose an adult survivor of childhood abuse has been deeply understood and validated by an effective therapist. Following this success with the therapist, the client risks disclosing this shame-laden secret to her husband and then her best friend, and they both respond affirmingly as well. Next, the client's profound wish is evoked that the abusing parent can now acknowledge or validate the reality of what occurred long ago or that the nonabusing parent, who could not be protective at the time, will now hear or believe the client. With expectations buoyed by successes with others, this client may address parents or other family members, only to encounter the same invalidation, scapegoating, and threats of ostracization that she received decades ago. Or, some clients may be profoundly discouraged to find that, unlike the therapist and some others in their lives, their marital partner cannot change or respond positively to this new behavior. Therapists must help clients examine their expectations of others before they try out new responses and must prepare clients by anticipating both how others are likely to respond and how the clients are likely to respond if they receive the old unwanted responses. Thus, in order to manage these potential setbacks and disappointments, therapists must prepare clients in the following ways:

1. By helping clients *realistically anticipate* how each new person is likely to respond to their changes. For example, if the client is a married woman, what will her husband probably say and do when she acts more assertively with him for the first time?

Client:

He's likely to smile at me like I'm a cute child or something, and then just change the topic as if I didn't even say anything—or didn't even matter.

2. By helping clients spell out in detail how they are likely to feel and what they are likely to say and do if they receive the old unwanted response to their healthier new behavior.

Client:

I think I'd feel ashamed, feel stupid for trying to stand up for myself, and just give up and withdraw inside—like I always have done.

3. By role-playing and providing new, more adaptive responses that, after rehearsed in therapy, clients can use at this discouraging moment, instead of repeating what they have always done in such circumstances.

Client:

I'm telling you something important right now and you are not taking me seriously. I don't like your smile, and I don't like feeling dismissed like this. Are you willing to listen seriously or should I stop now?

It's very likely that clients will continue to have experiences with others that reenact problematic relational patterns, confirm pathogenic beliefs about themselves and others, and evoke the familiar, painful feelings. By helping clients walk through these three steps, however, the therapist can help them find ways of changing their own internal and interpersonal responses—even when others are unable to change. For example, clients can learn that they are not to blame for, or deserving of, the way they were treated—even though the family still cannot affirm them. The married woman in our example learns that she can still advocate for herself and does not have to comply with her husband's dismissal and that she can establish new relationships in which her limits or opinions can be respected—even though her spouse cannot do this. In this process, not only does clients' current behavior change, but, more importantly, for enduring change to be sustained, clients' internal beliefs and responses are changing and taking on a more authentic form.

Thus, although the reenactments that occur as the client tries out new ways of responding with others can be deeply disappointing, they also provide an opportunity to further work through and resolve the client's conflict. Through each setback or crisis, the therapist and client gain further awareness of the client's conflicted relational patterns and deeper access to the painful feelings that accompany them. Working together, therapist and client recognize how the client's old patterns and expectations are being evoked again in this particular relationship and can observe how old response patterns or interpersonal coping patterns are still being activated. What is important about this process is that, in each successive encounter with the generic conflict, clients still have the opportunity to be understood and to receive a more satisfying response from the therapist, and thereby realize that they can change their own responses with problematic others and change within themselves. By repeatedly enacting a more satisfying interpersonal solution with the therapist, clients are able to feel more empowered and to adopt more effective ways of responding with some individuals beyond the therapy setting, even though others persist in their old patterns. Having developed healthier beliefs about themselves, others, and how relationships can be, clients will also begin to establish new, more affirming relationships that encompass these important changes. In other words, clients' resolution does not rest on changes in the parent, spouse, or others, as most clients believe when they enter treatment, but on how they change and respond to themselves and others in current interactions.

It is not always easy for therapists to provide a corrective emotional experience, however. Although clients are improving during the working-through period, they will still be struggling intensely with their central conflicts. Eliciting maneuvers, testing behavior, and transference distortions that fit the therapist into the old relational templates will still have to be negotiated. Especially when clients are distressed or feeling vulnerable, therapists must be able to accept and work through transference distortions, such as accusations that the therapist "doesn't really care about me," "is trying to control me," "isn't strong enough to handle me," "needs too much from me," or "will take advantage of me."

Furthermore, most clients will become discouraged at times by the repetitious working-through process. Seeing that the same old conflict is confronting them again, clients may feel that nothing has changed, that it is futile to keep trying, and that they should consider dropping out of therapy. At these critical moments, therapists must provide clients with a relationship that offers them the promise of change and must highlight the changes that have occurred. Therapists can make no guarantees of change, of course, and they cannot assume responsibility for the clients' motivation to continue. However, therapists must actively reach out and extend their care and concern to the clients. When clients feel discouraged, therapists must communicate that there can be another way, point out possible instances where such other ways have occurred, and convey that they remain committed to working with the clients, even though they know how frustrated or discouraged the clients are feeling right now. It is therapists' belief in the potential for change and their personal commitment to helping this particular client change that pulls clients through these expectable crisis points in the working-through phase.

As an illustration of the working-through phase, we now consider a case example in which the client changes in her relationship with the therapist and then, with the therapist's assistance, is able to extend that change to other conflict areas in her life. The episode in which clients' change first emerges—the *critical incident*—may be a dramatic, transference-laden conflict. In most cases, however, as in this example, it is a subtle aspect of the interaction with the therapist and may go unnoticed unless the therapist, informed by well-prepared working hypotheses, has been watching for such a development.

Our case example concerns Tracy, who habitually responded to others in a compliant, pleasing manner. Tracy always went along with others; she acted as if her wishes and feelings didn't matter. Tracy said that she didn't believe that others would ever be interested in listening to her or doing what she wanted to do. Her therapist linked Tracy's compliant behavior to her presenting symptom of crying spells and went on to explore Tracy's low self-esteem and feelings of worthlessness.

In addition, the therapist encouraged Tracy to respond differently in their relationship; the therapist showed that she cared about what Tracy had to say and invited Tracy to express her own opinions in therapy—even if that meant disagreeing with the therapist. Two weeks later, significant new behavior emerged in the therapeutic relationship:

Tracy:
You're right. I guess I never have felt very good about myself.

Therapist:
From other things you've said, it seems to me that your mother just wasn't there for you as a child. You didn't get the support you needed from her. Maybe we should explore that further.

Tracy:
No, I don't really want to talk about her. I remember her more fondly, but my father was pretty harsh with us.

Therapist:
You just said no, you don't want to talk about your mother, and that I was wrong about her.

Tracy:

I'm sorry; sure we can talk about her. What do you want to know?

Therapist:

Forget your mother; you just disagreed with me! You just told me you didn't want to do what I wanted to do and suggested what you thought would be better. That's great!

Tracy:

What?

Therapist:

You just did what we have been talking about. You expressed your own opinion, said what you thought, disagreed with me. I am so happy for you!

Tracy:

Aren't you mad at me? Didn't that hurt your feelings?

Therapist:

Oh no, we can disagree and still be close. That was a brave thing you just did.

Tracy:

I guess maybe it is OK for me to say what I think sometimes?

Therapist:

Sure it is. I care about what you think. I want to know what you feel, and I want to do things your way, too.

Tracy:

(tearing) But you're safe. It's easy to do that with you. You're not like other people.

Therapist:

Yes, I am "safe," and it is easier to do things like that with me than with other people. But you have changed to be able to do that with me; you've grown. You couldn't do that before, you know.

Tracy:

Yeah, that's right.

Therapist:

And if you can value yourself enough to express what you think in here with me, then you can begin to do that out there with other people as well.

Tracy:

Do you really think so?

Therapist:

Yes, I'm sure you can. If you can do it with me, you can do it with them.

Tracy:

Oh, I would love to be able to do this with my boyfriend.

Therapist:

Tell me about how this usually plays out with your boyfriend?

Tracy:

(describes the usual scenario)

Therapist:
OK, what do you want to be able to say to him instead?

Tracy:
(*describes an alternative scenario*)

Therapist:
Would it be helpful to you if we role-play that together? We might be able to work out together where the problems might develop if you try that with him.

Why is the therapist so excited about such a seemingly small change? Is this single manifestation of a new behavior all that is necessary for Tracy to resolve her problems? No, but this highly significant event brings her to the working-through phase of therapy. Within the safe orbit of the therapeutic relationship, Tracy was able to try out a new behavior that taps into her generic conflict and arouses intense anxiety. For Tracy, adopting such a seemingly insignificant self-assertive behavior arouses lifelong feelings of worthlessness. At the same time, she is discarding the primary means she has developed to protect herself from these painful feelings—that is, to go along with others. In order to resolve her conflict, Tracy will have to repeatedly experience the same sequence:

1. Her old relational pattern—having to go along—is aroused.
2. She tries out a new and more adaptive response with the therapist, which is to assert her own preference.
3. She receives a different and more satisfying response from the therapist: The therapist affirms Tracy's initiative.
4. She generalizes the new behavior to others beyond the therapy setting, by considering how she typically behaves with significant others in her life and, specifically, what she might want to say and do instead the next time she is in this situation with her boyfriend.

This repetitious reworking of the new behavior and the old conflict is called *working through.*

When Tracy returned for her next session, she made no mention of this incident with the therapist and said nothing about what went on with her boyfriend. Recognizing this as resistance, the therapist waited for the opportunity to acknowledge what had occurred between them the week before and to reassure Tracy that she was supportive of her new assertive response.

Therapist:
We haven't talked yet today about the important new way that you responded to me last week. Did you have any thoughts about that during the week?

Tracy:
No, not really.

Therapist:

Fine, but I just wanted to say again how happy I was for you. It felt very good to me to see you express your own opinion and be able to disagree with me.

Evidently, this reassurance must have given Tracy the permission she needed. The following week, Tracy was able to adopt the same type of assertive behavior with her boyfriend for the first time. This successful experience, in turn, encouraged her to confront the same issue in another more threatening relationship—with her father.

Tracy:

(*beaming*) Guess what! I did it! I've been dying to tell you all week!

Therapist:

Ha! What did you do?

Tracy:

I did what we talked about with my boyfriend. He was talking about something he thought we should do, and I disagreed with him and told him what I thought we should do instead. It felt great, and I don't think he minded either.

Therapist:

Good for you! You're on your way.

Tracy:

I can't believe how easy it was. I wish it could be that easy with my father. He's always putting me down when I say something.

Therapist:

Maybe you're ready to start changing your relationship with your father as well. How do you respond to him when he puts you down?

Tracy:

I get real quiet when he does that. I feel like crying, but I don't. I wish I could just tell him that I don't like it and I want him to stop.

Therapist:

Yeah, confronting him at the time he does that and setting limits with him would certainly change your relationship. It would be saying to him—and to yourself—that you value who you are and what you have to say and you're not going to let him hurt you like that anymore.

Tracy:

But I could never do that. I'd like to, but I just couldn't.

Therapist:

You haven't been able to do that in the past, but you have been changing in here with me, and now with your boyfriend, too. So maybe you can do something different with him as well. What holds you back from setting limits with him? What are you most afraid of?

Tracy:

He wouldn't take me seriously; he'd just laugh at me.

Therapist:

How would that make you feel?

Tracy:

Worthless. (*begins crying*)

Therapist:

So that's where that awful feeling comes from.

Tracy:

Yeah. (*long pause*) It makes me mad, too.

Therapist:

Of course it does! It hurts you very much when he diminishes you like that, and you have every right to be angry!

Tracy:

It's not fair! I don't want to let him do that anymore. It's not good for me.

Therapist:

That's right, it's not good for you, and you don't deserve it. Now that you can see that clearly, maybe you don't have to go along with it the way you used to.

Tracy:

What can I do?

Therapist:

What would you like to do the next time he does that?

Tracy:

Well, maybe I can tell him to stop it—tell him that I just don't want to be treated like that anymore.

Therapist:

I think that would be a strong and appropriate response on your part. But before you try it, let's explore further how that conversation might play out. Maybe we can identify the trouble spots and role-play some responses to help you through them.

In ever-widening circles such as these, clients confront and work through the same relational patterns in the different spheres of their lives.

Notice that the critical incident in this case was very subtle: When the therapist suggested that Tracy talk about her mother, Tracy said she didn't want to. Therapists could easily miss such slight shifts in behavior unless they have conceptualized the client's dynamics and hypothesized how the client's conflict might be reenacted in the therapeutic relationship. If Tracy's therapist hadn't *anticipated* that Tracy would tend to comply with her, and if the therapist hadn't been alert for any budding sign of self-direction or assertiveness from Tracy, the opportunity for change would most likely have been lost; that is, the therapist might have responded by focusing on the content (by saying, for example, "But it's important to look at your relationship with your mother because . . . ") rather than on the process (Tracy's self-assertion). In that case, Tracy would have complied with the therapist and dutifully talked about her relationship with her mother. The interpersonal process between Tracy and her therapist would then recapitulate her maladaptive relational pattern, and no change would result.

When working with a client who seems unable to change, the therapist may eventually grow discouraged and become critical toward, or disengaged from, the client. In those circumstances, the therapist should consider whether the therapeutic interaction is subtly reenacting the client's maladaptive relational patterns. That would explain why the client is unable to change, even though the therapist and client continue to talk about adopting more assertive behaviors with others, the therapist sincerely supports and encourages this, and the client learns useful skills that should facilitate the new behavior. Such subtle reenactments along the process dimension are predictable. If therapists utilize process notes (Appendix A) and generate working hypotheses (Appendix B), they can get an idea of how each particular client's conflict may be reenacted in the therapeutic relationship. By anticipating what to listen and watch for, therapists will be able to recognize such reenactments more quickly and respond to them more effectively.

Finally, when clients are generalizing change to relationships outside the therapy setting, they will at times experience nonaffirming responses. For example, Tracy's father may not respond as well as her boyfriend did. Therapists can use this experience as an opportunity to help clients see that some relationships can be different and some can't—and that nonvalidating responses from certain people do not constitute evidence of the clients' lack of inherent worth. In this way, the working-through phase focuses on both interpersonal and intrapersonal change and growth.

From Present Conflicts, through Family-of-Origin Work, and on to Future Plans

In the working-through phase, the therapeutic action is primarily in the present; the significant events are current interactions with the therapist that enact resolutions of problematic relational patterns. Simultaneously, these events propel intrapersonal reworkings of clients' old templates, which will be critical in assisting the clients to assimilate this new, more authentic, and more satisfying way of relating. These corrective experiences are then generalized, as clients try out with others the new ways of relating they have found with the therapist, which are now also congruent with the new ways in which they perceive themselves and how relationships can be. As we have seen, clients will have both successes and disappointments; some individuals will welcome the clients' changes, whereas others continue along well-worn, problematic lines. As clients explore the potential for change in current relationships and come to terms more realistically with the possibilities and limits, two new subphases in the working-through process may emerge. First, clients, on their own, often start to look back in order to better understand the formative experiences that shaped these problems they are resolving. In the first part of this section, we examine therapeutic guidelines to help with this family-of-origin work. Second, as clients develop a more meaningful narrative that helps them better understand their life and how they have become who they are, they also begin to look ahead: to think more about what they want their life to be in the future; and to reformulate life plans to better fit the person they are becoming. This subphase, often ushered in by the emergence of the Dream, will be explored in the second part of this section.

Family-of-origin work It can be liberating for clients to explore the familial interactions and developmental experiences that shaped their current conflicts. As they gain understanding of their childhood dilemmas, they become more accepting of the choices, compromises, and adaptations they have had to fashion in their lives. They also begin to feel empathy—both for themselves and for the personal limitations or difficult life circumstances that led their caregivers to respond in ways that may have been hurtful or problematic. Clearly, this developmental perspective can be profoundly enriching. However, in most cases, and especially early in treatment, it is not very productive for therapists to try and lead clients back in time, make historical interpretations, or attempt links between current relational patterns and historical family relationships.

Therapist:
>   I keep hearing the same theme: that what you are complaining about with your spouse seems in some ways to be similar to what used to occur with your father.

In contrast, from a shared or mutual exploration of familial and developmental experiences, therapists can indeed learn a great deal about the corrective experiences that clients need in treatment, and that will allow therapists to begin generating fruitful working hypotheses and case formulations. For example, the therapist might ask:

>   What was it like to be a child in your family?
>   Tell me what your parent would say and do when you were successful or acted stronger. What was the look on your mother's face when you showed her that spelling award?
>   Tell me about your parents' marriage. Can you bring it to life for me?

Early in treatment, this information will inform therapists about the formative experiences that shaped the client's relational templates. This provides therapists with invaluable guidelines to begin formulating the kinds of responses that are likely to reenact problematic relational patterns for this client and the types of therapeutic responses that are more likely to be reparative for this client, to create greater safety, to pass the client's tests, to disconfirm pathogenic aspects of the client's self-image and dysfunctional expectations of others, and to provide a corrective emotional experience. Early in treatment, however, most clients will not be able to make meaningful bridges between current and past relationships from this initial exploration initiated by the therapist. *Usually, when the counselor highlights seemingly obvious connections, the relational patterns do not come alive for clients until they have experienced a new and more satisfying response to these old relational patterns with the therapist.* Later in treatment, many clients will spontaneously lead the therapist back to developmental experiences. These client-initiated explorations, which often occur immediately after clients have a corrective emotional experience or have successfully adopted a new response with others, continue throughout the working-through period. They are enlivening for clients and routinely lead to productive behavior change in current relationships. Thus, unlike other counseling theories, the interpersonal process approach does not suggest that behavior change leads to insight or, conversely, that insight leads to behavior change. Although both of these change processes occur at times, a different mechanism of change is proposed here: Both meaningful insight and sustainable behavior change result from or immediately follow clients' new or corrective relational experience with the therapist.

As clients make their own meaningful links between formative and current relationships, they often ask the therapist what they should do about their current relationships with family members. We now examine two broad guidelines to help clients with this family-of-origin work. First, harkening back to the internal focus for change described in Chapter 4, we see how clients must change how *they* respond to old problematic relational patterns in their current interactions with family members. Second, we look at the concept of *grief work*, in which clients mourn, and come to terms with, what they have missed developmentally. Working in both of these areas results in an interpersonal and internal resolution of family dynamics.

Clients who struggle with more serious, pervasive, or long-standing problems usually have family-of-origin work to do. Some clients do not wish to discuss problems from the past with caregivers or family members; in this case, therapists should not pressure the clients to do so. However, some clients will want to acknowledge past conflicts with parents and other family members and try to resolve them. It is enormously gratifying—and a powerful impetus for change—when such rapprochements succeed. As we have already emphasized, however, such current resolutions of historical problems often do not occur. Occasionally, parents have grown and changed but, in many cases—especially with more serious problems—the client will often receive the same invalidating, blaming, or hurtful responses as in childhood. This is especially painful when the client's unacknowledged motivation for talking with the parent is a wish to get the parental love, affirmation, or protection that the client has always longed to receive. If the therapist has not prepared the client for this potential disappointment, the client may experience profound feelings of helplessness and hopelessness. Further, the client often will be breaking family rules by addressing conflicts with parents directly, and in many cases punitive homeostatic mechanisms will be set in motion. For example, when Tracy confronted her father about always putting her down, he again disparaged her—just as he had for the past 20 years. In addition, her mother threatened to disown her for being "disrespectful" and "ungrateful," and her depressed and obese sister telephoned and tried to make her feel guilty for "hurting Daddy" and "stirring up trouble"! Thus, the problem is not simply the parent's mistreating the child. Instead, more serious consequences—such as the self-hatred, bulimia and dysthymia that Tracy suffered—result because the parent and extended family system typically make the adult offspring again feel responsible for or deserving of the parent's original and current mistreatment.

Clients usually feel hopeless and blame themselves when parental caregivers and the broader family system cannot change. Although these clients often go on to act helpless in other current relationships, in actuality they are not. Their interpersonal resolution does not rest on the parent changing, but on how they change and respond in current interactions with living parents or, equally important, in their ongoing internal relationship with deceased caregivers. In order for clients to make enduring changes in deep-seated relational conflicts, they must change their own responses in these prototypic interactions. For example, Tracy's father was too limited to be able to hear her concern and to talk with her about the problem. However, Tracy stopped going along with her father's disparagement of her. Instead, Tracy began setting limits with her father and bravely metacommunicated with him, by saying: "I keep asking you to 'Stop it' when you put me down, and you laugh at me and keep on doing it. I can't make you stop, but I don't have to pretend that it's OK anymore. It hurts me and I don't like it." Although her father stayed in the same disparaging mode and again made fun of her, things changed profoundly for Tracy when she changed how she responded to him. As she was able to sustain this stronger stance toward him and the family system over the next few months, she felt less insecure and intimidated with them and others in her life than she had ever been. Her long-standing feelings of worthlessness and depression largely disappeared, and her intermittent battles with bulimia improved. These far-reaching changes occurred for Tracy even though her father—and the broader family system—could not change. A critical aspect of this was that Tracy's attitude toward herself changed; these changes were facilitated by her increasing internal focus and corrective emotional experiences with her therapist and supportive others.

Thus, clients change, in part, by changing how they respond in current problematic interactions with parental caregivers, other family members, and significant others in their lives. Clients also change by doing grief work—coming to terms with the feelings left in them by hurtful parent-child interactions. Clients do not need to discuss with parents and other caregivers problems in historical relationships, but they do need to stop disavowing and acknowledge to themselves what was legitimately wrong, set limits, stop their own participation in any current or ongoing mistreatment, and grieve for what they have missed. Regardless of whether formative attachment figures can change—and whether they are living or deceased—clients can resolve these conflicts through their own internal or personal work.

Testing the reality of new possibilities with family members is important and may or may not lead to improved relations. Often, aging caregivers cannot change; even when they do, grown offspring still need to acknowledge what was wrong in the past and mourn what they missed developmentally. The same model applies again: The therapist must provide a supportive holding environment that allows clients to fully integrate the sadness, anger, shame, and other feelings that were too threatening or unacceptable to fully experience, contain, and integrate before. The therapist must also help clients come to terms with both the good news and the bad news about their situation. The good news is that, through grief work, clients can disconfirm the pathogenic belief that they were in some way to blame for their rejection, mistreatment, or parentification. The bad news is that clients' developmental needs—such as the need for secure affectionate ties—were not adequately met then, and these same developmental needs cannot now be met in current adult relationships (for example, with a spouse, friend, or therapist). Only after such losses or deprivation can be acknowledged as real, fully experienced, and integrated (that is, mourned rather than disavowed) can clients go on for the first time to get adult versions of their needs met in current relationships. As long as developmental needs are operating unacknowledged, clients are unable to get legitimate adult needs met in current relationships and to feel sustained.

Two countertransference propensities can prevent therapists from helping clients achieve resolution of such family-of-origin work. First, because of their own splitting defenses, therapists sometimes do not have the breadth to help clients accept both what was good and what was problematic in their development. Therapists' countertransference propensities lead them to make one of two errors. Some therapists characteristically want to deny or downplay the extent and continuing influence of real and painful experiences.

Therapist:
That was then. You need to forget about what happened in the past. Let's try to do something about the problems you're having now.

Other therapists only want to criticize or blame the caregivers and ignore the strengths and contributions that were present in the relationship.

Therapist:
Your father was stupid and sick to do that to you and your sister!

Instead, therapists need to affirm the full reality of what went wrong in formative relations, while appreciating the (internalized) child's need to maintain whatever ties can be preserved. Except in extreme circumstances, therapists should discourage the client from ending all contact with caregivers. When a client breaks off all communication with family members, splitting defenses are usually operating, and internal resolution or lasting change is unlikely. In such cases, the therapist will be idealized, and clients then will be compelled to compulsively recreate the split-off bad side of their conflicts in other relationships or to make themselves "bad." Therapists will be far more effective—and clients will be better able to stay with this anxiety-arousing and guilt-inducing material—when counselors can acknowledge the bad news without taking the parental figure away.

Therapist:
I can see how much it hurt you when he did that.

VERSUS

Therapist:
I can't believe he would do that to you. He's a Nazi!

The second countertransference propensity that interferes with family-of-origin work is that therapists often want to bypass the work of mourning. When clients acknowledge the ways in which they were hurt or missed what they wanted or needed, painful feelings are aroused. Therapists may find it difficult to help clients come to terms with those feelings. Most clients do not want to end the hope—or fantasy—of getting what they missed; some may even become angry when therapists acknowledge these limitations. Therapists' reluctance to address these issues will be reinforced if, as routinely occurs, such material arouses therapists' own unfinished family-of-origin and grief work. However, if the therapist tries to bypass the work of mourning and move too quickly to a problem-solving approach—for example, by looking for substitute supportive relationships— clients will not be able to internalize much from these practical interventions. Only after original losses have been acknowledged, felt, and integrated (that is, grieved for in a supportive holding environment with the therapist) can clients open up their adult emotional needs and feel fulfilled or sustained by new relationships in a way they have not been able to risk before. As clients improve in these very significant ways, they become able to leave the past behind, to turn the page, and to begin looking ahead to a different future.

Next, we examine the emergence of the Dream and the second subphase of the working-through period.

The Dream   Clients make adaptive behavioral changes and feel subjectively better as they work through their conflicts. As they improve, the focus of therapy moves away from developmental conflicts and current interpersonal problems and gravitate toward plans for the future and existential issues regarding personal choices: How can clients create more meaning in their lives, live more authentically, and make the most of the time and relationships they have? This exciting transition from psycho- pathology, neurosis, and conflict to health, growth, and creativity is signaled by the emergence of the Dream.

In *The Seasons of a Man's Life*, Levinson (1978) describes the profound influence of the Dream in shaping the structure of adult life and the course of personality development. As Levinson points out, the Dream does not refer to casual waking or sleeping dreams but, in the largest sense, to the kind of life that one wants to lead. At first, the Dream may be poorly articulated and tenuously connected to reality, but it holds imagined possibilities of self in the adult world that generate excitement and vitality. It is the central issue in Martin Luther King's historic "I Have a Dream" speech or in Delmore Schwartz's story, "In Dreams Begin Responsibilities."

The Dream has roots in the grandiose and unrealistic hero fantasies of adolescence, but it is more than that. In early adulthood, the Dream still has the quality of a vision—of who and how we would like to be in the world. In this way, the Dream is inspiring and sustaining for the individual, even though it may be mundane to others. For example, individuals have Dreams to be a good husband or wife, a respected community leader, an ethical attorney, a skilled craftsperson, a medical doctor who relieves suffering, a caring teacher, a successful but honest businessperson, an artist, or a spiritual leader. *If the individual is to have purpose and a sense of being alive, occupational and marital choices must incorporate some aspects of the Dream. When the Dream has been abandoned or set aside, life will not be infused with vitality and meaning, even though the person may be successful.*

Up until the working-through phase of therapy, therapists are usually responding to the despair of broken dreams, the disillusionment of unfulfilling dreams, and the cynicism of abandoned dreams. In many cases, the life-infusing Dream that Levinson has articulated has not yet been addressed in therapy. In the working-through phase, however, as clients are resolving their conflicts and emerging as healthier individuals, therapists must help clients to articulate and renew their Dream and to find ways of incorporating aspects of their Dream in their everyday lives. To achieve this, therapists must first be able to differentiate the Dream from the broken dreams that underlie clients' presenting symptoms. These broken dreams often reflect failed attempts to rise above their conflicts and become special, as discussed in Chapter 7.

In the process of working through, clients gradually recognize their compensatory strivings to be special and to rise above their conflicts by pleasing others (moving toward), achieving success and power (moving against), or becoming safely aloof and cynically superior (moving away) and progressively relinquish these interpersonal coping styles. By doing so, they stop blocking the expression of their generic conflict. This presents the opportunity for clients' conflicts to be expressed and resolved within the therapeutic relationship. However, therapists must anticipate that, when clients with more problematic developmental histories relinquish these attempts to rise above their conflicts, they may experience a sense of devastating failure and a total loss of self-esteem. Thus, one important component of helping clients resolve their conflicts is to replace these defensive, grandiose strivings with clients' own attainable, yet sustaining, Dream.

Arthur Miller's (1949) Pulitzer Prize–winning play, *Death of a Salesman*, poignantly illustrates how the neurotic striving to rise above conflicts inevitably fails, how it can lock clients in their generic conflicts for a lifetime, and how it stifles the Dream. The protagonist, Willy Loman, has adopted a moving-toward interpersonal style to cope with his profound feelings of inadequacy and shame. With his smile and his freshly shined shoes, Willy pathetically strives not just to be liked but to be well-liked. In a brilliant exposition of multigenerational family roles and relationships, Miller shows how Willy's grandiose strivings to rise above his shame-based conflicts by being well-liked are also acted out through his oldest son, Biff. Willy exaggerates Biff's accomplishments as magnificent, excuses or ignores his shortcomings, and never responds to the reality of his son's actual feelings, needs, or wishes.

Near the end of the play, both father's and son's defensive strivings to rise above fail. Willy is fired from his job—a crushing humiliation that shatters his myth of being well liked and his lifelong coping strategy of winning approval. That same day, Biff fails to secure an important business opportunity that his father has encouraged. With this setback, Biff realizes that he has been living out his father's myth that he would become a fabulously successful entrepreneur. In the closing scene, father and son respond very differently to their respective crises.

Biff is able to relinquish the unrealistic script of superachievement and success that his father has demanded of him. In the last scene, Biff begs his father to release him from this grandiose role, to "speak the truth" in the family for the first time, to acknowledge that he is just another regular guy—nothing more, nothing less—and to let that be good enough. However, Willy cannot relinquish his own rising-above defense and, therefore, cannot grant Biff's plea to "let me off the hook." Instead, Willy alternately rejects Biff and idealizes him as magnificent.

Biff finally recognizes the conflict and interpersonal process that he and his father have been enacting all of their lives. By directly confronting his father, Biff is able to change his part of their conflict. Although this confrontation does not lead to an interpersonal resolution with Willy, it does lead to an internal resolution for Biff. Soon after the confrontation, Biff decides to leave home. He says that he now knows who he is and what he wants to do with his life. He is going to pursue his own Dream: to leave the city, move out west, and work outdoors where he can feel the sun on his back and smell the grass. Whereas Biff could relinquish the neurotic strivings to be special and allow his own attainable Dream to emerge, Willy could not.

In one last attempt to change their relationship and have a genuine closeness, Biff reaches out and puts his arms around his father. Horrified, Willy recoils in fear and disgust. Although his lifelong obsession has been to win the approval and respect of others, he must rigidly reject exactly what he has always wanted from his son. To accept the love that Biff was offering—to get what he wants—would reveal Willy's own unacceptable generic conflict—his unmet needs to be respected and wanted, his profound sense of being unlovable, and his intense feelings of inadequacy and shame. With no understanding of what these disavowed feelings are about and no supportive relationship to help him contain this flooding, his emotions would be intolerable and overwhelming. Thus, Willy is now trapped in his conflict: He is no longer able to maintain his own (or Biff's) defensive strivings to rise above, and he is unable to let his true feelings and concerns emerge. Desperately bound in that unresolvable conflict, Willy's only solution is to end his own life.

As clients' neurotic strivings to be special and rise above their conflicts fail, their generic conflicts are revealed. Without a holding environment to help them contain their feelings, some clients—like Willy—feel hopeless and see suicide as the only way out. In contrast, if the therapist can provide a different and more satisfying response to the conflict, clients can begin to come to terms with these feelings in a new way, resolve their conflicts, and change. As clients begin to improve during the working-through phase, the Dream will often emerge.

In order to help clients change, therapists must support and encourage the Dream. They can do this in many ways. First, therapists must listen for the material or issues that genuinely enliven clients, engage their intrinsic interest, and bring them pleasure. By attending closely to what clients really want to do and inquiring about what feels right for them, therapists can help clients discover and articulate their Dream. Some clients' Dream is completely unarticulated and undeveloped, because their caregivers never accepted their feelings or responded to their interests. These clients, with a less developed sense of self, are unaware not only of what they want to do in life but, more basically, even of what they like or dislike. Therapists can help these clients discover and successively articulate their feelings, preferences, interests, values, and ultimately their Dream by repeatedly

- Asking clients to attend to what they are experiencing at that instant
- Orienting clients to listen to themselves and simply be aware of what they want to do in different situations—even if they cannot act on their wishes (for example, to realize "I'm not enjoying this right now")
- Being interested in and entering into clients' subjective experience or worldview

- Calling attention to, and trying to participate in, what seems to interest or hold meaning for clients
- Acknowledging and taking overt pleasure in what clients do well
- Encouraging clients to act on their own internal preferences when possible (for example, even something as simple as saying, "I'm sleepy. I'm going to take a nap")

Although many clients have a more developed self than this, some will need to begin attending to their own internal experience in these very basic ways. It is very rewarding for the therapist to help clients differentiate a self by responding— perhaps for the first time in their lives—to their own feelings, perceptions, and interests.

As clients begin to experience their own inner life more fully, and the therapist responds affirmingly, their Dream will start to emerge. At this point, the therapist can do several things to help clients bring aspects of their Dream into their current lives. For example, the therapist must provide a safe practicing sphere in which clients can discuss and explore various possibilities of their Dream without having to become committed prematurely to any action. The therapist can also help clients tailor their Dream to reality and find ways to express aspects of their Dream in their everyday lives. Moreover, the therapist can help clients identify training or education routes that will facilitate realization of their Dream and must also give clients permission to develop close relationships with others who can inform them and act as mentors in their chosen fields and pursuits.

Finally, clients will confront the same generic conflicts in exploring their Dream that they have experienced in other aspects of their lives. Thus, the therapist will also have to help clients work through their old conflicts that are aroused by pursuing their Dream. For example, the therapist will have to help clients differentiate between their Dream and their interpersonal coping style and defensive strivings to rise above (for example, strivings to feel secure by being loved by everyone; to gain a sense of personal adequacy by becoming wealthy and powerful; or to escape criticism or rejection by being safely aloof and superior). *Progress will not be simple or unimpeded for many clients, however, because pursuing what they really want to do or getting what they want may threaten object ties and arouse separation guilt and separation anxiety.* However, as clients successively work through these conflicts and find that they do have a right to their own life, they can follow through on realistic plans for attaining aspects of their Dream, and therapy begins evolving to a natural conclusion.

Termination

Termination is an important and distinct phase of therapy that must be negotiated thoughtfully. Ending the therapeutic relationship will almost always be of great significance to clients. The way in which this separation experience is resolved is so important that it influences how well clients will be able to resolve future conflicts, losses, and endings in their lives. It also helps determine whether clients leave therapy with a greater sense of self-efficacy and the ability to manage their own lives more successfully. Therapists often underestimate how powerful an experience termination is for clients. It holds the potential either to undo or to confirm and extend the changes that have come about in therapy. Thus, therapists must be prepared to utilize the further potential for change that becomes available as therapists and clients prepare to end their relationship.

Early in their training, counselors often ask, "How will I know when it's time to end therapy?" It is time to end when clients have achieved significant relief from their symptoms, can respond more flexibly and adaptively in current situations rather than cataloging all experience into predisposed categories, and have begun to take steps toward promising new directions in their lives. Therapists know that clients are ready to terminate when they have converging reports of client change from three different sources: (1) when clients report that they consistently feel better, can respond in more adaptive ways to old conflict situations, and find themselves capable of new responses that were not available to them before; (2) when clients can consistently respond to the therapist in new, more direct, egalitarian, and reality-based ways that do not enact their old interpersonal coping styles or maladaptive relational patterns; and (3) when clients' significant others give them feedback that they are different or make comments such as, "You never used to do that before." With this convergence of perceptions that important changes have occurred, it is time for therapy to end. Before going on to discuss successful termination, we must distinguish between two types of endings: natural and unnatural.

The termination sequence just described is a natural ending, because clients' work is finished. One of the therapist's primary goals in these natural endings is to affirm both sides of clients' feelings about ending: to take pleasure in clients' independence and actively support their movement out on their own; but also to let clients know that the therapist will accept their need for help or contact in the future. Thus, in contrast to clients' old templates, therapists can clarify that they will not be disappointed or burdened if clients want to send a letter to stay in touch, call to share important life events such as marriages and births, or recontact the therapist for help if they have problems in the future. Many clients did not have permission to be both related and individuated at the same time: Leaving signified a complete loss of emotional support and connectedness, and staying meant not having their own voice, views, or feelings distinct from those of the people they were with. Thus, the termination phase provides these clients with an important opportunity to further resolve the separateness-relatedness dialectic and can be a potent corrective emotional experience. Assured of the therapist's support for both sides of their feelings, clients can further internalize the therapist and keep all that she or he has offered, solidify their own sense of self and personal efficacy, and successfully end the relationship.

Unfortunately, most graduate-student therapists do not have the satisfaction of seeing many cases evolve to a natural close. Instead, therapy ends before the client is finished, because the graduate student must move on to another placement or because the school year ends. Sometimes the therapist and client have made substantial gains at the point when external constraints demand termination; at other times, the treatment process may be only in midstream. In either case, significant therapeutic gains can still be made. To accomplish this more difficult task, however, it is essential that the therapist and client address and work through the complex issues that an unnatural or externally imposed ending usually arouses. Because graduate-student therapists often have to initiate terminations before clients are ready, we will focus on unnatural endings in the discussion that follows. It is necessary to examine unnatural or imposed endings closely, because they routinely evoke sad, angry, and distancing reactions in the client, which, in turn, activate guilt and other countertransference reactions in the therapist that can undo the significant gains that have been made in a therapeutic relationship.

Accepting That the Relationship Must End

It is usually difficult for both the therapist and the client to end their relationship. In many cases, bringing up the topic of termination will arouse emotional responses that are far more intense and difficult than either the therapist or the client has anticipated. As a result, therapists and clients regularly collude to deny the reality of the impending ending and to avoid the difficult feelings it arouses for both of them. When therapists and clients stumble in this way and do not confront the ending squarely, their interpersonal process will often recapitulate clients' generic conflicts and prevent clients from resolving their conflicts as fully as they could have.

The single most important guideline for negotiating a successful termination is to unambiguously acknowledge the reality of the ending.

Therapist:
> We have four more sessions left before the school year ends and we have
> to stop. Let's talk about what this ending is going to mean for both of us.

The therapist and client can then discuss all of the client's positive and negative reactions about the ending, especially the client's emotional reactions toward the therapist. Although the therapist and client may both want to avoid this topic, the therapist cannot let that happen. Once the therapist and client mutually decide that the client is ready to terminate, or that the therapist must terminate because of outside constraints, they need to set a specific date for the final session.

Because the therapist and client need time to work through their ending, the final session must be scheduled at least two weeks hence.

In order to have a successful ending, the therapist and client mutually establish a final date and then explore together the client's reactions to ending their relationship. In some cases, it will be difficult for the therapist to set a specific termination date and discuss the ending, because this arouses the therapist's own unresolved conflicts over this ending—or more commonly, endings in general. In addition to the therapist's own countertransference propensities, which are especially likely to be evoked around terminations, setting a clear termination date will often bring up the client's presenting symptoms and central conflicts again. In order to further resolve their conflicts, some clients may need to work through again, in the termination phase, aspects of the same issues that have been dealt with before. In other words, when termination becomes a reality, some clients will temporarily retreat from or undo the changes they have made. For all of these reasons, therapists need to acknowledge the termination forthrightly, count down the final sessions, and repeatedly invite clients to discuss their reactions to the termination:

> After today, we will have only three sessions left. How is it for you to hear me say that?
>
> We have only two sessions left now, and I think it's important that we talk together about our ending. What comes up for you when I acknowledge that soon we are going to have to stop working together?
>
> Next week will be our last session. We've accomplished some things together, but problems remain and our work is unfinished. What do you find yourself thinking about our relationship and the time we have spent together?
>
> This is the last time we are going to be able to meet. I'm feeling sad about that, and wondering what kind of feelings you're having.

This type of countdown precludes any ambiguity that clients may have about the ending. For example, it will prevent the client from asking, at the end of the last session, "Will I be seeing you next week?" This approach also keeps the therapist from acting out and avoiding the separation as well—for example, by saying at the end of the last session, "Oh, I see our time is up. Well, I guess that's it." When therapists do not address the termination directly, they are usually acting out their own separation anxieties: Their own unresolved feelings from problematic endings in the past—parental divorces, deaths, and other unanticipated or unwanted leavings—are being reevoked, because they never had the holding environment they needed to come to terms with their own feelings of loss. Such countertransference reactions are especially problematic because they often dovetail with clients' issues: Many clients are also struggling with their own unresolved conflicts over separation or are avoiding the topic because they sense it is difficult for the therapist. Unrecognized or unresolved conflicts over endings constitute one of the most common types of countertransference, and, when they occur, therapists need to seek consultation from a supervisor or colleague.

Because both clients' transference distortions and therapists' countertransference reactions are especially likely to occur around unnatural terminations and at the end of time-limited treatments, we must explore these issues further. Both therapists and clients often have experienced painful endings with significant others as "just happening to them." In many cases, they

- Were not prepared in advance for the separation
- Did not understand when or why this particular ending was occurring
- Were not able to participate in the leave-taking by discussing it with the departing person or saying good-bye

Such problematic endings have left many counselors and clients feeling powerless and out of control in regard to some of the most important experiences in their lives. In contrast, the approach suggested here enhances clients' efficacy by allowing them to be active, informed participants in the ending. Clients know when the ending is occurring; have the safety to express their sadness, anger or disappointment about it; and have the opportunity to say good-bye. Thus, in order to achieve a successful leave-taking the therapist must (1) acknowledge whatever therapeutic work remains unfinished; (2) nondefensively tolerate, affirm, and help to contain the client's sad, angry, and disappointed feelings, as well as other guilt-inducing responses that may occur; and (3) explore how this termination may be evoking other unwanted or unexpected endings in the client's past. Why is this so important? Especially with unnatural endings, it often will seem to the client that everything has now gone bad, and this relationship has just failed in the same way that others have in the past. Let's elaborate this expectable problem more fully.

As the old saying goes, children are supposed to grow up and put their feet under their own table, but it's hard to get up from the table when you're still hungry. The emotional deprivation and longing for fulfillment that many clients bring to therapy become ignited when treatment ends prematurely or before they feel ready to stop. Clients' relational templates may be such that their anger and disappointment—or even their sense of abandonment and betrayal over unmet attachment needs—will be directed toward the counselor, no matter how clearly the treatment parameters were explained at the beginning of treatment. Clients' angry outbursts can be difficult to hear:

You're just like my mother. You're abandoning me, too.
Our relationship never meant anything to me, anyway.
Nothing changed for me in therapy. This was just a waste of time.

The success of treatment in these cases depends in part on counselors' ability to remain nondefensive and tolerate clients' protests—to accept the clients' genuine experience and remain responsive. When, because of their own templates, counselors cannot tolerate the clients' accusations, they become defensive and feel guilty. In that case, they may make ineffective attempts to talk clients out of their feelings:

Therapist:
> But you *have* changed. You're not as depressed as you used to be, and now you can do _____, and before you couldn't do that.

Instead of these futile attempts, therapists need to acknowledge clients' feelings:

Therapist:
> You're really angry at me as our relationship is ending, and I respect those important feelings. Right now, it just seems like nothing meaningful has occurred and that nothing has gotten better for you.

By accepting clients' anger and protests in these ways—without fully agreeing that their factual content is correct—therapists give clients a response that doesn't match their relational templates. In turn, this affirmation of their current frustration or disappointment often enables clients to connect their strong emotional reactions to the developmental figures who originally disappointed or left them. Thus, what therapists often find hardest about unnatural endings—and about time-limited therapy—is managing their own guilt or defensiveness in the face of clients' anger over the ending. By drawing on the guidelines outlined here, however, therapists can help clients see how this ending is in fact different in some important ways from other unwanted endings in the clients' past, even though it may feel as if the same problematic pattern is occurring. The therapist can help make the termination of therapy different from past problematic endings—and, in doing so, help resolve this unfinished grief work from the past—by

- Telling the client about the ending in advance
- Inviting the client to share his or her angry, disappointed, or sad feelings and being able to accept and affirm those feelings without becoming defensive
- Talking with the client about the ending and the meaning it holds for both of them
- Validating the client's experience by acknowledging the ways that this ending does indeed evoke other conflicted or unwanted endings
- Making sure that the therapist and client say good-bye to each other

In other past endings that have been problematic, clients were not able to have such experiences with the departing person. Although most clients with significant conflicts over endings cannot recognize these differences initially, the therapist offers clients an opportunity to resolve long-standing problems by gradually helping clients differentiate this type of mutually acknowledged ending from other incomplete or unsatisfying endings with which they have had to cope in the past. In this way, successfully managing the termination becomes the prototype for resolving conflicts and dealing with future losses.

In the face of clients' anger, withdrawal, or accusations that their termination is just like prior losses and disappointments, therapists, too, may lose sight of the real differences that exist between this termination and clients' problematic past endings. In that case, therapists may accept the clients' blame, feel guilty, and avoid the ending—or invalidate the clients' sad, disappointed, or angry feelings by trying to talk clients out of them. Ironically, in that case, therapists make the clients' accusations come true: They reenact the clients' past conflicted endings. Therapists do better by acknowledging the similarity of the clients' feelings at termination and at past separations and discussing with clients how they can make this termination different from such problematic partings.

During the termination period, clients may have a diverse range of sad, hopeful, angry, excited, disappointed, and grateful feelings toward the therapist. As therapists track clients' reactions to the ending, they will often see how the clients' central conflict is simultaneously being aroused in two contexts. First, the clients' primary feelings about termination (anxiety, anger, guilt, helplessness, betrayal) may *match* the clients' feelings about the formative relationship in which the generic conflict originally arose. Second, clients' reaction toward termination may also match the feelings that were aroused in the crisis situation that originally brought the clients to treatment. By responding affirmingly to the clients' feelings, and clarifying their connection to these two sources, where appropriate, the therapist is helping them further resolve their central conflicts.

Finally, another useful intervention for the termination phase is to go through a Review-Predict-Practice sequence with all clients. Therapists *review* progress, accomplishments, successful changes and transitions, and unfinished issues with clients. Especially important, therapists help clients *predict* the developmental tasks or anxiety-arousing relationship issues that are likely to evoke their old dysfunctional patterns in the future. It is essential to anticipate these potentially problematic situations or relational scenarios as specifically as possible. As illustrated with Tracy, earlier in this chapter, the therapist might ask, "What is he likely to say, how is that going to make you feel, and what are you most likely to say and do at that moment?" *Practice* in dealing with these issues or patterns also prepares clients to succeed on their own. Therapists can provide this by role-playing effective and ineffective responses to the threatening or unwanted interactions that clients are likely to face in the future.

To sum up, the therapist must address the client's feelings about ending the relationship and help the client work them through. This is difficult to do, however, when both the therapist and the client wish to avoid the ending. The therapist and client have become close through their work together, and the relationship holds significant meaning for both of them. In order to avoid the conflicts that are activated by the termination, clients—and therapists, too—may try to deny the reality of the ending. Some clients may try to devalue the significance of their relationship with the therapist or to diminish the importance of the work that has been accomplished. Other clients will become symptomatic again and anxiously communicate that they are still too troubled to terminate and make it on their own. A few clients will ask the therapist if they can become friends. These and other common responses during the termination phase keep clients from facing current and past losses, preclude clients from receiving a new and more reparative response from the therapist, and keep clients from being able to internalize the therapist and integrate the gains they have made in treatment.

Ending the Relationship

Although some clients will reexperience their central conflicts at the time of termination, many will not. Both clients and therapists still need to talk about the ending, however. For example, therapists can talk with the client about the different ways in which they have seen the client change over the course of therapy. Therapists can also acknowledge the limitations of their relationship and the unfinished issues that the client will need to continue to work with on his or her own. Recollections of close moments, awkward misunderstandings, risks ventured, anxiety-arousing confrontations, and humorous incidents can be shared. This is also a time when therapists can reveal their own feelings about the client: what they have learned about themselves and life from the client; how the client has changed, enriched, or influenced them; ways in which therapists have cared about, have enjoyed, or will miss the client; and how therapists will remember the client.

Finally, in natural and unnatural endings, termination brings the therapist and client back to the separateness-relatedness dialectic again. In the case of natural endings, the therapist must give clients permission to leave; clients must know that the therapist enjoys their success, is pleased by their independence, and takes pleasure in seeing them become committed to new relationships. Other clients need to be reassured that they can have future contact with the therapist if necessary. Clients will be reassured to know that, whenever possible, the therapist is available to help them with future crises, should the need arise. Knowing that the therapist is available, if needed, but also supports their decision to leave, clients are able to internalize the therapist and the help the therapist has provided. This type of natural termination supports clients' own strivings for individuation and mastery in a supportive relational context, and helps clients claim their own self-efficacy and authentic voice.

As we have seen, graduate-student therapists often cannot be available in this way, because endings with their clients are of necessity unnatural. In these cases, the finality of the ending must be honestly acknowledged, and the therapist must respond to all of the client's emotional reactions. The unfinished business that remains must also be honestly acknowledged, and the therapist may need to help the client find a new therapist to finish the work that was just begun. If the therapist can do this, most clients will be able to accept the limitations of the relationship without having to reject the therapist or undo the work that was accomplished. When the client can keep inside what he or she has gained with the therapist, and the therapist can appreciate what he or she has learned from and given to the client, the relationship they have shared, although incomplete, has been successful and productive.

Closing

This book has followed the course of therapy from beginning to end. Just as clients still have unresolved conflicts when treatment ends, many questions and concerns about therapy will remain unanswered for therapists-in-training as this book draws to a close. Complex psychological issues and interpersonal processes have been introduced here, but counselors will need further reading and supervision to help with the exceptions, nuances, and complications that occur in every therapeutic relationship. Despite these limitations, the reader has learned essential guidelines for working with the process dimension and the relational aspects of psychotherapy. The conceptual framework presented here should enable beginning therapists to function more effectively and autonomously as professionals and to better assimilate other training experiences from different theoretical perspectives.

In closing, the central theme of this text is that, at every stage of development, growth and change occur within the context of significant emotional relationships. Accordingly, therapists are encouraged to regard themselves and the relationships they provide their clients as their most effective means of helping clients change.

Suggestions for Further Reading

1. The Dream is an important psychological dimension in people's lives that has received too little attention. The reader is encouraged to read Daniel Levinson's groundbreaking book, *The Seasons of a Man's Life* (New York: Ballantine, 1978). See especially pages 91–109.

2. As we saw in this chapter, a dramatic illustration of Horney's concepts of rising above and interpersonal coping styles can be found in Arthur Miller's play *Death of a Salesman* (New York: Penguin, 1949).

3. Useful information about identifying clients' generic conflicts and how they can be reactivated during the termination phase of treatment can be found in J. Mann's *Time-Limited Psychotherapy* (Cambridge: Harvard University Press, 1973).

4. Student therapists who have found the ideas presented here to be helpful are especially likely to enjoy *Between Therapist and Client* by M. Kahn (New York: Freeman, 1997) and *How Psychotherapy Works* by J. Weiss (New York: Guilford, 1993). For more experienced therapists, *Relational Concepts in Psychoanalysis* by Stephen Mitchell (Cambridge: Harvard University Press, 1988) and *Coping with Conflict: Supervising Counselors and Psychotherapists* by W. Mueller & B. Kell (New York: Appleton-Century-Crofts, 1972) both provide an informative presentation of interpersonally oriented psychotherapy. Also enjoyable is the introduction to Irving Yalom's *Love's Executioner* (New York: Basic Books, 1989).

Process Notes

Name: _____ Date: _____

1. Client Behavior: What is the client doing in this session?

What were the principal concerns expressed and new issues raised? What were the primary emotions experienced and characteristic relational themes or interactions evidenced? What interactions or situations tend to make this client anxious, and what are the interpersonal coping strategies the client employs to defend against these anxieties? How do I and others typically react to these coping strategies or eliciting maneuvers?

_____

_____

_____

_____

2. Therapeutic Relationship: What can I learn from our current interaction?

How did I feel about myself during this session, and what were my personal reactions toward the client? Do my emotional reactions typify the reactions of others toward the client (for example, angry, bored, frustrated, protective), or do they suggest my own personal issues or countertransference? Hypothesize several ways the current interaction may be similar to (reenacting) and different from (resolving) the core relational patterns that the client has been struggling with in other relationships.

_____

_____

_____

_____

3. Treatment Plans: What are my goals?

What issues do I need to explore further in order to understand this client better and clarify the focus of treatment?

How are the maladaptive relational patterns in the client's life—and the dysfunctional beliefs and conflicted feelings that accompany them—linked to his or her presenting problems and interpersonal coping strategies? Following from this, what are the specific relational experiences I need to provide for this client in order to enact the corrective emotional experiences necessary for change? Have there been any significant changes in the client's life situation or therapy to alter my treatment plans, or have I missed any factors (including cultural ones) that impact my treatment plan?

---

4. Intervention Strategies: Which interventions were effective in this session, and which were ineffective?

What interpersonal processes and intervention techniques did I use to try and reach my goals? (Relevant techniques might include, for example, highlighting repetitive relational patterns; reflecting predominant affect; validating subjective worldview; improving parenting skills; linking presenting problems to current interaction with therapist; clarifying automatic thoughts when distressed; and role-playing.) How did the client respond to these interventions? Were these techniques and interventions congruent with the corrective interpersonal process I was trying to enact with this client? With this understanding, what can I do next time to differentiate myself and our process from the relational patterns that have characterized other problematic relationships and to create greater interpersonal safety for this client?

Case Formulation Guidelines

It seems that nothing can make a new counselor feel anxious and inadequate more quickly than having to write a case formulation! It is hard to do, but keep trying and your efforts soon will be rewarded with more productive sessions and changing clients. The following guidelines will prove useful.

1. Formulation of the problems Summarize the client's presenting problems and suggest why this client is seeking treatment now. What has the client done in the past to address and resolve these problems? Are there any significant personal, familial, or cultural factors that make it difficult for this client to acknowledge having problems or to ask for help? How have sociocultural factors (such as race, religion, socioeconomic status, gender, and sexual orientation) affected the client and his or her experience with significant others? Provide a diagnosis with your rationale.

2. Treatment focus Clarify the maladaptive relational patterns—and the dysfunctional beliefs and conflicted feelings that accompany them—that recur in the client's presenting problems, in the client's current and past significant relationships, and in the client's relationship with you. For example, consider the client's internal working models or relational templates: (1) What does the client want or wish from others (for example, to be cared for)? (2) What does the client expect from others (for example, to be criticized or rejected)? (3) What is the client's experience of self in relationship to others (for example, a sense of being inadequate or burdensome)? (4) What are the conflicting affective states that typically result (for example, contempt and shame; or affection and security)? Thinking in terms of repetitive relational patterns and internal working models, identify the interpersonal coping strategies this client has typically used to manage his or her anxieties (for example, complying with others, withdrawing from others, trying to control others). What do these interpersonal coping strategies tend to elicit from the therapist and others? Try to provide a conceptual focus for treatment by integrating your formulations of this client's problems into two or three themes that capture what you think is wrong and needs to change.

3. Developmental context   How did the client's problems originally come about? Reveal the attachment history (for example, a self-absorbed and unresponsive caregiver), parenting practices (for example, a nurturant caregiver who was, however, unable to set limits and enforce discipline), familial roles (for example, a scapegoated or parentified child), family systems functioning and communication patterns (for example, the nature of the marital relationship, alliances, and coalitions), and trauma or other developmental experiences that may have shaped current problems. When you do not have sufficient developmental information, formulate working hypotheses about potential developmental experiences that may have transpired. Suggest how these developmental experiences (individual, familial, cultural) could interact with current stressors, crises, and social supports to shape the client's subjective worldview and current conflicts. Try to identify the client's pathogenic beliefs about himself or herself and faulty expectations of others.

4. Therapeutic process   What has counseling focused on to date, and how have you and the client interacted together? How do you tend to feel toward the client, and is this similiar or different from how others tend to feel toward the client? How does the client bring his or her conflicts into the therapeutic relationship? More specifically, consider several ways in which the therapeutic process you are enacting with the client may be different from, and similar to, the core relational patterns that the client has been struggling with in other relationships. What does the client tend to elicit from you and others? How does the client test (that is, confirm or disconfirm) core beliefs and expectations with you? How does the client tend to distort or misperceive you?

5. Goals and interventions   Where do you want to go with this client and how do you plan to get there? (For example, you might want to give a sexual abuse survivor the experience of secure boundaries; to convey a sense of being valued; or to support efficacy by encouraging the client's own initiative.) Focusing on shared goals that seem realistic to you and are experienced by the client as his or her own, utilize your interpersonal and developmental formulations of the client's problems to clarify what you are trying to accomplish right now, your intermediate goals, and your long-term goals for this client. Present your intervention strategy and how you plan to use yourself and the relationship you provide this client to reach these goals. Include the sequence of issues to be addressed and specific interventions or techniques to be employed. (Relevant interventions might include, for example, interpersonal feedback about the impact the client has on others, exploring the client's reactions to treatment or the therapist, cognitive reframing, educational input, process comments, affective containment, and interpretations.) Try to articulate as specifically as you can the interpersonal responses or experiences this particular client will need from you in order to receive a corrective emotional experience and change. (For example, for a client with detached parents, you would actively help him generate possible solutions; in contrast, for a client with intrusive or controlling parents, you would minimize suggestions, homework, and directions and validate all expressions of the client's own initiative.) Evaluate the appropriateness of the goals set and interventions used, given the client's personal, familial, and cultural context.

6. Impediments to change   If the client terminates prematurely or treatment is not successful, what factors could contribute? Identify the specific situations, relational scenarios, or affects that are likely to trigger the client's anxiety or diminish his or her sense of safety. How will the client express and/or defend against anxiety in his or her relationship with you (for example, by withdrawing, becoming critical, or complying)? Why have these anxieties and threats made sense in terms of the client's worldview or previous experience, and how were these coping strategies adaptive and necessary in the past? Suggest how the client's old relational scenarios are most likely to be reenacted in the therapeutic relationship, and how the client's conflicts or interpersonal style may interact with your own issues or countertransference propensities.

BIBLIOGRAPHY

Alexander, F., & French, T. M. (1980). *Psychoanalytic therapy: Principles and applications.* University of Nebraska Press.

Anchin, J., & Kiesler, D. (1982). *Handbook of interpersonal psychotherapy.* New York: Pergamon Press.

Anderson, C. M., & Steward, S. (1983). *Mastering resistance: A practical guide to family therapy.* New York: Guilford.

Bandura, A. (1977). Self efficacy: Toward a unifying theory of behavioral change. *Psychological Review, 84*(2), 191–215.

Bandura, A. (1982). Self-efficacy mechanisms in human agency. *American Psychologist, 37*, 122–147.

Baumrind, D. (1967). Child care practices anteceding three patterns of preschool behavior. *Genetic Psychology Monographs, 75*, 43–88.

Baumrind, D. (1971). Current patterns of parental authority. *Developmental Psychology Monograph, 4*(Whole No. 1, Pt. 2).

Baumrind, D. (1983). Familial antecedents of social competence in young children. *Psychological Bulletin, 94*(1), 132–142.

Beck, A. (1976). *Cognitive therapy and the emotional disorders.* New York: International Universities Press.

Beck, A., & Freeman, A. (1990). *Cognitive therapy of the personality disorders.* New York: Guilford.

Benjamin, L. (1995). *Interpersonal diagnosis and treatment of personality disorder* (2nd ed.). New York: Guilford.

Bergin, A., & Garfield, S. (Eds.). (1994). *Handbook of psychotherapy and behavior change* (4th ed.). New York: Wiley.

Boszormenyi-Nagy, I., & Spark, G. (1973). *Invisible loyalties: Reciprocity in inter-generational family therapy.* New York: Harper & Row.

Bowen, M. (1966). The use of family theory in clinical practice. *Comprehensive Psychiatry, 7*, 345–376.

Bowlby, J. (1988). *A secure base.* New York: Basic Books.

Bugental, J. (1965). *The search for authenticity: An existential-analytic approach to psychotherapy.* New York: Holt, Rinehart & Winston.

Carkhuff, R. (1969). *Helping and human relations: A primer for lay and professional helpers* (Vols. 1 and 2). New York: Holt, Rinehart & Winston.

Davanloo, H. (Ed.). (1980). *Short-term dynamic psychotherapy.* New York: Aronson.

Elkind, D. *Ties that stress.* Cambridge, MA: Harvard University Press.

Engel, L., & Ferguson, T. (1990). *Hidden guilt.* New York: Pocket Books.

Erikson, E. (1968). *Childhood and society* (2nd ed.). New York: Norton.

Farley, J. (1979, January). Family separation-individuation tolerance: A developmental conceptualization of the nuclear family. *Journal of Marital and Family Therapy*, 61–67.

Field, T. (1996). Attachment and separation in young children. *Annual Review of Psychology, 47*, 541–561.

Fraiberg, S., Adelson, E., & Shapiro, V. (1975). Ghosts in the nursery: A psychoanalytical approach to the problems of impaired mother-infant relationships. *Journal of the American Academy of Child Psychiatry, 14*, 387–421.

Frankl, V. (1963). *Man's search for meaning.* New York: Pocket Books.

Fromm, E. (1982). *Escape from freedom.* New York: Avon.

Fromm-Reichmann, F. (1960). *Principles of intensive psychotherapy.* Chicago: University of Chicago Press.

Gelso, C., & Carter, J. (1985). The relationship in counseling and psychotherapy: Components, consequences, and theoretical antecedents. *The Counseling Psychologist, 13*, 155–243.

Gill, M., & Muslin, H. (1976). Early interpretations of transference. *Journal of the American Psychoanalytic Association, 24*, 779–794.

Gilligan, C. (1982). *In a different voice: Psychological theory and women's development.* Cambridge: Harvard University Press.

Goldenberg, I., & Goldenberg, H. (1996). *Family therapy: An overview* (4th ed.). Pacific Grove, CA: Brooks/Cole.

Greenberg, J., & Mitchell, S. (1983). *Object relations in psychoanalytic theory.* Cambridge: Harvard University Press.

Greenson, R. (1967). *The technique and practice of psychoanalysis* (Vol. 1). New York: International Universities Press.

Guggenbuhl-Craig, A. (1971). *Power in the helping professions.* New York: Springer.

Haley, J. (1967). Toward a theory of pathological systems. In G. H. Zuk & I. Boszor-menyi-Nagy (Eds.), *Family therapy and disturbed families.* Palo Alto, CA: Science and Behavior Books.

Haley, J. (1980). *Leaving home: The therapy of disturbed young people.* New York: McGraw-Hill.

Hartman, E. (1978, October). Using eco-maps and genograms in family therapy. *Social Casework*, 464–476.

Horney, K. (1966). *Our inner conflicts.* New York: Norton.

Horney, K. (1970). *Neurosis and human growth.* New York: Norton.

Horowitz, M., Marmar, C., Krupnick, J., Wilner, N., Kaltreider, N., & Wallerstein, R. (1984). *Personality styles and brief psychotherapy.* New York: Basic Books.

Jordan, J., Kaplan, A., Miller, J., Stiver, P., & Surrey, J. (1991). *Women's growth in connection* (pp. 126–184). New York: Guilford.

Kafka, F. (1966). *Letter to his father.* New York: Schocken Books.

Kahn, M. (1997). *Between therapist and client.* New York: W. H. Freeman.

Kaiser, H. (1965). *Effective psychotherapy.* New York: Free Press.

Karen, R. (1994). *Becoming attached.* New York: Time/Warner.

Kell, B., & Mueller, W. (1966). *Impact and change: A study of counseling relationships.* Englewood Cliffs, NJ: Prentice-Hall.

Kelly, G. (1963). *The psychology of personal constructs.* New York: Norton.

Kiesler, D. (1988). *Therapeutic metacommunication: Therapist impact disclosure as feedback in psychotherapy.* Palo Alto, CA: Consulting Psychologists Press.

Kiesler, D., & Van Denburg, T. (1993). Therapeutic impact disclosure: A last taboo in psychoanalytic theory and practice. *Clinical Psychology & Psychotherapy, 1*(1), 3–13.

Klerman, G. L., Weissman, M., Rounsaville, B., & Chevron, E. (1984). *Interpersonal psychotherapy of depression*. New York: Basic Books.

Kohut, H. (1971). *The analysis of the self*. New York: International Universities Press.

Kohut, H. (1977). *The restoration of the self*. New York: International Universities Press.

Laing, R. D., & Esterson, A. (1970). *Sanity, madness and the family*. Middlesex, England: Penguin.

Lazarus, A. (1973). Multimodal behavior therapy: Treating the BASIC I.D. *Journal of Nervous and Mental Disease, 156*, 404–411.

Levenson, E. (1982). Language and healing. In S. Slip (Ed.), *Curative factors in dynamic psychotherapy*. New York: McGraw-Hill.

Levenson, H. (1995). *Time-limited dynamic therapy*. New York: Basic Books.

Levinson, D. (1978). *The seasons of a man's life*. New York: Ballantine.

Lewis, H. (1971). *Shame and guilt in neurosis*. New York: International Universities Press.

Luborsky, L., & Marks, D. (1991). Short-term supportive-expressive psychoanalytic psychotherapy. In P. Crits-Cristoph & J. Barber (Eds.), *Handbook of short-term dynamic psychotherapy* (pp. 110–136). New York: Basic Books.

Luborsky, L., & DeRubeis, R. (1984). The use of psychotherapy treatment manuals: A small revolution in psychotherapy research style. *Clinical Psychology Review, 4*, 5–14.

Mahler, M., Pine, F., & Bergman, A. (1975). *The psychological birth of the human infant: Symbiosis and individuation*. New York: Basic Books.

Malan, D. H. (1976). *The frontier of brief psychotherapy: An example of the convergence of research and clinical practice*. New York: Plenum.

Mann, J. (1973). *Time-limited psychotherapy*. Cambridge, MA: Harvard University Press.

Mann, J., & Goldman, R. (1982). *A casebook in time-limited psychotherapy*. New York: McGraw-Hill.

Masterson, J. (1972). *Treatment of the borderline adolescent: A developmental approach*. New York: Wiley.

Masterson, J. (1976). *Psychotherapy of the borderline adult: A developmental approach*. New York: Brunner/Mazel.

May, R. (1977). *The meaning of anxiety*. New York: Norton.

May, R., Angel, E., & Ellenberger, H. (1958). *Existence: A new dimension in psychology and psychiatry*. New York: Basic Books.

McCarthy, P. (1982). Differential effects of counselor self-referent responses and counselor status. *Journal of Counseling Psychology, 29*, 125–131.

McCarthy, P., & Betz, N. (1978). Differential effects of self-disclosing versus self-involving counselor statements. *Journal of Counseling Psychology, 25*, 252–256.

McClure, F., & Teyber, E. (1996). *Child and adolescent therapy: A multicultural-relational approach*. Fort Worth, TX: Harcourt Brace Jovanovich.

Miller, A. (1949). *Death of a salesman*. New York: Penguin.

Miller, A. (1981). *Prisoners of childhood*. New York: Basic Books.

Miller, A. (1984). *For your own good*. New York: Farrar, Straus & Giroux.

Mills, J., Bauer, G., & Miars, R. (1989). Use of transference in short-term dynamic psychotherapy. *Psychotherapy, 26*, 338–343.

Minuchin, S. (1974). *Families and family therapy*. Cambridge: Harvard University Press.

Mitchell, S. (1988). *Relational concepts in psychoanalysis*. Cambridge: Harvard University Press.

Mueller, W., & Kell, B. (1972). *Coping with conflict: Supervising counselors and psychotherapists*. New York: Appleton-Century-Crofts.

Olson, D. H., McCubbin, H. I., Barnes, H., Larsen, A., Muxen, M., & Wilson, M. (1983). *Families: What makes them work*. Newbury Park, CA: Sage.

Passons, W. (1975). *Gestalt approaches in counseling.* New York: Holt, Rinehart & Winston.

Pinderhughes, H. (1989). *Understanding race, ethnicity and power.* New York: Free Press.

Rocklin R., & Levitt, D. (1987). Those who broke the cycle: Therapy with nonabusive adults who were physically abused as children. *Psychotherapy, 24*(4), 769–778.

Rogers, C. (1951). *Client-centered therapy.* Boston: Houghton Mifflin.

Rogers, C. (1959). A theory of therapy, personality and interpersonal relationships as developed in the client-centered framework. In S. Koch (Ed.), *Psychology: A study of science:* Vol. 3. *Formulations of the person and the social context* (pp. 184–256). New York: McGraw-Hill.

Satir, V. (1967). *Conjoint family therapy.* Palo Alto, CA: Science and Behavior Books.

Schatzman, M. (1973). *Soul murder: Persecution in the family.* New York: Random House.

Seligman, M. (1975). *Helplessness.* New York: W. H. Freeman.

"Shame." (1995). [Special Issue]. *American Behavioral Scientist, 38*(8).

Sifneos, P. (1987). *Short-term dynamic psychotherapy: Evaluation and technique* (2nd ed.). New York: Plenum.

Solomon, M. (1973). A developmental, conceptual premise for family therapy. *Family Process, 12*(2), 179–188.

Speight, S., Myers, L., Cox, C., & Highlen, P. (1991). A redefinition of multicultural counseling. *Journal of Counseling and Development, 70*(1), 29–36.

Springmann, R. (1986). Countertransference: Clarifications in supervision. *Contemporary Psychoanalysis, 22,* 252–277.

Steinbeck, J. (1952). *East of Eden.* New York: Viking Press.

Stierlin, H. (1972). *Separating parents and adolescents.* New York: Quadrangle.

Strupp, H. H. (1980). Success and failure in time-limited psychotherapy. *Archives of General Psychiatry, 37,* 595–613, 708–716, 831–841, 947–954.

Strupp, H., & Binder, J. (1984). *Psychotherapy in a new key: A guide to time-limited dynamic psychotherapy.* New York: Basic Books.

Strupp, H., & Hadley, S. (1979). Specific versus nonspecific factors in psychotherapy: A controlled study of outcome. *Archives of General Psychiatry, 36,* 1125–1136.

Sullivan, H. S. (1968). *The interpersonal theory of psychiatry.* New York: Norton.

Sullivan, H. S. (1970). *The psychiatric interview.* New York: Norton.

Teyber, E. (1981). Structural family relations: A review. *Family Therapy, 1,* 39–48.

Teyber, E. (1983). Effects of the parental coalition on adolescent emancipation from the family. *Journal of Marital and Family Therapy, 9,* 89–99.

Teyber, E. (1992). *Helping children cope with divorce: A practical guide for parents.* San Francisco: Jossey-Bass.

Turock, A. (1980). Immediacy in counseling: Recognizing clients' unspoken message. *Personnel and Guidance Journal, 59,* 168–172.

Wachtel, P. (1982). *Psychoanalysis and behavior therapy: Toward an integration.* New York: Basic Books.

Wachtel, P. (1993). *Therapeutic communication: Principles and effective practice.* New York: Guilford.

Walborn, F. (1996). *Process variables: Four common elements of counseling and psychotherapy.* Pacific Grove, CA: Brooks/Cole.

Weiss, J. (1993). *How psychotherapy works.* New York: Guilford.

Weiss, J., & Sampson, H. (1986). *The psychoanalytic process: Theory, clinical observation and empirical research.* New York: Guilford.

Wenar, C. (1994). *Psychopathology from infancy through adolescence: A developmental approach* (3rd ed.). New York: Random House.

Wheelis, A. (1974). *How people change*. New York: Harper & Row.

White, R. (1959). Motivation reconsidered: The concept of competence, *Psychology Review, 66,* 297–333.

Winnicott, D. W. (1965). Ego distortion in terms of true and false self. In *The maturational process and the facilitating environment*. New York: International Universities Press.

Woodward, B. (1984). *Wired*. New York: Simon & Schuster.

Yalom, I. (1975). *The theory and practice of group psychotherapy* (2nd ed.). New York: Basic Books.

Yalom, I. (1981). *Existential psychotherapy*. New York: Basic Books.

NAME INDEX

Adelson, E., 58
Alexander, F., 19
Anchin, J., 30, 295
Anderson, C. M., 91

Bandura, A., 77, 78, 103, 104, 122
Bateson, G., 10
Bauer, G., 241
Baumrind, D., 174
Beck, A., 44, 50 51, 199, 225, 227
Benjamin, L., 5, 28
Bergin, A., 272
Bergman, A., 14, 257
Betz, N., 55
Binder, J., 6, 28, 50, 60, 199, 224
Block-Lewis, H., 137
Boszormenyi-Nagy, I., 11, 14, 60, 173, 190
Bowen, M., 6, 11, 60, 87, 173, 189, 190
Bowlby, J., 10, 30, 42, 44, 58, 130, 172
Bugental, J., 158

Carkhuff, R., 6, 54
Carter, J., 35
Cashdan, S., 55, 295
Chevron, E., 28
Cox, C., 84

Davanloo, H., 6, 20, 27, 50
DeRubeis, R., 50

Egan, G., 158
Elkind, D., 191
Engel, L., 257

Erikson, E., 6, 164, 172
Esterson, A., 43

Fairbairn, W., 10
Farley, J., 172, 191
Ferguson, T., 257
Field, T., 130
Fraiberg, S., 58
Frankl, V., 122
Freeman, A., 50, 199, 225, 228
French, T., 19
Freud, S., 5, 10
Fromm, E., 6, 177, 228
Fromm-Reichmann, F., 6, 7, 20

Garfield, S., 272
Gelso, C., 35
Gill, M., 250
Gilligan, C., 59
Goldenberg, H., 10
Goldenberg, I., 10
Goldman, R., 27
Greenberg, J., 10, 30
Greenson, R., 34, 60, 90
Guggenbuhl-Craig, A., 158
Guntrip, H., 10

Hadley, S., 54, 150
Haley, J., 10, 39, 60, 250, 257
Highlen, P., 84
Horney, K., 6, 7, 193–195, 199, 203, 224, 228, 326
Horowitz, M., 28

SUBJECT INDEX

TO THE OWNER OF THIS BOOK:

I hope that you have found *Interpersonal Process in Psychotherapy: A Relational Approach* useful. So that this book can be improved in a future edition, would you take the time to complete this sheet and return it? Thank you.

School and address: ⎯⎯⎯⎯⎯⎯⎯⎯⎯⎯⎯⎯⎯⎯⎯⎯⎯⎯⎯⎯⎯⎯⎯⎯⎯⎯

Department: ⎯⎯⎯⎯⎯⎯⎯⎯⎯⎯⎯⎯⎯⎯⎯⎯⎯⎯⎯⎯⎯⎯⎯⎯⎯⎯⎯⎯⎯

Instructor's name: ⎯⎯⎯⎯⎯⎯⎯⎯⎯⎯⎯⎯⎯⎯⎯⎯⎯⎯⎯⎯⎯⎯⎯⎯⎯⎯

1. What I like most about this book is: ⎯⎯⎯⎯⎯⎯⎯⎯⎯⎯⎯⎯⎯⎯⎯⎯

⎯⎯⎯⎯⎯⎯⎯⎯⎯⎯⎯⎯⎯⎯⎯⎯⎯⎯⎯⎯⎯⎯⎯⎯⎯⎯⎯⎯⎯⎯⎯⎯⎯⎯⎯⎯⎯

⎯⎯⎯⎯⎯⎯⎯⎯⎯⎯⎯⎯⎯⎯⎯⎯⎯⎯⎯⎯⎯⎯⎯⎯⎯⎯⎯⎯⎯⎯⎯⎯⎯⎯⎯⎯⎯

2. What I like least about this book is: ⎯⎯⎯⎯⎯⎯⎯⎯⎯⎯⎯⎯⎯⎯⎯⎯

⎯⎯⎯⎯⎯⎯⎯⎯⎯⎯⎯⎯⎯⎯⎯⎯⎯⎯⎯⎯⎯⎯⎯⎯⎯⎯⎯⎯⎯⎯⎯⎯⎯⎯⎯⎯⎯

⎯⎯⎯⎯⎯⎯⎯⎯⎯⎯⎯⎯⎯⎯⎯⎯⎯⎯⎯⎯⎯⎯⎯⎯⎯⎯⎯⎯⎯⎯⎯⎯⎯⎯⎯⎯⎯

3. My general reaction to this book is: ⎯⎯⎯⎯⎯⎯⎯⎯⎯⎯⎯⎯⎯⎯⎯⎯

⎯⎯⎯⎯⎯⎯⎯⎯⎯⎯⎯⎯⎯⎯⎯⎯⎯⎯⎯⎯⎯⎯⎯⎯⎯⎯⎯⎯⎯⎯⎯⎯⎯⎯⎯⎯⎯

4. The name of the course in which I used this book is: ⎯⎯⎯⎯⎯⎯⎯

⎯⎯⎯⎯⎯⎯⎯⎯⎯⎯⎯⎯⎯⎯⎯⎯⎯⎯⎯⎯⎯⎯⎯⎯⎯⎯⎯⎯⎯⎯⎯⎯⎯⎯⎯⎯⎯

5. Were all of the chapters of the book assigned for you to read? ⎯⎯⎯

 If not, which ones weren't? ⎯⎯⎯⎯⎯⎯⎯⎯⎯⎯⎯⎯⎯⎯⎯⎯⎯⎯⎯

6. In the space below, or on a separate sheet of paper, please write specific suggestions for improving this book and anything else you'd care to share about your experience in using the book.

⎯⎯⎯⎯⎯⎯⎯⎯⎯⎯⎯⎯⎯⎯⎯⎯⎯⎯⎯⎯⎯⎯⎯⎯⎯⎯⎯⎯⎯⎯⎯⎯⎯⎯⎯⎯⎯

⎯⎯⎯⎯⎯⎯⎯⎯⎯⎯⎯⎯⎯⎯⎯⎯⎯⎯⎯⎯⎯⎯⎯⎯⎯⎯⎯⎯⎯⎯⎯⎯⎯⎯⎯⎯⎯

⎯⎯⎯⎯⎯⎯⎯⎯⎯⎯⎯⎯⎯⎯⎯⎯⎯⎯⎯⎯⎯⎯⎯⎯⎯⎯⎯⎯⎯⎯⎯⎯⎯⎯⎯⎯⎯

⎯⎯⎯⎯⎯⎯⎯⎯⎯⎯⎯⎯⎯⎯⎯⎯⎯⎯⎯⎯⎯⎯⎯⎯⎯⎯⎯⎯⎯⎯⎯⎯⎯⎯⎯⎯⎯

⎯⎯⎯⎯⎯⎯⎯⎯⎯⎯⎯⎯⎯⎯⎯⎯⎯⎯⎯⎯⎯⎯⎯⎯⎯⎯⎯⎯⎯⎯⎯⎯⎯⎯⎯⎯⎯

Optional:

Your name: _____ Date: _____

May Brooks/Cole quote you, either in promotion for *Interpersonal Process in Psychotherapy: A Relational Approach* or in future publishing ventures?

Yes: _____ No: _____

Sincerely,

Edward Teyber

FOLD HERE

NO POSTAGE
NECESSARY
IF MAILED
IN THE
UNITED STATES

BUSINESS REPLY MAIL

FIRST CLASS   PERMIT NO. 358   PACIFIC GROVE, CA

POSTAGE WILL BE PAID BY ADDRESSEE

ATT: *Edward Teyber*

Brooks/Cole Publishing Company
511 Forest Lodge Road
Pacific Grove, California 93950-9968

FOLD HERE

Brooks/Cole Publishing is dedicated to publishing quality books for the helping professions. If you would like to learn more about our publications, please use this mailer to request our catalogue.

Name: _____

Street Address: _____

City, State, and Zip: _____

FOLD HERE

FOLD HERE